LIQUID CRYSTALS FOR ELECTRONIC DEVICES

LIQUID CRYSTALS FOR ELECTRONIC DEVICES

Edward L. Williams

NOYES DATA CORPORATION

Park Ridge, New Jersey London, England

1975

Library of Congress Catalog Card Number: 75-754
ISBN: 0-8155-0567-1
Printed in the United States

Published in the United States of America by
Noyes Data Corporation
Noyes Building, Park Ridge, New Jersey 07656

FOREWORD

The detailed, descriptive information in this book is based on U.S. patents relating to the chemistry and practical preparation of liquid crystals.

This book serves a double purpose in that it supplies detailed technical information and can be used as a guide to the U.S. patent literature in this field. By indicating all the information that is significant, and eliminating legal jargon and juristic phraseology, this book presents an advanced, technically oriented review of the manufacture of liquid crystals.

The U.S. patent literature is the largest and most comprehensive collection of technical information in the world. There is more practical, commercial, timely process information assembled here than is available from any other source. The technical information obtained from a patent is extremely reliable and comprehensive; sufficient information must be included to avoid rejection for "insufficient disclosure." These patents include practically all of those issued on the subject in the United States during the period under review; there has been no bias in the selection of patents for inclusion.

The patent literature covers a substantial amount of information not available in the journal literature. The patent literature is a prime source of basic commercially useful information. This information is overlooked by those who rely primarily on the periodical journal literature. It is realized that there is a lag between a patent application on a new process development and the granting of a patent, but it is felt that this may roughly parallel or even anticipate the lag in putting that development into commercial practice.

Many of these patents are being utilized commercially. Whether used or not, they offer opportunities for technological transfer. Also, a major purpose of this book is to describe the number of technical possibilities available, which may open up profitable areas to research and development. The information contained in this book will allow you to establish a sound background before launching into research in this field.

Advanced composition and production methods developed by Noyes Data are employed to bring our new durably bound books to you in a minimum of time. Special techniques are used to close the gap between "manuscript" and "completed book." Industrial technology is progressing so rapidly that time-honored conventional typesetting, binding and shipping methods are no longer suitable. We have bypassed the delays in the conventional book publishing cycle and provide the user with an effective and convenient means of reviewing up-to-date information in depth.

The Table of Contents is organized in such a way as to serve as a subject index. Other indexes by company, inventor and patent number help in providing easy access to the information contained in this book.

15 Reasons Why the U.S. Patent Office Literature Is Important to You —

1. The U.S. patent literature is the largest and most comprehensive collection of technical information in the world. There is more practical commercial process information assembled here than is available from any other source.

2. The technical information obtained from the patent literature is extremely comprehensive; sufficient information must be included to avoid rejection for "insufficient disclosure."

3. The patent literature is a prime source of basic commercially utilizable information. This information is overlooked by those who rely primarily on the periodical journal literature.

4. An important feature of the patent literature is that it can serve to avoid duplication of research and development.

5. Patents, unlike periodical literature, are bound by definition to contain new information, data and ideas.

6. It can serve as a source of new ideas in a different but related field, and may be outside the patent protection offered the original invention.

7. Since claims are narrowly defined, much valuable information is included that may be outside the legal protection afforded by the claims.

8. Patents discuss the difficulties associated with previous research, development or production techniques, and offer a specific method of overcoming problems. This gives clues to current process information that has not been published in periodicals or books.

9. Can aid in process design by providing a selection of alternate techniques. A powerful research and engineering tool.

10. Obtain licenses — many U.S. chemical patents have not been developed commercially.

11. Patents provide an excellent starting point for the next investigator.

12. Frequently, innovations derived from research are first disclosed in the patent literature, prior to coverage in the periodical literature.

13. Patents offer a most valuable method of keeping abreast of latest technologies, serving an individual's own "current awareness" program.

14. Copies of U.S. patents are easily obtained from the U.S. Patent Office at 50¢ a copy.

15. It is a creative source of ideas for those with imagination.

CONTENTS AND SUBJECT INDEX

INTRODUCTION 1

CHOLESTERIC COMPOUNDS 3
Compositions 3
Nematic and Potentially Cholesteric Materials Combination 3
Multicomponent Cholesteric Mixtures 6
Hysteretic Compositions 7
Glassy Optically Negative Materials 10
Addition of Phototropic Material 15
Cholesteryl Erucyl Carbonate as Stabilizer 16
Cholesteryl p-Nonylphenyl Carbonate as Stabilizer 17
Extended Mesomorphic Ranges with Low Temperature Stability 19
Ultraviolet Radiation Stabilizer 20
Encapsulation Techniques 21
Continuous Polymeric Matrix 21
Transparent Top Coating over Capsule Wall 30
Dispersions of Gelatin-Gum Arabic Capsules 32
Mixtures of Nematic and Cholesteric Liquid Crystals 34
Barrier Coatings 36
Sulfonated Polymers of Acrylamides 36
Cross-Linked Poly(Vinyl Alcohol) 38
Casein Glue 40
Acrylic Resin 41

APPLICATIONS FOR CHOLESTERIC CRYSTALS 43
Temperature Sensing Devices 43
Skin Patch with Encapsulated Crystals 43
Measuring Skin Temperature 43
Temperature-Limit Detection 47
Clinical Thermometers 47
Optical Filters 52
Notch Filters—Opposite Rotary Sense 52
Notch Filters—Same Intrinsic Rotary Sense 58
Band Pass Filter 62
Tuning Method for Plural Filters 63
Bandwidth Narrowing 68
Induced Optical Activity 68
Display and Imaging Systems 69

Electrically Controlled Imaging Systems 69
Matrix-Addressed Display 79
High Frequency Color Display System 83
Series of Alphanumeric Images 89
Variable Color Display Using Induced Deformational Stresses 92
Uniform Electrical Holding Field 97
AC Electrical Fields 105
Display of Infrared Beams Using Encapsulated Crystals 105
Optical Storage Capability 111
Electric Field Sensitive Devices 115
Undisturbed Image on a Disturbed Background 116
Molecular Orientation Technique for Xerography Process 120
Erasable and Reusable Imaging System 123
Ultraviolet Light 124
Other Applications and Devices 126
Decorative Iridescent Compositions 126
Tunable Liquid Crystal-Dye Laser 127
Analytical Methods 129
Detection of X-Ray Radiation Using Iodine Compounds 131
Conductive Ink Resistor for Heating 133
Detection System for Surface Flaws 135
Light Modulation System 135

NEMATIC COMPOUNDS 137
Aromatic Azoxy Compounds 137
Substituted Azoxybenzene Compounds 137
Azo Compounds Stable to Electrochemical Degradation 140
Azoxybenzene Compounds 143
Schiff Bases 145
p-[N-(p-Methoxybenzylidene)Amino] Phenyl 2-Ethylhexanoate 145
N-(o-Hydroxy-p-Methoxybenzylidene)-p-n-Butylaniline 146
Alkoxy and Acyloxybenzylidene Anils 148
N-(p-Alkanoyloxybenzylidene)-p'-n-Butylaniline 150
N-(p-n-Alkoxycarbonyloxybenzylidene)-p'-n-Butoxyaniline 151
N-(p-2-Methoxyethoxybenzylidene)-p-n-Butylaniline 151
Mixtures and Derivatives of p-Methoxybenzylidene-p'-n-Butylaniline 152
p-Methoxyformyloxybenzylidene-p-n-Butylaniline 156
N-(p-n-Alkoxycarbonyloxybenzylidene)-p'-Aminophenyl Acetate 161
p-(Anisalamino)Phenyl Acetates 162
Olefinic-Substituted Benzylidene Structures 164
N-(p-n-Alkanoyloxybenzylidene)-p'-Aminophenyl Acetate 165
Aromatic Esters 165
Aryl Ester Mixtures 165
Positive Dielectric Anisotropy 167
Alkyl-Stilbene Compounds 172
Multicomponent Mixtures 173
Low Voltage Crystal Display Mixture 173
For Transparency Control 174
Ternary Nematic Mixtures 180
For Dynamic Scattering Mode at Room Temperature 180
Nematic-Cholesteric Blend with Positive Anisotropy 182
Alkoxybenzylidene Anils and Asymmetric Cholesteric Compounds 187
Lyotropic Nematic Liquid Crystals 188
Ternary and Binary Mixtures 189
p-Alkoxyphenyl Esters of Capronyloxybenzoic Acid 191
Aligning and Doping Techniques 193
Hexadecyltrimethyl-Ammonium Bromide 193
Hexadecylpyridinium Bromide 194
Long Chain Acids, Ureas and Silanes 195

Surface Treatment for Orientation Purposes 196
Polyamide Resins 197
Oriented Light Scattering 200
Parallel Grooves in Contact Surfaces 200
Inert Coatings on Substrates 204
Silane Derivative Chemisorbed on Substrate 205
Squarylium Dye or Tetracyano-p-Benzoquinone 206
Complex Ions Added for Improved Dynamic Scattering 207
Additives to Reduce the Turn-Off Time 210

NEMATIC AND SMECTIC LIQUID CRYSTAL DISPLAY SYSTEMS AND DEVICES 212
Nematic 212
Rear Projection Screen 212
Pleochroic Dye for Color Display 214
Electrooptical Elements 217
Fast Self-Quenching 221
Electronically Tunable Birefringence 221
Thermally Addressed Display Cell for Optics Equipment 224
Kerr Effect Cells 227
Scintillation-Free Image from Coherent Light Source 230
Storage Device Employing CdS Photoconductor Layer 232
Electrooptic Shutter Having Thin Glass Layer 240
Electroluminescent Element 243
Light Valve Using Nematic and Cholesteric Mixtures 243
Smectic 245
Smectic-Nematic Liquid Crystals 245
Infrared Laser Beam 247
Variable-Tilt Compounds for Temperature Indicator 251
Memory System 252

COMPANY INDEX 259
INVENTOR INDEX 260
U.S. PATENT NUMBER INDEX 262

INTRODUCTION

Liquid crystals may loosely be defined as substances that have some properties ordinarily thought to be characteristic of liquids and some properties ordinarily thought to be characteristic of solids. These materials cannot be considered as rare; it has been estimated that one out of every 200 new compounds synthesized in the laboratory has liquid crystalline characteristics. Mechanically, liquid crystals resemble liquids, and have viscosities which range from that of runny glue to glassy solids. Some liquid crystals show unusual rheological properties in that the measured value of the viscosity is a strong function of the direction of flow of the substance. However, it is the optical properties of liquid crystals that most resemble those of crystalline solids. For example, some liquid crystalline substances scatter light in symmetrical patterns and the color of the reflected light is a function of the angle at which it is observed.

One way of looking at the difference between liquids, solids and liquid crystals is to consider the degrees of freedom of the molecules within the three states. In the liquid state the molecules are mobile in three directions and can rotate about three axes perpendicular to one another. In a crystal, the molecules are fixed and rotations are not possible. In liquid crystals, molecules are mobile in two directions and can rotate about one axis or are mobile in three directions and can rotate about one axis.

Liquid crystals are usually divided into three classes: smectic, nematic and cholesteric. In smectic liquid crystals, the molecules are arranged in layers with their long axes approximately normal to the plane of the layers. The molecules can move in two directions (in their own plane) and can rotate about one axis. The molecules within a layer may be regularly spaced as in a crystal or may be randomly distributed. The planes can slide easily over each other. The term "smectic" comes from a Greek work meaning grease or slime and was applied to this type of liquid crystal because of its slippery feeling.

In nematic substances the only restriction on the arrangement of molecules is that they preserve a parallel or nearly parallel position with respect to each other. The molecules have mobility in three directions with rotation about one axis being permitted. This arrangement is analogous to a long box filled with round pencils which can roll around and slide back and forth, but remain parallel to one another in the direction of the long axis of the pencil. The term "nematic" comes from a Greek word meaning thread and was applied to this type of liquid crystal because of the thread-like lines visible under microscopic observation. In cholesteric materials the molecules are arranged in parallel layers about 2000 A thick, in contrast to smectic materials where the layer thickness is about the length of a molecule, about 20 A.

Each molecule in the cholesteric state essentially lies in a plane a few Angstroms thick. Very many such planes stack upon one another to form a several thousand Angstrom thick layer. Side chains project upward from the plane of each molecule and some hydrogen atoms extend below. Because of this the direction of the long axis of a molecule in a given plane is slightly and systematically rotated from the direction of the axis of the molecules in adjacent planes.

The result is a helical structure which shows great optical activity because the layer distance of several thousand Angstroms corresponds to optical wavelengths. The term "cholesteric" is applied to this class of liquid crystal because the first examples of it were derivatives of the substance, cholesterol. Cholesterol itself does not have liquid crystalline properties because it lacks the necessary side chains.

Liquid crystals have many odd and interesting physical properties which may be exploited to practical advantage. A striking example is given by the optical properties such as reflection, birefringence, circular dichroism, optical rotation and color. These properties of a liquid crystal can all be very sensitive to changes in its environment such as temperature, pressure, presence of an electric or magnetic field or the presence of chemical vapors. The reason for this sensitivity is the weakness of the intermolecular forces responsible for the structure of the material. This weakness allows changes in the structure to occur under the influence of even very mild changes in environment. Perhaps the most widely exploited optical property of liquid crystals is the color sensitivity of cholesteric liquid crystals.

In the liquid phase these materials are colorless, but on cooling to the liquid crystalline phase or mesophase they go through a range of colors, appearing first violet, then blue, green, yellow, red and finally colorless again. The rate of change of color with temperature change varies from material to material. For a given material each color corresponds to an exact temperature. Hundreds of cholesteric materials, both pure and mixtures, are known and have been used to measure temperatures ranging from -20° to $250^\circ C$.

New uses for liquid crystals are rapidly emerging in all areas of applications-related-technology. They are currently being employed for wristwatch faces, disposable thermometers, breast tumor detection, digital readout for many instruments and minicomputers, optical filters, solvents for NMR spectroscopy, and for many toy and decorative materials. Liquid crystals may one day be used to provide a thin screen television display. These systems, requiring very low voltage and power, are desirable for use with integrated circuits in complex addressing schemes that would replace the conventional kinescope.

This book covers over 120 processes as described in the U.S. patent literature. Most of the products marketed today as well as the many future applications for liquid crystal compositions and devices are described in detail.

CHOLESTERIC COMPOUNDS

COMPOSITIONS

Nematic and Potentially Cholesteric Materials Combination

F.D. Saeva and R.L. Schank; U.S. Patent 3,819,531; June 25, 1974; assigned to Xerox Corporation describe liquid crystalline compositions having the optical properties of the cholesteric mesophase. The compositions comprise at least one nematic liquid crystalline material and at least one potentially cholesteric material which has a molecular structure which is similar to that of a nematic liquid crystal.

Cholesteric liquid crystals have the property that when the propagation direction of plane polarized or unpolarized light is along the helical axis thereof, i.e., when the light enters in a direction perpendicular to the long axes of the molecules, (neglecting absorption considerations), this light is essentially unaffected in transmission through thin films of such liquid crystals except for a wavelength band centered about some wavelength λ_0 where $\lambda_0 = 2np$ with n representing the index of refraction of the liquid crystal substance and p the pitch or repetition distance of the helical structure.

The bandwidth $\Delta\lambda_0$ of this wavelength band centered about λ_0 will typically be of the order of about $\lambda_0/14$. For light of a wavelength λ_0, the cholesteric liquid crystal, under these conditions, exhibits selective reflection of the light such that approximately 50% of the light is reflected and approximately 50% is transmitted, assuming negligible absorption which is usually the case, with both the reflected and transmitted beams being approximately circularly polarized in opposite directions.

For light having wavelengths around λ_0 but not at λ_0 the same effect is present but not as pronounced. The transmitted light is not circularly polarized but is instead elliptically polarized. The cholesteric liquid crystals which exhibit this property of selective reflection of light in a region centered around some wavelength λ_0 are said to be in the Grandjean or "disturbed" texture. If λ_0 is in the visible region of the spectrum the liquid crystalline film appears to have the color corresponding to λ_0 and if λ_0 is outside the visible spectral region the film appears colorless.

Depending upon the intrinsic rotary sense of the helix, i.e., whether it is right-handed or left-handed, the light that is transmitted in the region about λ_0 is either right-hand circularly polarized light (RHCPL) of left-hand circularly polarized light (LHCPL). The transmitted light is circularly polarized with the same sense of polarization as that intrinsic to the helix. Thus, a cholesteric liquid crystal having an intrinsic helical structure which is left-handed in sense will transmit LHCPL and one having a helical structure which is right-handed in sense will transmit RHCPL.

These cholesteric liquid crystal substances will be identified, in order to conform with popular convention, by the kind of light which is reflected at λ_0. When a film is said to be right-handed, it is meant that it reflects RHCPL, and when a film is said to be left-handed, it is meant that it reflects LHCPL.

A right-handed cholesteric liquid crystal substance transmits LHCPL essentially completely at λ_0 whereas the same substance reflects almost completely RHCPL. Conversely a left-handed film is almost transparent to RHCPL at λ_0 and reflects LHCPL. Since plane polarized or unpolarized light contain equal amounts of RHCPL and LHCPL, a cholesteric liquid crystal film is approximately 50% transmitting at λ_0 for these sources when the liquid crystal is in its Grandjean texture.

A further unique optical property of optically negative liquid crystal films is that contrary to the normal situation when light is reflected, such as by a mirror, where the sense of the circular polarization of the reflected light is reversed, this same phenomenon does not occur with light reflected by these liquid crystal films. The sense of the circular polarization of light reflected from these liquid crystal substances is not reversed but rather remains the same as it was before it came into contact with the liquid crystal substance.

For example, if RHCPL having a wavelength λ_0 is directed at a right-hand film having $\lambda_0 = 2np$ it is substantially completely reflected and, after reflection, remains RHCPL. If the same light were to be directed on a mirror the reflected light would be LHCPL.

It has been found that considerable versatility can be achieved with respect to cholesteric liquid crystals by mixing together combinations of right-handed and left-handed cholesteric liquid crystals. In such a mixture there is a composition at which the right-handed and left-handed components nullify each other to provide an infinite pitch. This technique also makes it possible to generally achieve a broader range of pitches than typically can be achieved by mixing together only right-handed or only left-handed materials. See *Proc. ACS Symposium on Ordered Fluids and Liquid Crystals*, Sept. 1969, page 463.

Many left-handed cholesteric liquid crystal materials are known; however to date only relatively few right-handed materials have been provided. Thus there exists a continuing need for optically negative liquid crystalline compositions, particularly those which are right-handed.

The process provides liquid crystalline compositions having the optical properties of the cholesteric mesophase which comprise at least one nematic liquid crystalline material and at least one potentially cholesteric material having a structure similar to that of a nematic liquid crystalline material. Thus, according to the process it is possible to form liquid crystalline compositions having optically negative characteristics by combining two materials, neither of which itself forms a cholesteric mesophase.

These compositions are typically prepared by heating a mixture of the individual components to the isotropic state, mixing the components thoroughly at this point and then allowing the mixture to cool. The existence of the optically negative characteristics can be determined by conventional microscopic analysis which shows, for example, cholesteric textures, a pitch band and extrinsic optical activity. Another technique for determining the existence of optically negative characteristics is described in U.S. Patent 3,780,307.

Any suitable nematic liquid crystalline material may be used in the compositions. Typical suitable nematic liquid crystalline materials include: p-azoxy-anisole, p-azoxyphenetole, p-butoxybenzoic acid, p-methoxycinnamic acid, butyl-p-anisylidene-p'-aminocinnamate, anisylidene-p-aminophenylacetate and p-ethoxybenzylamino-α-methylcinnamic acid.

Any suitable potentially cholesteric optically active material may be used in the compositions. These materials typically are chiral as single molecules because of molecular asymmetry or dissymmetry (racemic or nonchiral materials typically will not produce the desired effect); should be soluble to at least some extent in the nematic liquid crystalline materials with which they are combined; and have a molecular structure which is rod-like, preferably one which posesses a relatively high degree of polarizability and conformational rigidity. Typical suitable potentially cholesteric optically active materials can be generally described as being represented by the general formulas

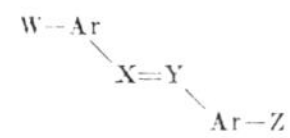

or W—Ar—Z or W—Z

where X and Y are chosen from the following radicals:

$$X = CH_2,\ Y = NH$$
$$X = Y = CH_2$$
$$X = Y = CH$$
$$X = CH,\ Y = N$$
$$X = Y = N$$
$$X = N,\ Y = N \rightarrow O$$
$$X = CH,\ Y = N \rightarrow O$$
$$(X = Y) = C = C$$

where W and Z can be the same or different radicals and can be a radical such as, alkenyl, alkynyl, alkoxy, halogen, alkyl, ester, nitro, cyano, amino, $(CH_3)_2N$, RCO_2, R—O—C, $RO(CH_2)_n$—O (where R = straight or branched alkyl), n is any integer and where Ar may be

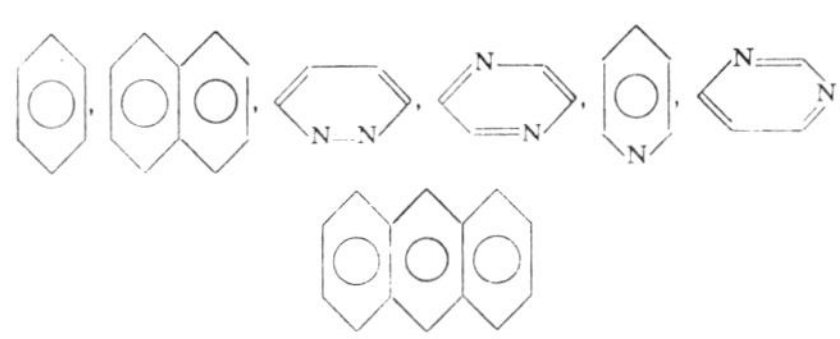

Example 1: In this and the following examples all parts and percentages are by weight unless otherwise specified. 0.01 mol of p-n-butylaniline and 0.01 mol of (S)-p-sec-amyloxy-benzaldehyde (previously prepared from the reaction of racemic (S)-sec-amyl alcohol with p-hydroxybenzaldehyde in the presence of base) are refluxed in 100 cc of absolute ethanol for about 1 hour after which the solvent is flash evaporated. The crude product is then distilled at 0.05 mm (204° to 205°C) to give a yellow isotropic liquid which is (S)-p-sec-amyloxy-benzaldehyde-p-n-butylaniline anil (1). Analysis of the compound shows 81.97% C, 9.13% H, and 4.16% N. $C_{22}H_{29}ON$ requires 81.68% C, 9.04% H and 4.33% N.

Compound 1 which does not possess any mesomorphic behavior is added to butyl (p-ethoxy-phenoxycarbonylphenyl) carbonate (BPC) to form a solution having a 10% concentration of Compound 1. The resulting mixture is heated at 100°C until it is completely isotropic and then mixed thoroughly while it is in this state.

The resulting supercooled mesomorphic material possesses the optical properties normally associated with the cholesteric mesophase, i.e., the existence of Grandjean and focal-conic cholesteric textures, a pitch band and extrinsic circular dichroism as determined by optical microscopy and by the use of a circular dichroism spectropolarimeter.

Example 2: Bis[(S)-sec-amyl] 4,4'-azoxycinnamate [Compound (2)] is prepared by the following technique. 0.10 mol of sec-amyl-4-nitrocinnamate [previously prepared from 4-nitrocinnamic acid chloride treated with (S)-sec-amyl alcohol in the presence of triethylamine] is suspended in 200 cc of water (the pH adjusted to 10 with ammonium hydroxide) along with 1.0 mol of sodium dithionite and stirred for about 2 hours at about 25°C. The crude hydroxylamine is collected by suction filtration and recrystallized from hexane.

0.05 mol of the hydroxylamine is then dissolved in 100 cc of absolute ethanol and the reaction mixture is refluxed for about 5 hours. The crude azoxy compound crystallizes from the reaction mixture on cooling and is collected by suction filtration and then recrystallized from absolute ethanol. Analysis of the compound shows 70.59% C, 7.04% H, 5.58% N. $C_{28}H_{34}N_2O_5$ requires 70.27% C, 7.16% H and 5.85% N.

Compound (2), which forms only a smectic mesophase between about 125° and 162.5°C, is added to 4,4-diheptyloxyazoxybenzene, a nematic material (Eastman Kodak) to make a solution having a 50% concentration of compound (2). The resulting mixture is heated at 170°C until it is completely isotropic and mixed thoroughly while it is in this state. The composition possesses the optical properties normally associated with cholesteric mesophases, i.e., cholesteric textures, a pitch band and extrinsic circular dichroism as determined by the techniques previously described.

The racemic form of Compound (1), described in Example 1, is added to 4,4-diheptyloxyazoxybenzene. The resulting composition does not possess the optical properties of the cholesteric mesophase.

Examples 3 through 10: The compositions formed in these examples possess the optical properties normally associated with the cholesteric mesophase, i.e., the existence of cholesteric textures, a pitch band and extrinsic circular dichroism as determined by the techniques previously described.

Example 3: A composition comprising about 5% p-(S)-sec-amyloxybenzylidene-p-butylaniline and about 95% BPC.

Example 4: A composition comprising about 5% (R)-2-octylethylcarbonate and about 95% BPC. The composition is found to be right-handed.

Example 5: A composition comprising about 5% of (S)-2-octylethylcarbonate and about 95% BPC. The composition is found to be left-handed.

Example 6: A composition comprising about 10% (S)-sec-amyl p-aminocinnamate and about 90% BPC. The composition is found to be right-handed.

Example 7: A composition comprising about 10% (R)-2-octyl p-aminobenzoate and about 90% BPC. The composition is found to be right-handed.

Example 8: A composition comprising about 10% (S)-2-octyl p-aminobenzoate and about 90% BPC. The composition is found to be left-handed.

Example 9: A composition comprising about 10% (S)-sec-amyl p-aminobenzoate and about 90% BPC. The composition is found to be right-handed.

Example 10: A composition comprising about 10% (S)-2-octyl p-aminocinnamate and about 90% BPC. The composition is found to be left-handed.

Multicomponent Cholesteric Mixtures

W.E. Woodmansee; U.S. Patent 3,441,513; April 29, 1969; assigned to The Boeing Company describes cholesteric liquid crystal materials exhibiting selective light scattering over characteristic temperature ranges having a first component providing color sensitivity at a relatively low temperature of cholesteryl oleate, a second component which narrows the temperature response to a small range of temperatures for the occurrence of the color phenomena selected from the group consisting of cholesteryl nonanoate, cholesteryl decanoate, cholesteryl octanoate, and cholesteryl para-nitrobenzoate; and a third component which adjusts the actual temperatures at which the color phenomena occur selected from the group consisting of cholesteryl acetate and cholesteryl propionate.

A fourth filler component providing increased visibility of the color patterns may be added selected from the group consisting of carbon, carbon black, black tempera powder, and finely divided metals and nonmetals. These cholesteric materials may be sandwiched between transparent plastic films for efficient handling and application.

Hysteretic Compositions

J.L. Fergason and N.N. Goldberg; U.S. Patent 3,529,156; September 15, 1970; assigned to Westinghouse Electric Corporation describe mixtures of compounds that possess a pronounced color change hysteresis after the transformation form the true liquid phase to the cholesteric phase. These hysteretic compositions may be employed as a thin film or layer over a dark background such as a pigmented plastic film to provide an information storing or recording device.

The film of the hysteretic cholesteric liquid crystalline composition will have a given color in the cholesteric phase but will turn colorless at a given clearing point, i.e., the temperature at which it has changed to the liquid phase. Upon cooling to a temperature below the clearing point, which may be the ambient temperature, the change to the cholesteric phase occurs but the return of color will be delayed for a significant period of time. During that time then, it constitutes a record of the information. As will become apparent hereinbelow, the clearing point of a basic hysteretic composition may be either raised or lowered by the addition of appropriate agents. The hysteretic color effect is also present after exposure to high intensity electrical fields and predetermined concentrations of certain vapors.

The hysteretic liquid crystalline cholesteric mixtures of this process contain (A) from about 15 to 40 percent by weight, of at least one halide selected from the group consisting of cholesteryl chloride, cholesteryl bromide, cholesteryl iodide, cholestanyl chloride, cholestanyl bromide and cholestanyl iodide and (B) at least about 30 percent by weight, of at least one compound selected from the group consisting of cholesteryl erucyl carbonate, cholestanyl erucyl carbonate, cholesteryl oleyl carbonate, cholestanyl oleyl carbonate, cholesteryl erucate, cholestanyl erucate, cholesteryl oleate and cholestanyl oleate.

Figure 1.1a (a cross-section of a device with a pigmented layer on one side of a plastic film and a thin layer of a hysteretic liquid crystal on the other side) shows a temperature recording or information storing device **10** in which the thin clear sheet of polyethylene terephthalate **11** has a thickness of 6 microns. The pigmented film **12** containing carbon black dispersed in a thin layer of polyvinyl alcohol has a thickness of 12 microns which is deposited on one side of the terephthalate film. Film **12** is applied as an aqueous solution or suspension and dried on the film **11**. A film of a hysteretic liquid crystalline cholesteric mixture **13** having a thickness of about 10 microns is deposited on the other side of the terephthalate base **11**. The film of a hysteretic liquid crystalline cholesteric mixture **13** is deposited by applying a 10% solution of the hysteretic mixture in petroleum ether onto the film **11** and evaporating the solvent.

In Figure 1.1b (a cross-sectional view of a body in contact with a recording device having a film of a hysteretic liquid crystalline composition irradiated with visible light and observable by a human eye), the recording device is placed so that the film **12** is in intimate contact with an aluminum plate or block **20**. The block **20** is adapted to be heated, as for example, by a resistance heater **21**, and is adapted to be cooled either by lower ambient temperatures after the heating has ceased or by cooling means such as the coil **22** containing circulating cold water. A thermocouple **23** is embedded in the plate **20** so that an accurate

FIGURE 1.1: HYSTERETIC CHOLESTERIC LIQUID CRYSTALLINE COMPOSITIONS AND DEVICES

a.

b.

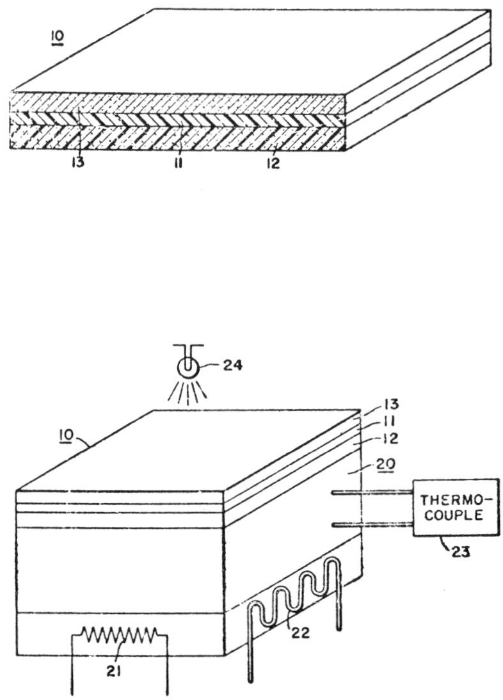

Source: J.L. Fergason and N.N. Goldberg; U.S. Patent 3,529,156; September 15, 1970

measurement of the temperatures of the block can be made. A source of white light **24** is provided so that the color of the layer of hysteretic film in the device will be visible to the naked eye. The visible color is, of course, a known phenomenon associated with the light scattering properties of the cholesteric phase. The following examples illustrate the process.

Examples 1 to 9: A series of recording devices was prepared, in the manner described above in conjunction with Figure 1.1a, using for film **13** a variety of simple admixtures or combinations of compounds in accordance with the constituent mixtures and composition ranges set forth above for the preparation of suitable hysteretic compositions. These simple devices were then placed in contact with the aluminum plate **20** of Figure 1.1b and heated until there was a transformation from the cholesteric phase to the liquid phase. The phase transformation was apparent as the layer was observed to change from a vivid color to a colorless layer.

When the layer was cooled, it reverted to the cholesteric phase but the color returned only after a significant delay in time. The compositions presented in Table 1 were found to be hysteretic and their room temperature colors, i.e., the color of the cholesteric phase, and their clearing points, i.e., the melt point or liquid transition point, were as indicated. In each case, at least a three minute time lag was observed in the color change, after the layer had been cooled to a temperature where the material was in the cholesteric phase. All of the percentages in the table below and those that appear thereafter are weight percentages unless otherwise specified.

TABLE 1: PROPERTIES OF HYSTERETIC MIXTURES

Example	Composition	Color of Cholesteric Phase	Clearing Point, °C.
1	30% cholesteryl chloride, 70% cholesteryl erucate	Red	37.5
2	20% cholesteryl chloride, 80% cholesteryl erucate	Green	38.2
3	15% cholesteryl chloride, 85% cholesteryl erucate	Blue	40
4	24% cholesteryl chloride, 76% cholesteryl oleyl carbonate	Green	45
5	20% cholesteryl chloride, 80% cholesteryl oleyl carbonate	Blue	45.4
6	30% cholesteryl chloride, 70% cholesteryl oleyl carbonate	Red	46
7	26% cholesteryl bromide, 74% cholesteryl oleyl carbonate	Green	43.8
8	26% cholesteryl bromide, 74% cholesteryl oleate	Green	41.1
9	26% cholesteryl chloride, 74% cholesteryl erucyl carbonate	Green	38.6

As is apparent from the data in Table 1, only a small variation in the clearing point occurs with varying amounts of compounds in a given combination of compounds. It has also been discovered, however, that significant changes in clearing points may be made by the addition of certain clearing point modulating agents. Up to 35% by weight, of (C) a clearing point elevating agent selected from the group of esters consisting of cholesteryl nonanoate, cholestanyl nonanoate, cholesteryl decanoate and cholestanyl decanoate may be included in the admixture of Compounds (A) and (B) to raise the clearing point temperature of the deposited cholesteric layer. The hysteretic effect is present in the layers even though the ester has been added to raise the clearing point.

However, the total amount of materials or compounds selected from (A) and (C) should not exceed 60 percent of the weight of the total mixture. As little as 1.0 percent of the esters in the group (C) will have noticeable and useful effect in elevating the clearing point temperature of a basic mixture of (A) and (B). In general, the addition of about 5 percent of any of the esters (C) will raise the clearing point about one degree centigrade.

Examples 10 to 13: Referring again to the device of Figure 1.1, mixtures containing the Compounds (A), (B) and the described esters (C) were employed to form the recording device **10** by depositing the mixture as a 10 micron thick layer **13** from a ten percent solution in petroleum ether onto one side of a 6 micron thick polyethylene terephthalate film **11**. The other side of the film **11** had the pigmented layer **12**.

Recording devices, with the compositions described in Table 2 providing the hysteretic liquid crystalline cholesteric layers, were tested in the apparatus of Figure 1.1b in the manner described in conjunction with the examples of Table 1. Again, at least a three minute time lag was observed in the color change after the layer was cooled below its clearing point. That the esters (C) have a pronounced and predictable effect in elevating the clearing point of a basic mixture of a hysteretic cholesteric composition of (A) and (B) is apparent from the data of Table 2.

In order to depress the clearing point of a given mixture of (A) and (B), up to about ten percent of another class of modulating agents (D), in this instance known as depressing agents, may be included in the mixture. Fatty compounds which dissolve the admixture (A) and (B) without crystallizing will function as clearing point depressing agents. Suitable fatty compounds for this purpose include the fatty acids, fatty esters and fatty alcohols, both saturated and unsaturated.

TABLE 2: ELEVATION OF CLEARING POINT

Example	Composition	Color of Cholesteric Phase	Clearing Point, °C.
10	28% cholesteryl chloride, 26% cholesteryl nonanoate, 46% cholesteryl oleyl carbonate	Red	51
11	28% cholesteryl chloride, 10% cholesteryl nonanoate, 62% cholesteryl oleyl carbonate	Red	44
12	28% cholesteryl chloride, 26% cholesteryl decanoate, 46% cholesteryl oleyl carbonate	Red	51
13	28% cholesteryl chloride, 10% cholesteryl decanoate, 62% cholesteryl oleyl carbonate	Red	44

Glassy Optically Negative Materials

L.B. Leder; U.S. Patent 3,789,225; January 29, 1974; assigned to Xerox Corporation describes a process for preparing various glassy optically negative liquid crystalline materials and a method for forming images on these materials. The glassy liquid crystals are relatively hard films which retain the optical properties of the normally softer liquid crystalline compositions. These liquid crystalline materials are imaged by flash exposure to a burst of imagewise activating radiation followed by relatively rapid cooling. The heat from the energy of the activating radiation typically causes a change in the pitch of the liquid crystal in the exposed areas of the film and the new pitch is retained by the rapid cooling.

The process of preparing glassy cholesteric liquid crystal materials comprises, generally speaking, heating the cholesteric liquid crystalline substance above its isotropic transition temperature, cooling the substance to some critical temperature within the cholesteric mesomorphic range, disturbing the substance while it is at this critical temperature and subsequently cooling the substance at a relatively rapid rate until it obtains a glassy, or substantially solid state.

Relatively thin films of these glassy cholesteric liquid crystalline materials are imaged by flash exposure to a very short burst of activating radiation in imagewise fashion. Typically, the pitch of the liquid crystalline material in the exposed areas of the film is caused to change and the new pitch is retained by relatively rapid cooling of the liquid crystal film. The imaging method provides a tendency to form clear images on a colored background.

Figure 1.2a shows flow chart showing the individual steps of this process; Figure 1.2b is schematic illustration of the practice of an imagining method of this process; and Figure 1.2c shows the absorption edges of the absorption spectra of a typical cholesteric liquid crystalline substance and other materials.

In the process shown in Figure 1.2a the cholesteric liquid crystalline substance is initially heated to a temperature above the isotropic transition temperature of the substance, i.e., the point at which the substance becomes a true isotropic liquid. This heating step may be carried out by any of various methods. Typically, a thin film or a particulate layer of the liquid crystalline substance is spread on some suitable substrate, e.g., a glass slide, which is then placed on a hot plate.

The liquid substance is then allowed to cool to some critical temperature within its mesomorphic temperature range, i.e., the temperature range over which the substance possesses the cholesteric liquid crystalline structure. This is typically accomplished by removing the source of heat used in the initial process step and allowing the substance to stand in air. Generally a time period of up to about 2 to 5 minutes is sufficient to cool the substance to the desired temperature.

It will be recognized that the critical temperature may be any which is within the temperature range in which the substance is in the cholesteric mesomorphic phase. Thus, the critical temperature may be chosen with respect to some desired color since, for normal viewing, the cholesteric liquid crystal substance exhibits various colors as the temperature of the substance changes. This is so because, as is well-known, λ for any cholesteric liquid crystalline substance, i.e., the center wavelength of the wavelength band reflected by the substance, is equal to 2np (where n is the refractive index of the substance and p is the pitch) and the pitch is responsive to and changes with the change in temperature.

Of course, the light reflected by the cholesteric liquid crystalline film will only be visible to the eye when λ is in the visible region of the spectrum. Where the λ value of the substance is not in the visible region the film will typically appear to be colorless. It will be obvious to those skilled in the art that the critical temperature to which the substance is cooled will vary with different cholesteric liquid crystalline substances since different substances are in the cholesteric mesomorphic state at varying temperatures.

When the cholesteric liquid crystalline substance is at the desired critical temperature it is then "disturbed." By disturbing the liquid crystal is meant exerting some external stimulus thereto, which may be a mechanical or electrical force, in order to change the substance from the focal-conic or "undisturbed" texture to the Grandjean or "disturbed" texture. Typically, most cholesteric liquid crystalline substances will cool to the focal-conic state unless disturbed by some means.

FIGURE 1.2: GLASSY LIQUID CRYSTALS

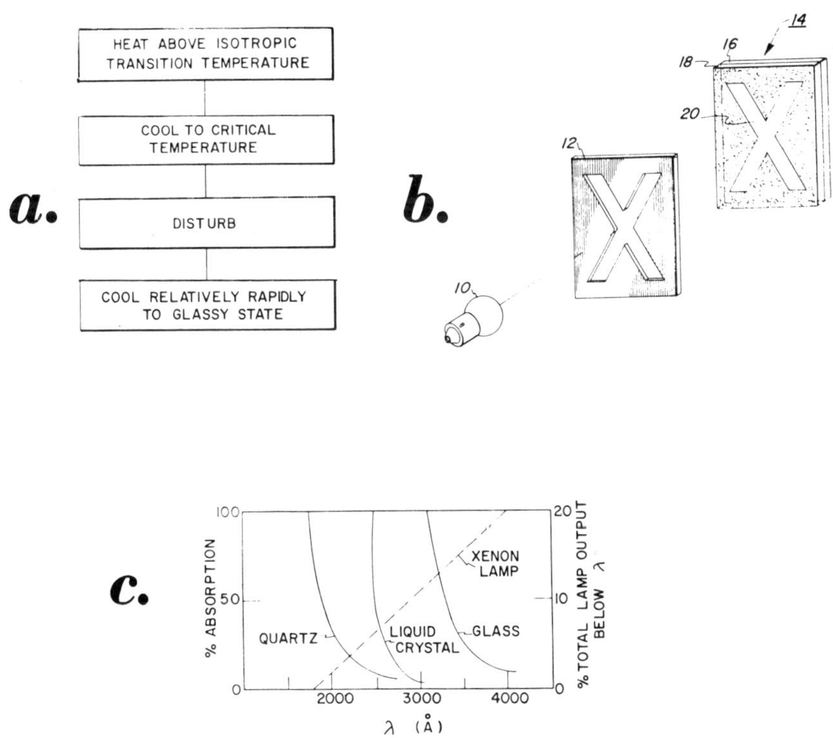

Source: L.B. Leder; U.S. Patent 3,789,225; January 29, 1974

The focal-conic texture state is predominantly characterized by its highly diffuse light scattering appearance caused by a distribution of small, birefringent domains. This texture has no single optic axis. The Grandjean texture is typically characterized by the selective reflection of incident light, at normal incidence, around a central wavelength λ where $\lambda = 2np$ as has been previously discussed. The Grandjean state is additionally characterized by optical activity for wavelengths of light away from λ, i.e., the plane of polarization of these wavelengths is rotated by some angle as they are transmitted. Where λ is in the visible spectrum the liquid crystalline composition film appears to have the color corresponding to λ. The Grandjean texture is sometimes referred to as the disturbed texture state.

Generally, therefore, the liquid crystalline substance may be disturbed by any means suitable for changing the substance from the focal-conic texture state to the Grandjean texture state including those which are presently known as well as those which may be discovered in the future. Known suitable methods by which the liquid crystalline substance can be disturbed include applying a shearing force to the surface of the liquid crystal film such as by sliding a glass plate across the surface thereof, mechanical vibration, high frequency ultrasonic vibration and, for appropriate cholesteric substances, applying an electric field thereto.

By disturbing the cholesteric liquid crystalline substance at the critical temperature it is theorized that the optical properties unique to the cholesteric state are thus retained in the substance when it is subsequently cooled. It is essential that the liquid crystalline substance be disturbed at the critical temperature for if this process step is carried out before the substance obtains the cholesteric mesomorphic state the desired optical properties will not be incorporated into the final product, and if done at a temperature below the cholesteric mesomorphic temperature range the substance will have crystallized and have lost its liquid crystalline properties.

The liquid crystalline substance is then cooled relatively rapidly until it adopts the glassy, or hard, form. This typically is accomplished by placing the substrate carrying the liquid crystal film on a cool surface and allowing it to stand in air until it becomes glassy. Of course, it will be obvious that the heating-disturbing-cooling cycle can also be carried out mechanically by utilizing temperature-controlled chambers and automatic disturbance devices.

The glassy cholesteric liquid crystalline materials made according to the process are stable at room temperature over long periods of time, e.g., up to a year or more and are not substantially affected by the presence of foreign stimuli. Thin films of these glassy liquid crystals can be utilized as imaging members.

Further these materials can be reprocessed according to the process described above thus providing reusable materials. Of course if the glassy liquid crystals are heated and subsequently allowed to cool to room temperature without being disturbed at a critical temperature they will crystallize and no longer possess any liquid crystalline properties.

The glassy cholesteric liquid crystalline films of the process can be utilized as imaging members according to the flash imaging method of the process. Figure 1.2b illustrates a typical example of the practice of the flash imaging method for forming images on glassy cholesteric liquid crystals. Referring to Figure 1.2b there is seen a high energy power source **10**, information-bearing member **12** and an imaging member, generally designated **14**, comprising substrate **16** and glassy cholesteric liquid crystalline film **18**.

According to the flash imaging method, imaging member **14** is exposed to an imagewise burst of activating radiation by directing activating radiation from energy source **10** through an information-bearing member **12**. An image **20** is thereby formed on the glassy liquid crystalline film **18** which typically will be a clear image on colored background areas. Alternately, by appropriate energy control an image of one color can be achieved on a background of the original color of the film.

The glassy cholesteric liquid crystalline film **18**, typically from about 1 to 10 mils in thickness, which can be any prepared by the process described above, is preferably arranged on a substrate **16** which may be any suitable material, transparent or opaque. The substrate functions as a support member for the glassy liquid crystal film and thus may be comprised of any suitable material having the necessary physical properties required for this purpose, such as for example, glass plates, metal sheets such as aluminum or the like, etc. Where the substrate is transparent to activating radiation exposure may be carried out from either side of the imaging member; of course, where the substrate material is opaque then exposure would have to be made from the liquid crystal film side of the imaging member as is illustrated.

When glassy cholesteric liquid crystalline films of the type described above are normally heated they do not cool back to the original highly colored state but instead change appearance due to crystallization. Crystallization results because the cooling times involved are sufficiently long to permit this process to occur.

However, when flash imaging of such glassy films is carried out by any means whereby the exposed areas of the films are heated very quickly the pitch of the liquid crystalline material in the exposed areas is changed and the relatively rapid cooling retains the new pitch (and/or color). The changes in the exposed areas are typically exhibited as a change in reflectivity or transmission or a change in color. Typically, the pitch of the liquid crystalline material in the exposed areas becomes smaller thus causing λ to be shifted toward the ultraviolet region of the electromagnetic spectrum. Therefore, the flash imaging method has a tendency to form relatively clear images on a colored background.

Alternatively, if sufficient energy is directed upon the film, the exposed areas thereof could be heated above the isotropic transition temperature of the material. Typically, these areas would crystallize upon cooling thereby losing their liquid crystalline characteristics. Thus, this procedure would also form clear images on a colored background.

Any suitable energy source which is capable of heating the cholesteric liquid crystalline material quickly may be utilized in the flash imaging method. Typical suitable energy sources include, for example, a Xenon flash lamp and other gas-filled flash lamps which provide ultraviolet wavelengths which can be absorbed by the cholesteric liquid crystalline materials and thus cause rapid heating thereof.

It has been found that for typical cholesteric liquid crystalline substances there is essentially no absorption of light at wavelengths longer than about 2500 A. However, at 2550 A and below there is strong absorption of light. Since a quartz envelope Xenon flash lamp has emission below 2500 A this component of the light is essentially totally absorbed by the glassy cholesteric liquid crystal film thereby typically causing a change in the pitch of liquid crystal material as well as a general smoothening of the film surface due to uniform heating and cooling.

That ultraviolet absorption causes the typical pitch change in the glassy cholesteric liquid crystal film according to the imaging method has been demonstrated experimentally by alternately placing a quartz plate and a glass plate over the film while exposing it to a burst of energy from a Xenon flash lamp. The absorption edges for glass and quartz relative to that of a typical cholesteric liquid crystal material, in this case, cholesteryl cinnamate, are shown in Figure 1.2c. Thus, it can be seen that the glass plate nullifies the imaging effect described above since it absorbs the ultraviolet radiation while the quartz plate does not alter the effect since it does not substantially absorb radiation of wavelengths which the liquid crystal substance absorbs.

It has been found that a burst of energy of approximately 1 joule/cm^2 in about 50 to 100 microseconds from a Xenon flash lamp is adequate to heat a glassy liquid crystal film sufficiently to cause imaging to occur. The distance which the energy source must be positioned from the film for imaging to occur will, of course, be dependent upon the source energy. Of course, it will be recognized that only a small portion of the energy from a Xenon flash lamp is activating, specifically that portion which is typically of a wavelength of about 2500 A or less. There is also seen in Figure 1.2c a plot of the total Xenon flash lamp output which is below the particular wavelengths indicated.

In a preferred form of the flash imaging method the glassy liquid crystalline film **18** has carbon black particles incorporated therein. Since carbon black absorbs radiation over a wide range of wavelengths, as has been discussed previously, the incorporation of such particles in the film increases the efficiency of the imaging method and further provides a wider latitude of choice with respect to the energy sources which can be employed.

The carbon black particles also provide a more contrasting image with respect to the background areas. Thus where carbon black is present in the liquid crystal film an energy source other than one which provides ultraviolet wavelengths can be used in the practice of the flash imaging method since the carbon black particles will absorb the wavelengths of radiation outside of the ultraviolet region and transfer heat to the liquid crystal thereby causing imaging to be effected.

Moreover where an energy source which provides an output containing wavelengths of radiation within the ultraviolet region as well as outside of this region, e.g., a Xenon flash lamp, is employed to effect imaging the presence of the carbon black in the liquid crystal film will result in a more efficient method since substantially all of the output from the energy source will be utilized. The following examples illustrate the process. All parts and percentages recited in the examples are by weight unless otherwise indicated.

Example 1: A thin film, approximately 1 to 2 mils thick, of a dicholesteryl carbonate (Eastman Organic Chemicals) is spread on a glass slide. The glass slide is then placed on a hot plate and heated until the temperature of the dicholesteryl carbonate is above its isotropic transition temperature which is about 220°C. At this point the material has a clear water-like appearance.

The slide is then removed from the hot plate and allowed to cool in air until the temperature of the dicholesteryl carbonate is within its cholesteric mesomorphic temperature range which is from about 175° to 220°C. When the material is in this temperature range it becomes cloudy in appearance. The dicholesteryl carbonate is then disturbed by applying a shearing force to the surface of the film with a second glass plate. The glass plate carrying the dicholesteryl chloride film is then removed to a relatively cool surface and allowed to cool to room temperature. A hard glassy dicholesteryl carbonate film which has a clear colorless appearance is obtained.

Example 2: A thin film of cholesteryl cinnamate (Eastman Organic Chemicals) is spread on a glass slide. The glass slide is then placed on a hot plate and heated until the temperature of the cholesteryl cinnamate is above its isotropic transition temperature which is about 214°C.

The slide is then removed from the hot plate and cooled in air until the cholesteryl cinnamate film reflects a green color. At this point the cholesteryl cinnamate is disturbed by applying a shearing force to the surface of the film with a second glass plate. The slide carrying the cholesteryl cinnamate film is then placed on a hot plate adjusted to a temperature less than the isotropic transition temperature but greater than the crystallization point which is about 141°C and allowed to come to equilibrium with the second temperature which is about 150°C. The plate is then removed to a cool surface and the cholesteryl cinnamate film is cooled relatively rapidly. A hard glassy film is obtained.

Example 3: The procedure described in Example 1 is repeated using cholesteryl propynyl carbonate (Eastman Organic Chemicals). Cholesteryl propynyl carbonate has a crystallization point of about 59°C and an isotropic transition temperature of about 90°C. Vitrification occurs in about 2 hours after the material has cooled to room temperature.

Example 4: The procedure described in Example 1 is repeated using p-nitrocholesteryl benzoate (Pfaltz & Bauer). This substance has a cholesteric mesomorphic temperature range of from 191° to 265°C. A similar result is obtained.

Addition of Phototropic Material

According to a process described by *N.N. Goldberg and J.L. Fergason; U.S. Patents 3,627,699; December 14, 1971 and 3,720,658; March 13, 1973; assigned to Westinghouse Electric Corporation* electromagnetic radiation in the frequency range of $10^{12.5}$ to 10^{17} cycles per second is detected, using a cholesteric liquid-crystal material to which there has been added, in the case of radiation of lower frequency, a suitable oil or oil-soluble dye, and in the case of radiation of higher frequency, a phototropic material such as beta-carotene or cholesteryl p-phenylazophenyl carbonate. The following examples illustrate the process.

Example 1: To a piece of polyethylene terephthalate film about 0.00025 inch thick, there is applied on one side a coating of black spray enamel. Then, on the other side of the film there is applied a liquid made by mixing into 500 milliliters of petroleum ether the following materials: 5 grams of Brilliant Green dye, 10 grams of oleyl cholesteryl carbonate, 10 grams of cholesteryl benzoate, and 25 grams of cholesteryl nonanoate. This is permitted to dry, to leave a stratum about 20 to 30 microns thick, on the one side of the film and at a temperature of about 50°C this film appears green. Maintained at such temperature, with electromagnetic radiation of infrared frequency of sufficient amplitude incident thereon, the film turned bluish-green.

Example 2: To a piece of polyethylene terephthalate film about 0.00025 inch thick, there is applied on one side a coating of black spray enamel. Then, to the other side of the film, there is applied a liquid made by mixing into 500 milliliters of petroleum ether the following materials: 5 grams of methyl orange dye, 10 grams of oleyl cholesteryl carbonate, ten grams of cholesteryl benzoate, and 25 grams of cholesteryl nonanoate. This is permitted to dry, to leave a stratum about 20 to 30 microns thick on the one side of the film, and at a temperature of about 20°C this film appears orange. Maintained at about such temperature, with electromagnetic radiation of infrared frequency of sufficient amplitude incident thereon, the film turns yellow or brown.

Example 3: To a piece of polyethylene terephthalate film about 0.00025 inch thick, there is applied on one side a coating of black spray enamel. Then, to the other side of the film there is applied a liquid made by mixing into 500 milliliters of a mixture of petroleum ether in chloroform the following materials: 30 grams of oleyl cholesteryl carbonate, 20 grams of cholestryl nonanoate, and 1 gram of cholesteryl paraphenylazophenyl carbonate. (Cholesteryl paraphenylazophenyl carbonate is a compound, prepared as indicated below in Example 4.) The liquid thus applied to the other side of the polyethylene terephthalate film is permitted to dry, leaving a stratum about 0.25 micron thick on the one side of the film.

After the solvent evaporates, a further layer of polyethylene terephthalate is placed over the liquid crystal material, in vacuum. The film thus prepared is exposed to ultraviolet radiation, using a filtered AH-4 mercury photoflood. Ordinary negatives may be used to place an image on the films for evaluation of gray scale and resolution. It is found that approximately 40 shades of discrimination are available, using this technique, and the spatial resolution is in excess of 20 lines per millimeter. The effects of the ultraviolet radiation on the liquid-crystal material are reversible. After being heated to about 150°C for thirty seconds, or being allowed to remain at 30°C for 8 hours, the film appeared as if it had not been exposed.

Example 4: Cholesteryl paraphenylazophenyl carbonate is prepared as follows. To a flask, there are added, dissolved in 75 milliliters of benzene, 22.5 grams (0.05 mol) of cholesteryl chloroformate. There are added 3.9 grams (0.05 mol) pyridine dissolved in 5 milliliters of benzene. Then, to the flask, there is added a suspension of 10.0 grams of paraphenylazophenol (0.05 mol) in benzene.

The materials in the flask are then heated to a reflux temperature and retained at the reflux temperature for 2 hours with stirring. The resultant reaction mixture was cooled and filtered to remove the precipitated pyridine hydrochloride. To the filtrate, methanol is added, causing the product, cholesteryl paraphenylazophenyl carbonate, to be precipitated from the solution. This precipitate is then recrystallized from the benzene-methanol solution in a yield of better than 90%, based upon the quantity of cholesteryl chloroformate charged. It exhibits a melting point range of 166° to 167.5°C and gives a bright, vivid color play at a temperature slightly above that.

Example 5: Example 4 was repeated, except that twice as much of the carbonate was used. The film thus obtained was slightly more sensitive to ultraviolet radiation, but the results are otherwise the same.

Cholesteryl Erucyl Carbonate as Stabilizer

N.N. Goldberg and J.L. Fergason; U.S. Patent 3,580,864; May 25, 1971; assigned to Westinghouse Electric Corporation have found that the stability of cholesteric-phase liquid-crystal compositions is improved by incorporating therein an amount, up to 100%, of cholesteryl erucyl carbonate, this compound is made by reaction of cholesteryl chloroformate with erucyl alcohol and pyridine. The carbonate does not crystallize to true solid in several months, whereas most other liquid-crystal compositions transform to true solid within several hours to several days. Moreover, the carbonate does itself exhibit a color play, so that it may be used in very large amounts in cholesteric-phase liquid-crystal compositions, without decreasing the vividness of the color changes exhibited thereby.

In the process, cholesteryl chloroformate, either as a reagent or formed in situ by reacting phosgene with cholesterol, is reacted with pyridine and erucyl alcohol, with pyridine hydrochloride being precipitated and removed by filtration. The carbonate is then isolated from the filtrate. The chief chemical reaction involved is that of the following equation:

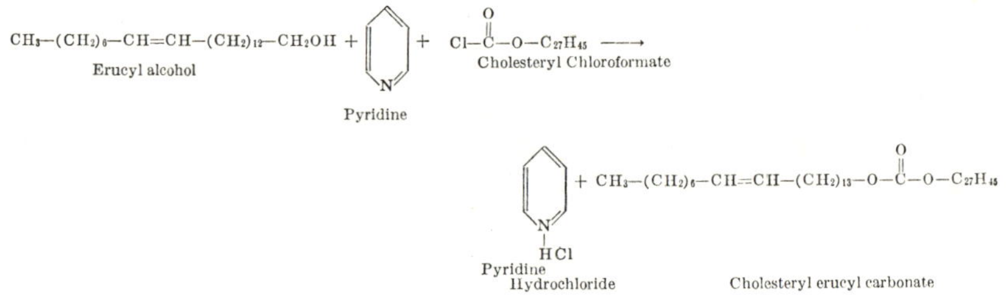

Examples 1 through 11: A three-necked glass flask having ground glass joints was fitted with a reflux condenser, a stirrer and an addition funnel. The flask is charged with 0.75 mol (336.9 grams) of cholesteryl chloroformate and 450 ml of benzene. Heat was applied until a complete solution was effected. In a separate flask, 0.76 mol (246.7 grams) of erucyl alcohol was added to 300 milliliters of benzene, and heat was applied until complete solution was effected. Through the addition funnel, the erucyl alcohol solution was then added to the cholesteryl chloroformate solution. Then, a solution of 0.75 mol (60.7 ml) of pyridine dissolved in 75 milliliters of benzene was added dropwise, through the addition funnel, to the contents of the flask. An exothermic reaction resulted, and a white solid was formed. After complete addition of pyridine, the contents of the flask were heated to and maintained at a reflux temperature for 1.0 hour.

The solid formed by the reaction, namely, pyridine hydrochloride, was vacuum-filtered by means of a Buchner funnel. The filtrate (a benzene solution containing the desired product) was treated with methanol, and an oily product came out of the solution. By means of a separatory funnel the oily (lower) layer was separated from the benzene-methanol layer. As prepared, the cholesteryl erucyl carbonate

$$(C_{27}H_{45}{-}O{-}\overset{\overset{\displaystyle O}{\|}}{C}{-}O{-}C_{22}H_{43})$$

exhibited a color-play temperature range of about 25.5° to 26.5°C, being green at about 26°C. After repeated washings with water, the color-play temperature range of the material was raised to 33.5° to 34.5°C. The product oil exhibited no tendency to crystallize to a true solid, even after prolonged periods of storage in a refrigerator. Under refrigeration, the material became grease-like in consistency. When used by itself, the compound exhibits a color-play within a narrow temperature range.

The carbonate is used by incorporating it, in amounts up to 100%, in cholesteric-phase liquid-crystal compositions. It may be mixed readily with any of a great number of compounds that are known to exhibit cholesteric-phase liquid crystals at temperatures about 20°C. Reference may be had to British Patent 1,041,490, lines 5-102 of page 4 for a comprehensive list of compounds suitable for use in making such compositions.

The carbonate is effective when used in amounts as small as about 5 weight percent of the stabilized composition, and it may be used in greater amounts, up to 100% by weight. In most instances, it will be desirable to use the compound by itself, or with relatively small amounts, up to about 30%, of other materials, but the possibility of using it in small amounts, such as 5 to 20 weight percent, simply to stabilize a composition whose color-play temperature range is in large part determined by the other materials present, is not to be overlooked. The following table presents specific examples that illustrate the use of the carbonate in cholesteric-phase liquid-crystal compositions:

Example	Cholesteryl erucyl carbonate, wt. percent	Other components, wt. percent	Color play temp. range, °C.
2	97.5	2.5 cholesteryl nonanoate [1]	27.1-28.2
3	95.0	5.0 cholesteryl nonanoate	27.6-28.6
4	95.0	5.0 cholesteryl crotonate [2]	26.6-27.9
5	85.0	5.0 cholesteryl nonanoate, 10.0 cholesteryl phenyl carbonate.[3]	23.5-25.2
6	75.0	15.0 cholesteryl phenyl carbonate, 10.0 cholesteryl nonanoate.	19-21.2
7	70.0	15.0 cholesteryl phenyl carbonate, 15.0 cholesteryl laurate.[4]	26-27.6
8	70.0	15.0 cholesteryl myristate,[5] 15.0 cholesteryl phenyl carbonate.	27.5-29.2
9	80.0	20.0 cholesteryl phenyl carbonate	18-19.9
10	85.0	15.0 cholesteryl phenyl carbonate	21.8-23.3
11	90.0	10.0 cholesteryl phenylethyl carbonate.[6]	19.5-21.2

[1] Melts at 79.3-80.0° C., clears at 91.8° C.
[2] Melts at 126-128° C., clears at 149° C.
[3] Melts at 124° C.
[4] Melts at 88° C.
[5] Melts at 83° C.
[6] Melts at 98-99.5° C.

The above compositions are made in the usual manner, namely, by mixing the named components in the amounts indicated with suitable solvent, such as benzene or petroleum ether, then forming the film by pouring the mixture onto a surface or substrate to permit the solvent to evaporate therefrom, leaving a stratum of about 1 to 50 microns thick. The compositions may also be encapsulated or enclosed in an appropriate translucent plastic capsule, the stratum being formed by a layer of the capsules. None of the compositions in the above table exhibited any true-solid formation within several months.

Cholesteryl p-Nonylphenyl Carbonate as Stabilizer

N.N. Goldberg; U.S. Patent 3,580,865; May 25, 1971; assigned to Westinghouse Electric Corporation has found that the stability of cholesteric-phase liquid-crystal compositions is improved by incorporating therein cholesteryl p-nonylphenol carbonate. This compound is made by reaction of cholesteryl chloroformate with p-nonylphenol and pyridine. The p-nonylphenyl carbonate does not crystallize to true solid in three years, but the corresponding p-octylphenyl carbonate starts crystallizing to true solid in three weeks.

In the process, cholesteryl chloroformate, either as a reagent or formed in situ by reacting phosgene with cholesterol, is reacted with pyridine and p-nonylphenol, with pyridine hydrochloride being precipitated and removed by filtration. The carbonate is then isolated from the filtrate. The chief chemical reaction involved is that of the following equation:

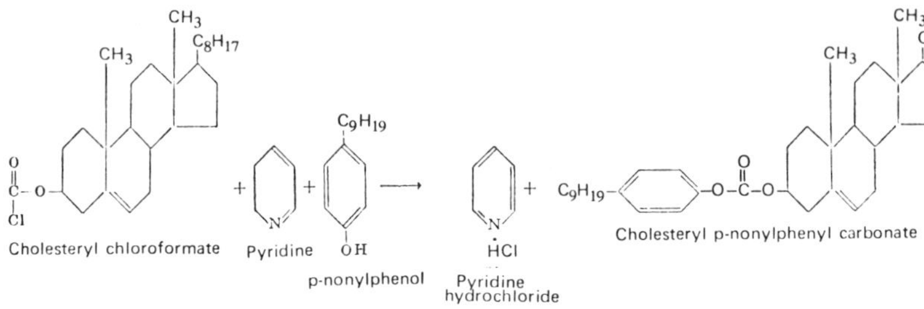

Examples 1 through 9: A three-necked glass flask having ground-glass joints is fitted with a reflux condenser, a stirrer, and an addition funnel. The flask is charged with one gram-mol of cholesteryl chloroformate (449.2 grams) dissolved in 800 milliliters of benzene. Through the addition funnel, there is slowly added with constant stirring a mixture of 1 gram-mol (220 grams) p-nonylphenol, 1 gram-mol (79.1 grams) pyridine, and 600 milliliters of benzene. After the addition is complete, the flask is heated to and maintained at reflux temperature for 1.0 hour. The reaction mixture is then cooled to room temperature.

During the addition and the subsequent heating, there is formed by the reaction of the equation cited above a solid, namely, pyridine hydrochloride, and it is removed from the reaction mixture by vacuum filtration. The filtrate comprises a benzene solution of the desired cholesteryl p-nonylphenyl carbonate, and the benzene may be removed from this solution with the use of heat and vacuum techniques to obtain, in good yield (about 95 to 98%) the desired carbonate, a very viscous pale yellow liquid.

It should be understood that chloroformates of other β-hydroxysterals, (other than cholesterol) may be substituted for the cholesteryl chloroformate of Example 1. For example stigmasteryl chloroformate may be substituted for the cholesteryl chloroformate.

The carbonate is used by incorporating it in cholesteric-phase liquid-crystal compositions. It may be mixed readily with any of a great number of compounds that are known to exhibit cholesteric-phase liquid crystals at temperatures above 45°C. Reference may be had to British Patent 1,041,490, lines 5-102 of page 4, for a comprehensive list of compounds suitable for use in making such compositions. The carbonate itself does not change color, and stabilized compositions containing great amounts of the carbonate accordingly are less vivid in their changes of color than compositions stabilized with less of the carbonate. The p-nonylphenyl carbonate is far more effective as a stabilizer against formation of true solid than its adjacent homolog, cholesteryl p-octylphenyl carbonate. The octylphenyl carbonate begins to crystallize to a true solid after three weeks, but the nonylphenyl carbonate shows no sign of undergoing such crystallization after three years. The following table illustrates the use of the carbonate in cholesteric-phase liquid-crystals.

	Percent					
Example	Cholesteryl p-Nonylphenyl Carbonate	Cholesteryl 2,4-Dichloro-benzoate	Cholesteryl 3,4-Dichloro-benzoate	Cholesteryl Nonanoate	Cholesteryl p-n-Butoxy-phenyl Carbonate	Color-Play Temperature Range (°C)
2	20	20	--	40	20	42-45
3	20	---	17.5	42.5	20	50-52
4	20	--	20	30	30	60-63
5	20	--	20	20	40	68-70
6	20	--	30	20	30	68-71
7	20	--	20	10	50	81-83
8	20	--	--	--	70	90-91
9	20	--	20	--	60	94-95

Cholesteryl 2,4-dichlorobenzoate melts at 128.5° to 129.5° and clears at 202°C. Cholesteryl 3,4-dichlorobenzoate melts at 150° to 157°C and clears at 185°C. Cholesteryl nonanoate melts at 79.3° to 80.0°C and clears at 180°C. Cholesteryl p-n-butoxyphenyl carbonate melts at 113° to 115°C and clears at 160°C.

The above compositions are made in the usual manner, namely, by mixing the named components in the amounts indicated with a suitable solvent, such as benzene or petroleum ether, and then forming a film by pouring the mixture onto a surface to permit the solvent to evaporate therefrom, leaving a stratum about 1 to 50 microns thick. The compositions may also be encapsulated in plastic capsules and the stratum may be formed from the capsules. None of the compositions in the above table exhibited any true-solid formation within a year, but control compositions, otherwise the same but with the carbonate omitted, exhibited true-solid formation within several hours to several days.

Extended Mesomorphic Ranges with Low Temperature Stability

W.E.L. Haas, J.B. Flannery, Jr., B. Mechlowitz and J.E. Adams; U.S. Patent 3,779,751; December 18, 1973; assigned to Xerox Corporation describe optically negative liquid crystalline compositions which are formed by mixing together cholesteric liquid crystalline materials with particular combinations of liquid crystalline compounds having a general formula

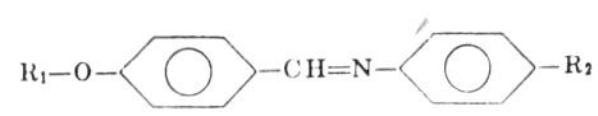

and which have a nematic mesophase which is stable at relatively low temperatures. Liquid crystalline compounds which belong to the class of compounds having the general formula

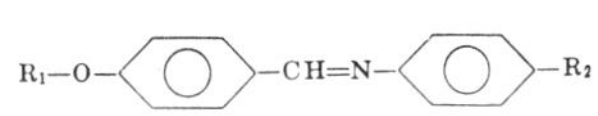

and where R_1 and R_2 are particular combinations of alkyl groups are listed in the following table.

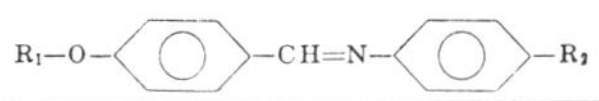

R_1	R_2	Smectic (° C.)	Nematic (° C.)	Name
C_3H_7...	C_4H_9	....	39-59	p-n-Propoxybenzylidene-p'-n-butylaniline.
C_4H_9...	C_4H_9	7-46	46-75	p-n-Butoxybenzylidene-p'-n-butylaniline.
C_5H_{11}...	C_4H_9	24-54	54-71	p-n-Pentyloxybenzylidene-p'-butylaniline.
C_6H_{13}...	C_4H_9	25-70	70-78	p-n-Hexyloxybenzylidene-p'-n-butylaniline.
C_7H_{15}...	C_4H_9	29-76	76-77.5	p-n-Heptyloxybenzylidene-p'-n-butylaniline.
C_4H_9...	CH_3	*(49)—66	66-68	p-n-Butoxybenzylidene-p'-toluidine.
C_4H_9...	C_2H_5	42-52	52-67	p-n-Butoxybenzylidene-p'-ethylaniline.
C_5H_{11}...	CH_3	35-45	45-64	p-n-Pentyloxybenzylidene-p'-toluidine

*Denotes monotropic transition.

It has been found that by utilizing some of the compounds listed in the above table and other members of the same generic class it is possible to form compositions which have significantly larger nematic mesomorphic temperature ranges than those of the individual compounds. These nematic liquid crystalline compositions are described in U.S. Patent 3,816,113. The liquid crystalline compositions described in the patent are: mixtures of p-methoxybenzylidene-p'-n-butylaniline ($R_1 = CH_3$ and $R_2 = C_4H_9$) (referred to as ABUTA for simplicity) and p-n-butoxybenzylidene-p'-n-butylaniline (BOBUTA); mixtures of p-ethoxybenzylidene-p'-n-butylaniline ($R_1 = C_2H_5$ and $R_2 = C_4H_9$) (EOBUTA) and ABUTA; mixtures of EOBUTA and BOBUTA; mixtures of p-n-butoxybenzylidene p'-ethylaniline (BOETHA) and ABUTA; mixtures of BOETHA and EOBUTA; mixtures of p-n-pentyloxybenzylidene-p'-n-butylaniline (PENTOBUTA) and ABUTA; and mixtures of p-n-hexyloxybenzylidene-p'-n-butylaniline (HEXOBUTA) and ABUTA. These compositions typically have nematic mesomorphic temperature ranges which are stable at room temperature or slightly below and in some cases significantly below, typically extending to below 0°C.

According to this process, it has been found that optically negative liquid crystalline compositions which have a cholesteric mesophase at, near or below room temperature can be formed by mixing together a suitable cholesteric liquid crystalline substance, or mixture of these substances, with the compositions listed above. In other words, each of the nematogenic compositions previously described can be mixed together with a sufficient amount of any suitable cholesteric liquid crystalline material, or combination thereof, to form liquid crystalline compositions which have optically negative characteristics. The optically negative compositions of the process have a cholesteric mesomorphic temperature range which is significantly larger than those of the individual components. The following examples illustrate the process. All parts and percentages listed are by weight unless otherwise specified.

All of the compositions listed in the following examples are prepared by placing the constituents in a vessel, heating above the isotropic transition temperatures of the constituents with mixing to ensure a homogeneous composition. The isotropic transition temperatures are determined by differential thermal analysis.

Example 1: A cholesteric liquid crystalline composition is prepared from 10% cholesteryl erucate, 54% ABUTA and 36% BOBUTA. The composition is cholesteric at room temperature and has an isotropic transition temperature of about 44°C.

Example 2: A cholesteric liquid crystalline composition is prepared from 20% cholesteryl erucate, 48% ABUTA and 32% BOBUTA. The composition is cholesteric at room temperature and has an isotropic transition temperature of about 44°C.

Example 3: A cholesteric liquid crystalline composition is prepared from 90% cholesteryl erucate, 6% ABUTA and 4% BOBUTA. The composition has an isotropic transition temperature of about 43°C.

Example 4: A cholesteric liquid crystalline composition is prepared from 10% cholesteryl erucate, 54% ABUTA and 36% EOBUTA. The composition is cholesteric at room temperature and has an isotropic transition temperature of about 46°C.

Ultraviolet Radiation Stabilizer

G.D. Dixon and L.C. Scala; U.S. Patent 3,656,909; April 18, 1972; assigned to Westinghouse Electric Corporation have found that damage to the color characteristics of cholesteric liquid crystals exposed to artificial or natural light is in great part due to the effect of ultraviolet irradiation, which could reduce the useful life of liquid crystal films as a detector element to less than 2 weeks. It has further been found that aromatic azo compounds Ar—N=N—Ar' and aromatic azoxy compounds

$$Ar-\underset{\downarrow \atop O}{N}=N-Ar'$$

having stable substitutions on one or both of the aryl (Ar or Ar') rings, when added to cholesteric liquid crystalline materials, materially decreased the deleterious effects caused by ultraviolet radiation. The azo and azoxy compounds useful in this process are those that while prolonging the useful life of liquid crystals, do not substantially alter their color response.

A variety of substituted aromatic azo and azoxy compounds were added to cholesteric liquid crystals to give sensor compositions. Among those found particularly suitable in increasing heat sensing life of the compositions were: 4-(phenylazo)phenol; 4-[(4'-ethoxyphenyl)azo]phenyl 10-undecenoate; 4-[(4'-ethoxy phenyl)azo]phenyl hexanoate and azoxydianisole.

The upper weight percent limit of the substituted aromatic azo and azoxy ultraviolet absorber stabilizers is limited by solubility in the cholesteric liquid crystalline materials to which they are added. The useful concentration range of stabilizer in the sensing composition (liquid crystal plus stabilizer) is between about 0.3 to 7 weight percent, with a preferred range between 1 to 6 weight percent, and an optimum range between 2 to 5 weight percent. Generally, at above about 6 weight percent stabilizer in the heat sensing compound (liquid crystal plus stabilizer), the stabilizer will crystallize out of the liquid crystal mixture. Generally, at below about 0.5 weight percent, the effect of the additive stabilizer in protecting the liquid crystal will not be very noticeable.

Degradation of the useful properties of cholesteric liquid crystals is due also to the influence of airborne particulate impurities. It was found that protecting a cholesteric liquid crystal material with certain clear colorless compatible polymeric materials increased the lifetime of the liquid crystal by preventing textural changes caused by atmospheric contaminants. Dust particles and such alighting on the surface of the liquid crystal without such protective packaging, would produce rippled areas of noncolor response surrounding the particles.

The protective packaging of the liquid crystals and ultraviolet absorber sensor compositions can also be accomplished by admixing the protective polymeric material in low molecular weight liquid form (average molecular weight of the polymer less than about 10,000). In this method the detector element will be cast or otherwise deposited on a substrate as a mixture containing: (1) a sensor composition of cholesteric liquid crystalline materials and generally an ultraviolet radiation stabilizer, in solution and (2) low molecular weight, liquid, chemically compatible, protecting polymeric material. Suitable low molecular weight, protecting polymeric materials would include silicone resins, polyisoprenes, polyvinyl chloride, polystyrene, polyacrylate, polycarbonate, polyesters and linear polyurethanes.

In all cases, the resulting cast detector element, whether in gel or solid form, will comprise a substantially homogeneous mixture of protective polymeric material (resin) and liquid crystalline material, with or without ultraviolet radiation absorbers. The protective polymeric material can comprise between about 1 to 20 weight percent of the resulting cast mixture after evaporation of solvents, with a preferred range of between about 3 to 10 weight percent. If below one weight percent of the evaporated substantially homogeneous mixture is the protective polymeric material, the protective effect is too small and if above twenty weight percent of the evaporated substantially homogeneous mixture is the protective polymeric material, the response colors are diluted.

ENCAPSULATION TECHNIQUES

Continuous Polymeric Matrix

A process described by *D. Churchill and J.V. Cartmell; U.S. Patents 3,734,597; May 22, 1973; 3,600,060; August 17, 1971; 3,578,844; May 18, 1971; assigned to The National Cash Register Company* is directed to an article of manufacture for utilizing an electric field to obtain an easily-discernible, stable, and optionally either comparatively permanent or readily heat-eraseable chromatic (color and/or reflectance, viz, reflective intensity) representation of comparable outline to the electric field.

Since the chromatic representation is stable, it can serve a storage (memory) function. One of the particularly advantageous features of the process is that the chromatic representation can be generated by a one-shot signal output, viz, an electronic computer output signal, a single sweep signal such as a radar scan, a display panel where information may be entered once for an extended period of storage such as an air terminal schedule display. Also, the chromatic representation in the color stable (storage) state is capable of representing a fixed body or state of information which can be retrieved and examined readily without requiring continuous electric field potential, e.g., continuous potential such as that obtained from an electron gun, e.g., cathode ray tube, using conventional electronic computer equipment.

The article comprises, in its essential components, a pair of spaced electrodes with a matrix-bound, droplet-containing cholesteric liquid crystalline member located in such a position that the electrodes impose an electric potential on the liquid crystal member. Usually the electrodes are positioned such that the first electrode (the electrode closer to the viewing agency, e.g., eyes of the viewer) and second electrode (the electrode more remote from the viewing agency, e.g., viewer's eyes) are in electric field-generating proximity to one another and the intermediate cholesteric liquid crystalline member can, but need not, be in contact with the first and/or second electrode. Both direct current and alternating current can be employed to produce the electric field.

The cholesteric liquid crystalline member can be composed of a single substantially continuous polymer matrix-bound, naked droplet or inclusion of cholesteric liquid crystalline material, but usually the member is composed of a plurality (and preferably a profusion) of individual, minute, naked droplets or inclusions of cholesteric liquid crystalline material.

Insofar as the large majority of droplets are concerned, the polymer matrix is continuous, that is to say that the internally located droplets are totally enveloped by a given portion of the matrix. Many of the droplets located close to the outer surface(s) of the matrix are likewise in a continuous matrix of polymer material. Some of the droplets are on the surface or a portion thereof has direct access to the open air through an opening(s) in the polymer matrix. The term "substantially continuous" accurately describes the polymer matrix which contains the vast majority of liquid crystal droplets. At least one of the materials in the cholesteric liquid crystalline phase must be chromatically responsive to an electric field of the requisite intensity.

Moreover, so long as the cholesteric liquid crystalline member is located intermediate (between) the electric field-imposing electrodes, the direction at which the field is applied to the member, viz, perpendicular or essentially parallel to the direction of viewing is immaterial.

While the first and second electrodes need not be (and usually are not) coextensive throughout their entire extent and the cholesteric liquid crystalline member need not be in direct contact with either or both electrodes; it is necessary for both the first and second electrodes to be close enough to one another and to the intermediate cholesteric liquid crystalline member to enable the formation of an electric field between the first and second electrodes.

For most purposes, the cholesteric liquid crystalline member contains a cholesteric liquid crystalline material which is chromatically responsive to an electric field of from about 10,000 to about 1,000,000 volts per centimeter of thickness of the cholesteric liquid crystalline droplet-containing member. By "chromatically responsive" is meant that the cholesteric liquid crystalline material (present in droplet form in the substantially continuous polymer matrix defining the cholesteric liquid crystalline member) must exhibit either an apparent change in color and/or reflective intensity (as viewed from incident white light) upon application of an electrical field of the above specified requisite intensity thereto.

Hence the term "chromatic response" is intended to include both changes (shift of wavelength) in color and changes in intensity of reflectance of the same color so that it appears different in color or shade of color from the previous color state(s). Thus, as used herein, a change in color is synonymous with a change in reflective intensity and vice versa.

In addition to protecting the droplets of cholesteric liquid crystal material, the substantially continuous polymeric matrix enables better adhesion between the cholesteric liquid crystal member and the first and second electrodes or comparable contiguous surfaces. Also, the presence of the polymeric matrix (which can be controlled to be of uniform and accurate thickness) serves to enhance the optical homogeneity of the chromatic representation (image) observed in the display device.

Various natural and synthetic film-forming polymeric materials can be employed to constitute the polymeric matrix, film or coating in which the individual droplets or inclusions of cholesteric liquid crystal material are located. Any transparent or substantially transparent film-forming polymeric material with adequate electrical insulation properties and which is soluble in a liquid which does not dissolve or substantially chemically affect the liquid crystal material adversely can be used.

Suitable representative film-forming polymeric materials for this purpose include the following: polyvinyl alcohol; gelatin, gum arabic, zein, a prolamine film former derived from the alcohol extraction of zea mays, a grain commonly called Indian corn; hydroxy ethyl cellulose; polyvinyl pyrrolidone; polyethylene oxide; copolymers of ethylene and maleic anhydride; copolymers of vinyl methyl ether and maleic anhydride; etc.

The cholesteric liquid crystals can be dispersed or positioned within the polymer matrix conveniently by emulsifying minute droplets of liquid crystal material in a dryable liquid solution of the film-forming polymeric material which is to constitute the polymer matrix. According to a preferred form of this process, the droplets of liquid crystal material are emulsified in an aqueous solution of film-forming polymer. Since an extremely small droplet size of cholesteric liquid crystal material can be maintained in an emulsion, coatings or films prepared therefrom allow a good optical resolution and have a smooth surface(s).

These characteristics enhance the optical or visual readout of display devices containing the droplets in the matrix. In general, individual droplet size can range from 0.5 to 50 microns, but usually the individual droplet size ranges from about 1 to 30 microns. Average droplet size can range from about 1 to 30 microns but usually ranges from about 5 to 20 microns. Films prepared by drying these emulsions, containing the minute individual liquid crystal droplets or inclusions, can be stained or tinted as desired to enable the polymer matrix to serve as a color filter for light traveling to and from the liquid crystal material.

Such a system can be used where a narrowing of the broad iridescent effect present in some liquid crystal materials in the second and third chromatic states is desired. Moreover, such films can also be pigmented slightly, but care should be exercised to avoid use of an excessive amount of pigment or other nonlight-reflecting material as it can diminish the color response and brilliance due to interference with incident and reflected light.

Comparative tests between polymer matrix-protected cholesteric liquid crystal material versus films of unprotected liquid crystal materials of identical composition reveal several important advantages for the films comprised of droplets of cholesteric liquid crystal material in a substantially continuous polymer matrix when employed in the articles of this process. One of the major advantages resides in the apparent prevention of crystallization of the liquid crystals and mixtures thereof, or at least a diminution in the tendency towards crystallization. Hence, many cholesteric liquid crystal materials and mixtures are normally solid at room temperature. These mixtures often supercool considerably so that crystallization is not immediate at ambient room temperatures.

However, they do crystallize within a day or a longer time period and must be reheated to the liquid crystal transition temperature range in order to be suitable for display, for example in chromatic image recognition use. The dispersion of these mixtures in the polymer matrix seems to inhibit crystallization, and the cholesteric liquid crystal materials when dispersed within the polymer matrix are observed to remain in the liquid crystal state for a longer period of time than "pure" films of the same materials (no polymer matrix).

Another advantage attainable in the articles of the process is the ability of the matrix-entrapped material to retain the third chromatic state for comparatively permanent periods of time versus the comparatively transitory retention of the third chromatic state in the case of unrestrained cholesteric liquid crystals.

A further and equally important practical advantage of use of the matrix-retained droplets of cholesteric liquid crystals is their ability to be used in multicomponent (multicolor) systems, viz, as coatings or incorporated in a matrix containing a plurality of different cholesteric liquid crystalline mixtures, each mixture giving a characteristic color throughout the three chromatic stages with the different mixtures being employed in close proximity so that a multicolored display is achieved in one, two or all three chromatic states. Moreover, polychromatic displays can be achieved by using a plurality of "switched" leads or contacts.

In one case, some switches can be turned on and the others left off which creates a polycolored effect using differently colored mixtures or one mixture. Another way to achieve such an effect is to use a plurality of mixtures each of which is responsive at a different level of intensity of field potential to exhibit a different apparent color or shade of color. Yet another way of achieving the polychromatic displays is to have some switches turned on and then turn a portion of them off so that some switches are on, and some switches are off (never having been turned on) some of which were previously on. This works well with a single cholesteric liquid crystal material, one mixture or a plurality of differently colored mixtures.

Figure 1.3 schematically illustrates the process where 1.3a shows a cross-sectional view through a liquid crystal member or component; 1.3b a cross-sectional view through one form of the process which illustrates a display device; 1.3c a cross-sectional view of an article illustrating an alternative display device; 1.3d a cross-sectional view of a vacuum tube display device where the field is supplied by electrons drawn to the surface of the liquid crystalline member; 1.3e a cross-sectional view through a cathode ray tube; 1.3f a cross-sectional view of an alternative form of cathode ray tube; and 1.3g a cross-sectional view of another device where both an electric field and thermal energy can be utilized to impose transient and/or comparatively permanent colored images on the same cholesteric liquid crystal member.

Figure 1.3a shows the presence of cholesteric liquid crystal member **3** comprised of a plurality of minute, individual, "naked" droplets or inclusions, **a**, of a cholesteric liquid crystal material confined within a substantially continuous matrix, **b**, of polymeric material. The droplets have a random shape and are reasonably uniformly distributed (dispersed) within the polymer matrix.

An optional thin, protective barrier film e.g., polyethylene glycol terephthalate, Mylar, polyester can be deposited on one or both major surfaces of liquid crystal member **3** to prevent deleterious solvation of surface located liquid crystal material, e.g., by a solvent component(s) such as may be contained in opaque lacquer film **4** of Figure 1.3b.

As shown in Figure 1.3b, transparent (glass or plastic) insulating protective layer **1** is directly in contact with transparent first electrode **2** which in turn is in intimate contact with the cholesteric liquid crystalline member **3**. The cholesteric liquid crystalline member **3** is composed of an array or profusion of minute droplets, **a**, of cholesteric liquid crystalline material retained within a substantially continuous polymeric matrix, **b**, with or without a protective barrier film on one or both sides thereof. In addition to enhancing the optical properties of the liquid crystalline member and protecting the droplets from deleterious exposure, the polymeric film, **b**, serves as an effective insulator between both of the electrodes.

Next to the cholesteric liquid crystalline member **3** is located a black lacquer insulating film **4** to enhance the observation of the chromatic changes in the liquid crystalline member. Conductive elements **5** are located intermediate between lower protective layer **6** (usually glass or plastic) and black insulating film **4**. While Figure 1.3b shows three such elements, it should be clearly understood that any desired number can be used and the configuration

FIGURE 1.3: PROCESS FOR PRODUCING COLOR STATE IN DISPLAY DEVICE

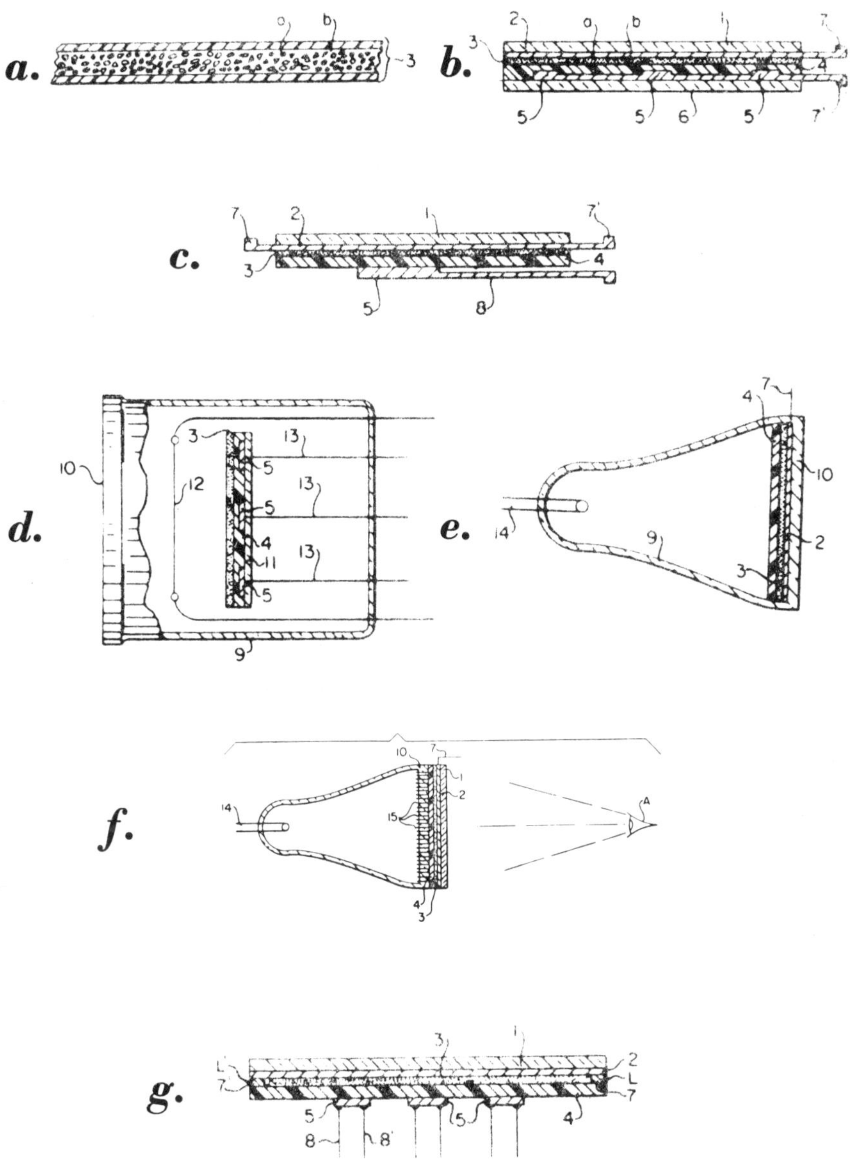

Source: D. Churchill and J.V. Cartmell; U.S. Patent 3,734,597; May 22, 1973

can be arranged in any desired display pattern, sequence or shape. Upper and lower electrodes (leads) **7** and **7'**, respectively allow the passage of an electric current from a suitable potential source (not shown) into electroconductive contact with both the first transparent electrode and the conductive elements **5**. Upon subjecting the liquid crystalline member to the electric field by applying an electric potential across the leads **7, 7'**; the chromatic representation viewed through the transparent layers of **1** and **2** will conform closely with the configuration defined by the field established between conductive elements **5** and the top electrode. That is to say that the configuration established by the elements **5** will be reproduced, but in different color or intensity of reflectance from the surrounding film areas when looking down through the transparent protective layer **1** and transparent first electrode **2**.

Upon opening the circuit, a further chromatic change takes place in the configuration areas wherein the cholesteric liquid crystalline material rapidly changes from the second chromatic state to the third chromatic state characterized by the ability of the cholesteric liquid crystalline member to retain a substantially permanent and different chromatic state for extended periods of time under normal atmospheric conditions of temperature and humidity. If desired, a sealer tape or potting compound can be applied to the outer peripheral edge(s) to aid in sealing the various layers against lateral and interfacial exposure to the atmosphere at the peripheral edge surfaces.

Figure 1.3c illustrates an alternative display device and is comprised of transparent glass plate **1** in contact with transparent electroconductive film **2** which in turn is in contact with liquid crystalline member **3**, which is composed of a profusion of minute cholesteric liquid crystal droplets or inclusions present in a retaining matrix of insulating polymeric film. Black insulating film **4** is deposited by coating on the cholesteric liquid crystalline member **3** throughout its entire extent. Electroconductive film **5** (as deposited in any desired configuration to constitute for example a printed circuit) is deposited in intimate contact with black insulating layer **4**. A conductor lead wire **8** is attached to or made integral with conductor film **5**. Leads, **7** and **7'**, are attached to bus bars in contact with the transparent tin oxide electroconductive film **2** at the left- and right-hand sides, respectively, thereof.

Upon application of potential across the upper transparent electrode **2** and the lower electroconductive film **5** through the intermediate cholesteric liquid crystalline member **3**, the visual chromatic image corresponding to the configuration of printed circuit electroconductive film **5** appears to the viewer through the protective layer **1** and transparent electroconductive film **2**, thereby exhibiting a colored image, the first, second (potential on) and third (potential off) chromatic states of which will depend upon the composition of the selected cholesteric liquid crystal material.

On passing a current through leads **7** and **7'** through the upper tin oxide electrode, the upper electrode can be heated to the isotropic transition temperature of the cholesteric liquid crystals thereby erasing the image produced during the third chromatic state. Upon cooling to ambient temperature, the liquid crystals then return to their original (first) chromatic state. Hence, it will be seen that this process allows storage of information until desired followed by erasure followed by imposition of new, e.g., updated information all accomplished electronically on a one-shot basis without the necessity for continuous electronic output during either storage or erasure.

Figure 1.3d illustrates an article of the process where the field is established by electrons attracted to the surface of the cholesteric liquid crystal-containing film. A matrix of conductive elements, **5**, is supported on insulator, e.g., fiber glass substrate in a vacuum tube **9** with a transparent viewing port or face plate **10**. The conductive elements are held at a positive potential by means of connecting the leads **13** to a potential source. A current is passed through thin, nontransparent resistive wire **12**, so that it is heated and emits electrons. The electrons are drawn to the surface of the liquid crystal film **3**, directly over the positively charged conductive elements. The field formed between the surface electrons and the conductive elements causes the liquid crystal to change color.

As shown here, three lead elements are used but it is possible to use any desired number and configuration of elements to describe the pattern desired. After the potential is removed from the elements, **5**, the pattern will remain until erased by heating to the isotropic transition temperature. This can be accomplished by means of any suitable energy source, e.g., heat lamp, heat wires or passing a current through the conductive elements.

The device of Figure 1.3e is a conventional cathode ray tube envelope **9** (CRT) with a transparent conductor **2** coated on the inner surface of the transparent face plate **10**. A film **3** of a profusion of minute droplets of cholesteric liquid crystal confined in a polymeric matrix is coated in direct contact with the transparent electrode **2** on the inner surface of the face plate, and a black insulator film **4** is coated on the liquid crystal film. The transparent electrode **2** is held at a positive potential via lead **7**, and electrons from the electron gun **14** are caused by suitable deflection techniques to impinge on the black insulator in the desired pattern.

The local fields set up between the electrons at the insulator and the front electrode cause the liquid crystal to change color in those areas forming a visual pattern when viewed from the front of the tube. The pattern is semipermanent and one sweep of the electron beam will suffice to establish a permanent image. This image can be erased by heating the liquid crystal above its isotropic transition temperature. A suitable way to do this is to pass a current through the transparent electrode as described in the previous illustration.

In the device of Figure 1.3f a conventional CRT is prepared with a multitude of conductive wires, **15** extending through the face plate **10** and flush with the outer surface on the face plate. A black insulating film **4** is formed on the face plate and the liquid crystal film member **3** is coated on the insulator. A transparent protective outer plate **1** upon which a transparent conductive electrode **2** has been deposited is brought in contact with the liquid crystal film with the conductive surface in closest proximity to the liquid crystal member. The transparent electrode **2** is held at a positive potential via lead **7**. The electrons from the electron gun **14** are deflected by any suitable conventional technique such that they describe a pattern on the inner surface of the CRT face plate. The electrons striking the wires **15** are drawn to the insulator film **4** and local fields are established across the liquid crystal film **3** between the charged wires **15** and the outer transparent electrode **2**. The liquid crystal in these areas will demonstrate a color change visible to the viewing agency, **A**.

The black insulating layer is often used to absorb light transmitted through the liquid crystal film and thereby increase the contrast of the liquid crystal colors. This light absorbing layer is not always necessary for adequate contrast in this type of a CRT display device however.

In the device of Figure 1.3g, transparent electroconductive coating **2**, e.g., tin oxide, is deposited on transparent protective element **1**, which also serves as a substrate or support for the coating. Bus bars **7**, **7'** are attached along the edges of coating **2** and have conductive leads **L**, **L'**, respectively, attached thereto. Opaque, e.g., black, nonreflective insulator film **4** has a layer or film **3** containing a profusion of minute droplets of cholesteric liquid crystal material confined within a polymeric matrix deposited thereon. Resistive conductor elements **5** can be deposited directly on insulator **4**, or on a lower supportive and protective layer such as layer **6** of Figure 1.3a.

The cholesteric liquid crystal member **3** is positioned in contiguous or closely spaced relationship to coating **2**. Each resistive element **5** has a pair of leads **8**, **8'** attached thereto, thus enabling each resistive element **5** to be operable separately or in groups, e.g., by appropriate conventional switching. Resistive elements **5** can be formed, or deposited, in any desired configuration or design. The device of Figure 1.3g has two primary modes of operation.

Field Effect Mode of Operation With Image Storage and Erasure: An electric field is applied across the liquid crystal film by a potential applied to the top conductive coating **2** and any one or more of the resistive elements **5**. The liquid crystal transforms to the second chromatic state in the area of electric field imposition. After the field is removed it assumes the third chromatic state in the areas over the resistive elements **5**. These areas can be selectively "erased" by passing a current through the rear resistive elements, thereby heating the liquid crystal to the isotropic melt temperature. On cooling, those areas which had been heated return to the first chromatic state and a semipermanent (stored) pattern is displayed from the areas over the elements which were not thermally erased.

When it is desired to change the stored image; the remaining resistive elements can be heated or the entire display unit can be heated by passing a current through the top conductive coating **2** to effect erasure. This mode of operation is more complex for dynamic display but has the advantage that the field can be applied to all the resistive elements at one time and individual switches to these elements are not required.

Thermal Mode of Operation With Storage Option: The device is operated thermally in the following manner. An electric current is passed through the resistive element **5** by means of separate pairs of leads **8, 8'** for each such resistive element. The heating of each element raises the temperature of the encapsulated liquid crystal directly above the element to the isotropic melting point. The heated portion becomes nonreflecting in contrast to the colored appearance of the unheated areas. When the current is interrupted, the element cools to ambient temperature and the liquid crystal returns to its initial colored (first chromatic) state. Thus different characters and designs (corresponding in shape and size to that of resistive elements **5**) can be displayed in a transient manner.

When it is desired to store an image; one of the leads **8, 8'** is connected to a high voltage source (not shown), and the other side of this source is connected to the transparent top electrode **2**, through either of leads **L, L'** and their associated bus bars **7, 7'**. A field generated across the liquid crystal film will cause a color change over the chosen resistive element **5**; and when the field is removed, the electric field-induced colored image will remain in the third chromatic state.

Erasure of this image can be accomplished readily by heating the liquid crystal member to the isotropic melting temperature either by heating only those elements **5**, which were used to generate the field or by passing a current through the top conductive coating **2**, by applying a potential across leads **L, L'**. On cooling, the liquid crystal film returns to its original (first) colored state.

Hence it will be apparent that the device of Figure 1.3g is, in effect, a display device having polychromatic capability for chromatically representing an optionally comparatively permanent (third chromatic state) yet thermally erasable configuration established by an electric field in optional conjunction with an independently operable yet thermally erasable transient, thermally operable, separate display means.

The device comprises a transparent support **1**; a transparent, electroconductive coating **2** on the support; conductive electrodes **7, 7'** in contact with the coating and operatively connected by leads **L, L'** to an electric field source; an insulator film **4** having a cholesteric liquid crystal member **3** in intimate proximity to the film and the coating. The device also has at least one electrically conductive resistive element **5** in thermally responsive association with the member, each resistive element having a separate pair **8, 8'** of conductive leads connected to a current source operable independently form the electric field source where the cholesteric liquid crystal is chromatically responsive to both an electric field and thermal energy. In the following examples all percents and parts are by weight.

Examples 1 through 4: Cholesteric members comprised of droplets of cholesteric liquid crystal material confined in substantially continuous solid polymer matrices are prepared using the below tabulated variety of specific matrix and liquid crystal materials. In each case 60 grams of cholesteric liquid crystal material is dispersed in 100 cubic centimeters of the aqueous polymer solution in a Waring blendor heated to 70°C by a heating jacket. Upon dispersion and formation of the desired emulsion droplet size (in the range of one to fifty microns), the various solutions, respectively, were coated onto glass which was coated previously with tin oxide (in conventional manner). The tin oxide coating is of a thickness corresponding to a resistivity of approximately 100 ohms per square. After drying black lacquer (Spray On No. 6223008), which contains carbon black in an insulating binder matrix, is sprayed over the exposed surface of the liquid crystal-containing polymer matrix, and dried.

Silver conductive paint is then applied to form an electrode on the black paint, and wires are attached to the silver electrode and the tin oxide coating and connected to an alternating current source of electric potential (although either alternating current or direct current can be utilized). The resulting basic configuration is essentially the same as those shown in Figures 1.3c and 1.3g.

Upon application of the alternating current potential, the reflectance from the area over the silver electrode when viewed with incident white light, shifts from green (first chromatic state) to blue (second chromatic state). After the electric field is removed, the reflectance does not return to the original green color, but instead assumes a grey appearance (third chromatic state), which is in very good chromatic and configurational contrast to the green background areas which are not subjected to the field.

On reapplication of the electric field, that area of the encapsulated cholesteric liquid crystalline member through which the field passes instantaneously turns to the blue color (second chromatic state). Repeated, extensive cycling of the current on and off clearly demonstrates that the cholesteric liquid crystal member possesses the ability to be cycled for extended periods without breakdown or loss of clearly discernible color contrast between the first, second and third chromatic states. The pertinent compositional make-up of the liquid crystal materials and matrix polymers used are tabulated below.

Example	Matrix Polymer	Polymer Concentration in Aqueous Solution Weight Percent	Cholesteric Liquid Crystal Mixtures Component	Weight Percent
1	Polyvinyl alcohol (Elvanol 7124)*	10	Cholesteryl nonanoate	70
			Cholesteryl chloride	25
			Cholesteryl cinnamate	5
2	Polyvinyl alcohol (Elvanol 5105)*	25	Cholesteryl nonanoate	38
			Oleyl cholesteryl carbonate	38
			Cholesteryl chloride	24
3	Polyvinyl alcohol (Elvanol 7005)*	12.5	Same as Ex. 2	
4	Gelatin (Gelatin AE-4)**	10	Same as Ex. 2	

*Du Pont polymers
**an acid extracted pigskin gelatin having a Bloom strength of 285 to 305 grams and an isoelectric point of pH 8 to 9

With further samples of the same articles produced as indicated herein, the information present in the display device (viz., the colored areas) on a different colored background produced by the electric potential) can be erased by application of heat to attain the isotropic melting point of the liquid crystal mixture in question, thus regenerating the entire film to its original green color (first chromatic state) when cooled again to room temperature. Another way of regenerating the film is to regenerate it selectively by application of the heat or an electric current of sufficient intensity to raise the temperature to the isotropic melting point of the liquid crystal mixture in selected areas only. In such a case, electrical current is passed through those resistive elements whose chromatic representation or image is to be erased, causing the elements to be heated to the isotropic transition temperature. Hence, the erasure of the chromatic image can be general (to regenerate the entire film) or selective to a particular area thereof.

The electroconductive force or potential levels required to secure the second chromatic state and erasure of the third chromatic state will depend mainly upon the thickness of the cholesteric liquid crystalline member and the spacing of the conducting electrodes. For example, erasure can be achieved readily for 10 to 100 micron thick cholesteric liquid crystalline members by applying a potential of 100 to 500 volts across the conductive elements to heat the elements to the isotropic melting point of a given liquid crystal mixture.

A variant of the article of Example 2 is prepared by applying the liquid crystal droplet-polymer matrix emulsion coating on "Mylar" film, painting the uncoated side of the Mylar film black with Spray On No. 6223008 and pressing the liquid crystal droplet-polymer matrix side of the film to the tin oxide-coated glass. The polyester Mylar film serves as an additional protective (barrier) film to protect the liquid crystal droplets, viz, additional to the polymer matrix. Of course, other equivalent protective polymeric materials can likewise be used. The use of such a protective film is preferable as it reduces the difficulty sometimes encountered in painting the silver of other electroconductive material directly on on the liquid crystal droplet-containing polymer matrix film.

While the cause of this difficulty is not entirely understood, it is believed due to some unconfined surface inclusions of liquid crystal material in the matrix film which allow a solvent component(s) in the silver paint to dissolve in the unprotected liquid crystals in the area(s) where such unconfined inclusions are present.

Example 5: The procedure of Example 1 is repeated except using a Mylar protective film and a mixture of cholesteric liquid crystalline materials containing 73 weight percent cholesteryl nonanoate and 27 weight percent cholesteryl chloride. Upon application of the electric field a color change of from red (first chromatic state) to dull grey or black (second chromatic state) was observed.

It should be noted here that the comparatively small amount of heat generated due to application of the electric field is insufficient to cause the liquid crystal to become liquid. Upon turning off the current, the color instantaneously changes from dull grey-black to a dull red (third chromatic state). This third chromatic state is sufficiently sharp in contrast to that of the first chromatic state to enable clear distinction thereof to the naked eye.

Transparent Top Coating over Capsule Wall

T.L. Hodson, J.V. Cartmell, D. Churchill and J.W. Jones; U.S. Patent 3,585,381; June 15, 1971; assigned to The National Cash Register Company describe display devices, having an encapsulated liquid-crystal member of enhanced color purity, color contrast and visual resolution due to an overlying essentially transparent top layer having a substantially smooth exterior surface. This essentially transparent layer has an index of refraction which approximates that of the capsule cell wall material and any polymeric or other binder used in conjunction with the encapsulated liquid crystals in the encapsulated liquid crystal layer of member. The advantages in color purity (optical spectral purity) and contrast attainable by the process are evident in the following detailed examples. All percents and parts are by weight unless noted otherwise.

Example 1: Topcoating — The advantages attainable by the process were illustrated by preparation of two comparative typical forms of display devices. One utilized the transparent, smooth-surfaced top layer in accordance with this process whereas the other had no such top layer structure. Both forms of display devices utilized the same electroconductive resistive elements to generate heat and were operated in the same way as thermal mode display devices. Thin-film tantalum-resistive elements were deposited onto a glass substrate by vacuum deposition in conventional manner. The substrate (with resistors) was then coated with a opaque black paint to aid in observing the liquid crystals upon heating.

A film of encapsulated liquid crystals with a polyvinyl alcohol binder was then coated over the black film. The encapsulated cholesteric liquid crystal layer was formulated (as specified below) to have a comparatively low color-temperature response, i.e., the scattered color being the same over a fairly wide range of temperature. The capsule of the article prepared in accordance with this process was then overcoated with a transparent acrylic polymer, whereas the comparison article (control) received no top layer.

In operation, the resistive elements were heated by an electric current from a 20 volt potential source and those regions of the encapsulated liquid crystal layer overlying the resistive elements were heated to the isotropic state which is transparent. At this point the black background undercoat becomes visible to yield a black image on a green colored substrate.

A liquid crystal mixture of 70 weight percent cholesteryl nonanoate, 25 weight percent cholesteryl chloride and 5 weight percent cholesteryl cinnamate was prepared and encapsulated in conventional manner using a standard two-way gelatin coacervation system. The encapsulation was conducted specifically as follows: into an aqueous solution of one weight part of acid-extracted pigskin gelatin (having a Bloom strength of 285 to 305 grams and an isoelectric point of pH 8 to 9) in 12.1 weight parts of distilled water at 55°C there were placed 13.7 weight parts of the liquid crystal melt.

The liquid crystal melt was milled with a shear agitator until the desired particle size was achieved, viz, from 15 to 30 microns. While the milling progressed, an aqueous solution of one weight part of gum arabic in 95.6 weight parts of distilled water was prepared in a separate container and maintained at a temperature of 55°C. When the desired particle size is achieved, the gelatin-liquid crystal emulsion is added slowly to the gum arabic solution. The pH is adjusted to 4.85 and the coacervate is permitted to cool to 27°C over a period of two and one-quarter hours. The resultant capsules are cooled to below 15°C and hardened with 0.5 weight parts of a 25 weight percent aqueous solution of glutaraldehyde for 12 to 15 hours. The resulting capsular slurry was then concentrated by decanting excess water.

Upon completion of encapsulation and subsequent concentration of the slurry, approximately one volume part of a ten weight percent solution of polyvinyl alcohol in water acts as a binder for the coating mixture forming the encapsulated cholesteric liquid crystal layer. The above emulsion coating was then coated on two glass plates previously provided with resistive heating elements which had in turn been overcoated with black paint (Krylon black lacquer) which was dried prior to application of the ELC layer.

Upon drying of the ELC layer a transparent top layer overcoat of a solution of 10 weight percent "Acryloid-B-72" dissolved in benzene was cast over the capsule coating to form the substantially smooth, essentially transparent top layer. Acryloid-B-72 is a methacrylic acid ester copolymeric material characterized by a viscosity of 480 to 640 centipoises in a 40 percent, by weight, solution in toluene at 30°C. The coating was applied to a wet thickness of approximately 10 mils by a conventional draw down technique. The polymer top layer dried to a clear, substantially smooth, transparent film.

Reflectance measurements were then conducted on both display devices at an angle of incidence of 45° while sensing normal to the plane of the encapsulated liquid crystal film surface. A white magnesium carbonate block was employed for reference. The below table indicates the value observed, and the test was conducted under the same conditions for both test specimens.

While the essentially transparent, substantially smooth polymer top layer has but little effect on the reflectance maximum of the colored state; a significant and quite visually observable improvement is noted for the contrast ratio. This improvement is believed to be due to elimination of undesirable light scattering from the capsule walls of the encapsulated cholesteric liquid crystals. Decreasing the reflectance band width results in a greater spectral purity and color contrast and hence visual resolution.

Value	Uncoated	Polymer Overcoat
Lambda max, colored (nanometers)*	515	521
Reflectance percent (versus magnesium carbonate)	33	36
One-half width of reflectance band (nanometers)	52	39
Contrast ratio colored/black	5	22

*Wavelength at which the maximum reflection is obtained

Example 2: Inverse Coating — A liquid crystal mixture containing 80 weight percent cholesteryl pelargonate, 15 weight percent cholesteryl chloride and 5 weight percent of oleyl cholesteryl carbonate was encapsulated using the same materials and in the same manner specified above in Example 1.

The encapsulated liquid crystal slurry was applied to one major (upper) surface of transparent Mylar (polyethylene glycol terephthalate) sheets having thicknesses ranging from approximately 0.5 to 20 mils. The encapsulated liquid crystals were coated uniformly on the upper surface of the Mylar substrate from an aqueous dispersion or slurry of 32 weight percent encapsulated liquid crystals, 64.2 weight percent water and 3.8 weight percent of a mixture of polyvinyl alcohol and polyvinyl acetate. The coating weight was 14 to 16 pounds per ream (475,000 square inches). This compares to an average thickness for the ELC layer of 35 to 60 microns.

The coating operation was conducted by screen printing in the following manner: The Mylar substrate to be printed upon was laid on a flat surface. A nylon screen, to which a stencilled vinyl back film had been adhered to give the desired printed configuration, mounted on a soft white pine frame, was positioned over the substrate. A supply of capsular ink was then poured on a blocked out region or border of the screen. A neoprene rubber squeegee was used to pull the supply of ink across the screen, at the same time pressing it through the open mesh of the screen. The screen was then lifted from the substrate leaving the capsular ink affixed to the Mylar.

After the capsular ink had dried, the process was repeated using a 4.4 weight percent black screen process ink to provide the necessary background for the encapsulated liquid crystals attached to the Mylar. Upon drying of the black opaque background layer, the encapsulated liquid crystal article was inverted thereby allowing a viewer to observe color changes occurring in the encapsulated liquid crystal layer(s) due to changes in temperature (or other equivalent color change inducing forms of energy) when viewed by incident white light through the transparent Mylar top layer. The transparent Mylar layer, originally serving as substrate for the deposition of the encapsulated liquid crystals, now serves as a brightness enhancing and spectral purity or color intensity improving top layer.

For the specific liquid crystal composition employed in the articles of this example, heating the encapsulated liquid crystal layer to a temperature of 30° to 31°C results in a red color when illuminated by incident white light at angles of incidence of 0° to 90° and viewed from approximately 90° from the horizontally located liquid crystal layer.

Dispersions of Gelatin-Gum Arabic Capsules

D. Churchill, J.V. Cartmell and R.E. Miller; U.S. Patents 3,697,297; October 10, 1972; and 3,732,119; May 8, 1973; both assigned to The National Cash Register Company describe a temperature-responsive display device or composition, comprising mesomorphic (liquid-crystal) material(s) contained in minute capsules as the temperature-responsive component.

The minute capsules, when placed in heat-conductive relation with a heat source together with an associated thermal gradient pattern to be displayed, exhibit an iridescence indicative of a certain characteristic temperature range, which temperature range varies according to the types and mixtures of mesomorphic materials. The capsule affords physical and chemical protection from various kinds of degradation and allows their use in practical and efficient thermal-gradient display systems. Following are specific examples of an encapsulating procedure and the mesomorphic materials used for preparing capsules of the temperature-responsive sensing system.

Example 1: 1.25 grams of acid-extracted pigskin gelatin, having a Bloom strength of 285 to 305 grams and an isoelectric point of pH 8 to 9, and 1.25 grams of gum arabic were stirred with 125 grams of distilled water at 55°C in a Waring blendor to yield a solution, which was at about pH 4.5. When the solution was formed, the pH was adjusted to 6.0 by the drop-by-drop addition of 20%, by weight, aqueous sodium hydroxide solution. To the above system was added a solution of 2 grams of cholesteryl propionate in 8 grams of cholesteryl oleate. The cholesterol derivative materials were emulsified in the Waring blendor vessel to an average droplet size of 10 to 50 microns, and the pH of the system was slowly reduced by the dropwise addition of 14%, by weight, aqueous acetic acid solution. Addition of acid was continued until the single, liquid-walled, capsules clustered to form aggregates having diameters of about 25 to 100 microns. At this final state, the system pH was about 5.

The entire liquid system was then chilled with an ice bath to below 10°C, while the agitation was continued. At 10°C, 0.6 milliliter of a 25%, by weight, aqueous solution of pentanediol, a chemical hardening agent for the gelatin, was added to the blender vessel, and the system was allowed to stir for about 12 hours while slowly returning to room temperature. At the end of that time, the capsule walls were firm and hardened, and the capsular system was poured through a wire mesh sieve having openings of 74 microns. That which passed the screen was suitable for coating the intended substrate. Capsules, along with the equilibrium liquid which passed through the sieve, were coated, using a drawdown applicator, to a wet thickness of more than 2 but less than 10 mils onto a blackened, 5 mil thick, polyethylene terephthalate film (Mylar) and then were dried in air at about 25°C. If temperature was raised at a point on the coated layer, concentric rings representing a temperature gradient pattern appeared outwardly from a central ring of blue, then spectrally in order through red, as the temperature gradient decreased.

As the temperature of the point was increased or decreased, the effect was evidenced by expansion or contraction of the rings. The system of Example 1 first exhibited the mesophase coloration transition at 25°C and continued to about 25° to 30°C. Systems of adjacent expanding and contracting rings intersected, and the resultant color was a function of the temperature of a given area.

Example 2: The encapsulation process of this example is identical with that of Example 1 except that a different mixture of cholesterol derivatives was employed. In this example, the mesomorphic materials were 5 grams of cholesteryl propionate and 5 grams of cholesteryl nonanoate to be substituted for the cholesteryl propionate and cholesteryl oleate used in Example 1. With regard to color change and transition temperatures, this system behaved much like the system of Example 1, the proportions being selected with the end in view of making the behavior of the system aligned with ordinary living-environment temperatures, as was also the objective of Example 1.

Example 3: In this example, two different cholesterol derivative mixtures were separately encapsulated by the method described in Example 1, and the capsules were coated, as a mixture, onto a blackened substrate film. The mesomorphic mixtures were (a) oleyl cholesteryl carbonate and cholesteryl nonanoate in a one-to-one ratio, and (b) oleyl cholesteryl carbonate and cholesteryl nonanoate in a one-to-three ratio, respectively. Maximum response to incident light (strong in 5500 angstrom components) was at 34.5°C in one system (a) and at 48.8°C in the other system (b).

The visual effect of such a coated mixture of capsules was much the same as that of the previous examples, except that the iridescence occurred in two sequences beginning at two different temperatures. In warming a test point on the coated layer, the first sequence of concentric rings of color appeared circling the point of heat application. When the spot on the coated layer was further heated above the first color advent temperature but below the color advent temperature of the second encapsulated material, the test point again appeared black with only a ring of color remaining, at that time, a fixed distance from the center of heat source, due to the static temperature gradient.

Upon further heating of the test point on the coated layer, a second set of color rings began to appear and move outwardly, but never reached the first rings, giving rise to a visual indication of the temperature at two different locations on the coated substrate. So much for point application of heat.

When the coated substrate film, as a whole, was heated, the film reflected changing colors, not in rings, but all over, colors first appearing due to one encapsulated material and then, on further heating, due to the second encapsulated material, the film appearing black for a time at temperatures intermediate to the different color advent temperatures. The color changes appeared in reverse order on cooling the film. If capsules of different mesomorphic materials which have overlapping color-change temperature ranges are combined, then the reflected colors will be combined to alter the visual appearance, depending on the relation of the different responses and the resolution of the capsular system. As discussed previously, smaller capsules yield systems with higher resolution.

Example 4: Capsules for use in this example were prepared by the technique of Example 1 and contained mixtures of oleyl cholesteryl carbonate (OCC) and cholesteryl nonanoate (CN) in proportions and having reflective characteristics using incident white light, as presented in the following table:

Capsule type	(OCC)/(CN), by weight	Advent temperature (degrees centigrade) for maximum 5,500 angstrom reflectance	Temperature range (degrees centigrade) for 6,000-4,500 angstrom reflectance
#1	25/75	48.8	48.5-49.8
#2	50/50	34.5	34.3-35.4
#3	65.2/34.8	28.3	28.1-29.1

These three encapsulated materials, #1, #2, and #3, were cast (as in Example 1) onto blackened film separately but in a pattern of squares and in such a way that the individual squares overlapped to form various couplet and triplet thicknesses. When this pattern was heated, each of the three encapsulated materials demonstrated the previously described color change phenomenon, wherein one color fringe traveled across single-material squares, two fringes appeared on the doubly overlapped layers, and three such color fringes were evidenced on portions of the pattern where all three of the mesomorphic materials were present.

The principal object of this example was to indicate that overlapped layers of the mesomorphic materials operate independently and that the layers are transparent to yield the dramatic iridescent effect even when overcoated with other encapsulated mesomorphic materials having different color-temperature characteristics.

Mixtures of Nematic and Cholesteric Liquid Crystals

J.V. Cartmell and D. Churchill; U.S. Patents 3,720,623; March 13, 1973; and 3,795,529; March 5, 1974; both assigned to The National Cash Register Company describe a mixture of cholesteric liquid crystal material such as cholesteryl nonanoate and nematic liquid crystal material such as methoxybenzylidene-p-n-butylaniline. The mixture can be encapsulated and employed in temperature-sensitive visual display devices. The addition of nematic liquid crystal material to cholesteric liquid crystal material enhances the brightness of films employing cholesteric liquid crystal material and prolongs the usable life of the films.

The proportion of nematic liquid crystal in the encapsulated mixture determines the color advent temperature, i.e., the temperature at which color first appears when the films are heated. Nematic liquid crystals greatly extend the latitude of the color advent temperature making it possible to formulate encapsulated cholesteric liquid crystal films which show color below 0°C. Such low temperature systems are not possible with standard mixtures containing oleyl cholesteryl carbonate in combination with other cholesteric compounds.

An additional feature of this process is the appearance of a thermal memory effect. If films of the encapsulated mixtures are heated to the isotropic state and cooled, a dark image of the heated area is evident. This image gradually disappears and the color returns over a period of several hours. This feature is useful if it is desired to know if a surface has been above a certain temperature within a period of several hours. Such a feature is described for unencapsulated mixtures containing high amounts of nematic liquid crystals employing the application of electric fields in *Laser Focus Magazine,* Volume 6, No. 9, pages 49-50 (September 1970). The following examples illustrate the process.

Example 1: 1.25 grams of acid-extracted pigskin gelatin, having a Bloom strength of 285 to 305 grams and an isoelectric point of pH 8 to 9, and 1.25 grams of gum arabic were stirred with 125 grams of distilled water at 55°C in a Waring blendor to yield a solution, which was at about pH 4.5. When the solution was formed, the pH was adjusted to 6.0 by the drop-by-drop addition of 20% by weight, aqueous sodium hydroxide solution. To the above system was added a solution of two grams of cholesteryl chloride in 8 grams of cholesteryl nonanoate and 1 gram of methoxybenzylidene-p-n-butylaniline.

The liquid crystal materials were emulsified in the Waring blendor vessel to an average droplet size of 5 to 30 microns, and the pH of the system was slowly reduced by the dropwise addition of 14% by weight, aqueous acetic acid solution. Addition of acid was continued until the polymeric material phased out of solution and began to deposit on the liquid crystal droplets. At this final state, the system pH was about 5.

The entire liquid system was then chilled with an ice bath to below 10°C, while the agitation was continued. At 10°C, 0.6 milliliter of a 25% by weight, aqueous solution of pentanediol, a chemical hardening agent for the gelatin, was added to the blendor vessel, and the the system was allowed to stir for about 12 hours while slowly returning to room temperature.

At the end of that time, the capsule walls were firm and hardened, and the capsular system was poured through a wire mesh sieve having openings of 74 microns. That which passed the the screen was suitable for coating the intended substrate. Capsules, along with the equilibrium liquid which passed through the sieve, were coated, using a drawndown applicator, to a wet thickness of more than 2 but less than 10 mils onto a blackened, 5 mil thick, polyethylene terephthalate film (Mylar) and then were dried in air at about 25°C. Note all percents are by weight unless otherwise specified.

Example 2: Other encapsulated mixtures of methoxybenzylidene-p-n-butylaniline (hereinafter referred to as MBBA) and cholesteric esters were prepared according to the general procedure described in Example 1. They are as follows: (1) 70% cholesteryl nonanoate, 25% cholesteryl chloride, 5% cholesteryl cinnamate, + 10% MBBA; (2) 95% cholesteryl nonanoate, 5% cholesteryl chloride, + 10% MBBA; (3) 79.2% cholesteryl nonanoate, 20.8% MBBA; (4) 81.8% cholesteryl nonanoate, 5.7% cholesteryl chloride, 12.5% MBBA; and (5) 72.0% cholesteryl nonanoate, 6.1% cholesteryl chloride, 18.4% oleyl cholesteryl carbonate, 3.5% MBBA.

The addition of nematic liquid crystals such as methoxybenzylidene-p-n-butylaniline to cholesteric liquid crystals lowers event temperatures, increases brightness of reflected colors, lengthens event temperature control and improves stability of the mixture of liquid crystals.

The addition of 1 to 25 weight percent MBBA to cholesteryl nonanoate controlled event temperatures from -5° to 60°C. The spectral range remained constant, i.e., 4°C for red-blue hues.

BARRIER COATINGS

Sulfonated Polymers of Acrylamides

R.C. Sutton; U.S. Patent 3,795,517; March 5, 1974; assigned to Eastman Kodak Company describes an improved barrier layer containing a polymerized blend of at least 50 mol percent of one or more α,β-ethylenically unsaturated amides, 0 to 50 mol percent of an α,β-ethylenically unsaturated sulfo ester, and 0 to 10 mol percent of a cross-linkable monomer for use in an electrooptical element containing a photoconductive layer, a liquid crystal film, and the barrier layer. The following examples illustrate the process.

Example 1: A 10 micron thick (dry thickness) photoconductive layer is prepared composed of polycarbonate resin, 4,4'-diethylamino-2,2'-dimethyltriphenylmethane organic photoconductor and a thiapyrylium sensitizing dye. The support for the layer is 4 mil poly(ethylene terephthalate), which is vacuum deposited with nickel to provide a conducting support. The resultant organic photoconductive film is used in each of the examples.

A control barrier layer (5 microns dry thickness) of cellulose nitrate is coated over the above described organic photoconductor film. The coating solution consists of the following:

	Grams
Parlodion (cellulose nitrate)	120
Methanol	1,320
n-Butanol	560
2-Methoxyethanol	100

A liquid crystalline film is then coated on the surface of the cellulose nitrate barrier layer. The composition of the liquid crystal film is set forth hereinafter.

The resultant element is tested for both shelf-life stability and imaging capabilities. Details and tabulated results follow the examples. Shelf-life stability is good. However, the resultant scattered light images are of low quality and a display cell containing the photoconductor-barrier layer-liquid crystal element of this example exhibits poor electrical response (i.e., undesirably high voltage pulses and long pulse times are required for image formation).

Example 2: A photoconductor-barrier layer-liquid crystal containing element similar to Example 1 is prepared, except that the control cellulose nitrate barrier layer is 3 microns thick. The element has good shelf stability and produces good images, but exhibits poor electrical response.

Example 3: A photoconductor-barrier layer-liquid crystal containing element is prepared similar to Example 1, except that the control cellulose nitrate barrier is 1.5 microns thick. This element has good shelf stability, and produces good quality images but does not exhibit good electrical response.

Example 4: A separate adhesion promoting subbing layer consisting of 60% poly(acrylic acid) and 40% poly(vinyl acetate) is coated on the organic photoconductor film described in Example 1 at 0.025 g/ft^2 from a 3% methanol solution. The resultant adhesion promoting layer has a dry thickness of about 0.2 micron. This serves as a base layer for the examples that follow. The adhesion promoting layer is optional and is not a good barrier layer itself because of its liquid crystal solvating properties.

Example 5: Poly(acrylamide-co-3-methacryloyloxypropane-1-sulfonic acid, sodium salt-co-2-acetoacetoxyethyl methacrylate) (weight ratio 90.25:4.75:5) is prepared by charging a 500 ml flask with 204 ml of water, 22.5 grams of acrylamide, 1.19 grams of 3-methacryloyloxypropane-1-sulfonic acid, sodium salt, and 22.5 grams of absolute ethanol containing 1.25 grams of 2-acetoacetoxyethyl methacrylate. The mixture is sparged with nitrogen for 20 minutes. The flask is immersed in a 60°C thermostated water bath. When the contents of the flask reach 60°C, 0.06 gram of 2,2'-azobis(2-methylpropionitrile) dissolved in 2 ml of acetone is added and the mixture is stirred under a nitrogen atmosphere for 16 hours. The bulk viscosity of the reaction mixture (25°C) at 11.1% solids content is 1,550 cp.

Another batch prepared by the same procedure but on a larger (5 gallon capacity) scale contained 11.8% solids and had a bulk viscosity of 3,300 cp. A sample of the large-scale batch was isolated by pouring 500 ml of the reaction mixture into 3 liters of isopropyl alcohol, collecting the precipitated polymer by filtration, washing with isopropyl alcohol, then with diethyl ether, and drying. The dry sample had a T_g of 184°C and an inherent viscosity of 1.51 measured at a concentration of 0.25 g/deciliter in 1.0 N NaCl solution at 25°C.

A coating of the polymer described immediately above is applied to the composite film structure (photoconductor plus adhesive layer) described in Example 4. The resultant coating is about 3 microns thick (dry thickness). A liquid crystal film is then coated on the surface of the above described polymer to form a photoconductor-barrier layer-liquid crystal element of the process. The resultant element has good shelf stability and exhibits excellent imaging quality and electrical response characteristics. See Tables 1 and 2.

Example 6: Poly(acrylamide-co-3-methacryloyloxypropane-1-sulfonic acid, sodium salt) (weight ratio 80:20) is prepared by charging a 5 gallon reactor with 12.8 kg of water, 1,440 grams of ethanol, 1,280 grams of acrylamide, and 320 grams of 3-methacryloyloxy-propane-1-sulfonic acid, sodium salt. The mixture is sparged with nitrogen gas for 20 min while the temperature is brought up to 60°C. When the contents of the reactor reaches 60°C, 25.6 ml of 30% hydrogen peroxide solution is added and the mixture is stirred under a nitrogen atmosphere for 16 hours.

Upon cooling (25°C) the reaction mixture has a bulk viscosity of 4,650 cp and a solids content of 11.1%. A sample isolated as in Example 1 has a T_g of 84°C and an inherent viscosity in 1.0 N NaCl (measured as in Example 5) of 1.75.

A coating of this polymer is applied to the composite film structure (photoconductor plus adhesive layer) described in Example 6. The resultant coating is about 3 microns thick (dry thickness). A liquid crystalline film is coated on the surface of the above described polymer to form a photoconductor-barrier layer-liquid crystal element of the process. The resultant element has good shelf stability and exhibits excellent imaging quality and electrical response characteristics. See Tables 1 and 2.

The evaluation for chemical stability (shelf life) for Examples 1 through 3 and 5 and 6 consists of maintaining the photoconductor-barrier layer structure in direct contact with the liquid crystal material at room temperature (23°C) for a specified period. At the termination of this period, the samples are visually examined for chemically caused defects.

To eliminate environmental contamination, the samples are protected with microscope slide cover glass. The barrier layer-containing elements of Examples 1 through 3 and 5 and 6 are tested for chemical stability using two different liquid crystal films. Data is given in Table 1 for the following liquid-crystal storage mixtures.

Storage Mix 1: 90% by weight N-(p-methoxybenzylidene)-p-butylaniline and 10% by weight cholesteryl oleyl carbonate;

Storage Mix 2: An interconverted mixture of 66.6 mol percent N-(p-valeryloxybenzylidene)-p-pentyloxycarbonyloxyaniline and 33.3 mol percent N-(p-acetoxybenzylidene)-p-methoxycarbonyloxyaniline, and small amounts of a cholesteryl ester and a quaternary salt comprising a quinolinium system with a long alkyl chain on the nitrogen.

In general, the shelf life is greatly extended by the use of storage mix 2, whereas with storage mix 1 only the thickest cellulose nitrate barrier coat has an appreciable lifetime. The results of the above described shelf life test are shown for Examples 1 through 3 and 5 and 6 in Table 1.

In addition, the shelf life test of Example 5 and 6 shown in Table 1 for storage mixture 2 is continued beyond 90 days for a total of 171 days. No visible damage to the photoconductor element coated with the barrier layer of Example 5 and 6 is observed.

TABLE 1: SHELF LIFE TEST RESULTS

Example	Storage Mix 1		Storage Mix 2	
	Test duration	Rating*	Test duration	Rating*
1 (control)	33 days	A	90 days	A
2 (control)	do	B	do	A
3 (control)	do	C	do	A
5	14 days	B	90 days	A
6	do	B	do	A
Organic photoconductive film with liquid crystal film and no barrier layer (control).	1 day	C	60 days	B

*Rating code:
A = Zero defects,
B = Minor damage,
C = Total destruction.

The imaging capabilities of the materials described in Examples 1 through 3, 5 and 6 are evaluated in a display device comprising a liquid crystal storage mixture sandwiched between conducting glass and the barrier-coated organic photoconductor elements of Examples 1 through 3, 5 and 6. With a voltage pulse applied across the structure, photocurrent causes imagewise dynamic scatter in the liquid crystal storage mixture. Liquid crystal storage mixture 2 is used for these studies. The cell area is 2 x 2 inches with a liquid crystal thickness of 12.7 microns. The photoconductor is illuminated with a 150 watt continuous xenon source providing 9×10^3 ergs/cm^2-sec for incident radiant energy. A line-copy positive transparency is used for imaging. Table 2 summarizes the electrical response parameters which provide a qualitative index of cell performance.

TABLE 2: ELECTRICAL RESPONSE AND IMAGE QUALITY TEST RESULTS

Example	Pulse amplitude, volts	Pulse duration, seconds	Image quality*
1 (control)	450	**5.0	5
2 (control)	450	**9.0	2
3 (control)	450	**2.0	1
5	300	0.4	1
6	480	1.5	1
Organic photoconductive film overcoated with liquid crystal film and having no barrier layer (control)	200	0.4	1

* Image quality rated 1 to 5; 1 is excellent; 5 is poor.
** Manually simulated pulse conditions.

Cross-Linked Poly(Vinyl Alcohol)

R.W. Stahr and H.G. Franke; U.S. Patent 3,795,516; March 5, 1974; assigned to Eastman Kodak Company describe a barrier layer containing poly(vinyl alcohol) for use in an electro-optical element containing a photoconductive layer, a liquid crystal film, and the barrier layer.

The polymeric barrier layer materials have the special advantages that they (1) possess optimum electrical and imaging properties, (2) effectively withstand solvent and chemical attack by the liquid crystal material, and (3) are relatively impermeable to a liquid crystal composition. The following examples illustrate the process.

Example 1: A 10 micron thick (dry thickness) photoconductive layer is prepared composed of polycarbonate resin, 4,4'-diethylamino-2,2'-dimethyltriphenylmethane organic photoconductor and a thiapyrylium sensitizing dye. The support for the layer is 4 mil poly(ethylene terephthalate) which is vacuum-deposited with nickel to provide a conducting support. The resultant organic photoconductive film is used in each of the examples.

A control barrier layer (5 microns dry thickness) of cellulose nitrate is coated over the above described organic photoconductor film. The coating solution consists of the following:

	Grams
Parlodion (cellulose nitrate)	120
Methanol	1,320
n-Butanol	560
2-Methoxyethanol	100

A liquid crystal film is then coated on the surface of the cellulose nitrate barrier layer. The composition of the liquid crystal film is set forth hereinafter.

The resultant element is tested for both shelf-life stability and imaging capabilities. Details and tabulated results follow the examples. Shelf-life stability is good. However, the resultant scattered light images are of low quality and a display cell containing the photoconductor-barrier layer-liquid crystal element of this example exhibits poor electrical response (i.e., undesirably high-voltage pulses and long pulse times are required for image formation.

Example 2: A photoconductor-barrier layer-liquid crystal-containing element similar to Example 1 is prepared, except that the control cellulose nitrate barrier layer is 3 microns thick. The element has good shelf stability and produces good images, but exhibits poor electrical response.

Example 3: A photoconductor-barrier layer-liquid crystal-containing element is prepared similar to Example 1, except that the control cellulose nitrate barrier is 1.5 microns thick. This element has good shelf stability and produces good quality images, but does not exhibit good electrical response.

Example 4: A separate adhesion-promoting subbing layer consisting of 60% poly(acrylic acid) and 40% poly(vinyl acetate) is coated on the organic photoconductor film described in Example 1 at 0.025 g/ft^2 from a 3% methanol solution. The resultant adhesion-promoting layer has a dry thickness of about 0.2 micron. This serves as a base layer for the examples that follow. The adhesion-promoting layer is optional and is not a good barrier layer itself because of its liquid crystal-solvating properties.

Example 5: A 2 micron thick coating (dry thickness) of a medium molecular weight poly(vinyl alcohol) (99% hydrolyzed, Elvanol 71-30) is applied from a 3% aqueous solution on the composite film structure (photoconductor plus adhesive layer) described in Example 4. A liquid crystal film is then coated on the surface of the above described poly(vinyl alcohol) coating to form a photoconductor-barrier layer-liquid crystal element of the process. The resultant element has good shelf-life stability; image quality is excellent and the element has very good electrical response.

Example 6: This example is similar to Example 5, except that a high molecular weight poly(vinyl alcohol) (87 to 89% hydrolyzed, Elvanol 50-42) is coated on the composite film structure described in Example 4. A liquid crystal film is then coated on the surface of the above described poly(vinyl alcohol) coating to form a photoconductor-barrier layer-liquid crystal element of the process. The resultant element has good shelf-life stability; image quality is excellent and electrical response is very good.

Example 7: This example is similar to Example 6, except that the poly(vinyl alcohol) is crosslinked with poly(silicic acid) as it is coated on the composite film structure of Example 4. The coating solution contains 200 ml of a 3 wt % aqueous solution of Elvanol 50-42, 160 ml 1:1 methanol-water mixture, 5 ml N HCl and 40 ml poly(silicic acid) solution which is prepared by mixing 100 parts by weight of tetraethyl orthosilicate with 70 parts by weight of ethyl alcohol and 22 parts of a 0.1 N HCl solution. The resultant dry cross-linked coating is comprised of about equal parts by weight of poly(vinyl alcohol) and poly-(silicic acid). A liquid crystal film is then coated on the surface of the above described cross-linked poly(vinyl alcohol) coating to form a photoconductor-barrier layer-liquid crystal element of the process. The resultant element has good shelf-life stability; image quality is excellent and the electrical response is very good.

Casein Glue

A process described by *F. Davis; U.S. Patent 3,619,254; November 9, 1971; assigned to Liquid Crystal Industries* provides thermosensing articles which can be prepared and then maintained for an indefinite time, as well as articles which will continue to function in a hostile environment. Such thermosensing articles comprise a base or a substrate, a layer of a cholesteric composition on the substrate, and a solid material coated on the cholesteric composition, the solid material being chemically inert to, and immiscible with, the cholesteric substance. Since the coating composition may itself be susceptible to attack by moisture or other ambient vapors and gases, in certain aspects of the process a protective overcoating of a second solid material is also employed.

The known cholesteric compositions are organic substances which are oily in nature. Attempts have been made to coat films of such compositions with various lacquers, varnishes, and the like, but such attempts radically alter or totally destroy the cholesteric compositions and their color play properties. Accordingly, because of the poor reproducibility of such thin films of coated cholesteric compositions, they have been unsuccessful for most practical applications.

It has been found according to this process that the incompatibility of aqueous coating substances and the cholesteric compositions can be overcome and that cholesteric compositions can be coated with a film which does not alter their properties. The substrate is coated with a film of cholesteric material, and then with a viscous liquid, solid-forming, lipophobic composition, this composition being immiscible with and inert to the cholesteric composition. After coating on the cholesteric composition, the lipophobic composition is rapidly given an initial set by evaporating the aqueous vehicle or polymerizing the solids with infrared radiation or conduction such as direct contact, a hot air blast, or the like and then the drying and setting of the composition to a solid is completed.

The immiscible coating for the cholesteric material is lipophobic, but it has been found that such materials provide a useful protective layer for the cholesteric composition despite the usual incompatibility of the two types of materials. The lipophobic coating is applied in the form of a viscous liquid which will form a solid material. The viscosity of such a liquid is higher than that of water, but it is desirably thin enough that it can readily be flowed onto the surface or used to dip, spray, or otherwise deposit a uniform coating on the cholesteric composition. The viscosity should not be so low as to spread unduly in the short setting time required. Setting time as used herein means the time required for the lipophobic coating to exhibit some initial stiffness by formation of an integument of dried or polymerized material on top of or stiffness within the layer.

The coating composition should be inert to, that is, not chemically reactive with, the cholesteric composition. The coating composition is also solid-forming, that is, it dries by loss of water or other lipophobic vehicle or polymerizes or cross-links to provide a material which is not liquid or vaporous. The coating, while solid, is preferably not rigid, but will exhibit at least as much flexibility as the substrate or substrate-base combination, so that the coating will not crack under any flexure which the substrate permits.

As noted above, the lipophobic coating material forms a solid by loss of water or other vehicle and/or by formation or cross-linking of a polymer. Solutions, suspensions, emulsions, or dispersions of the solids or the solid-forming materials can be applied to the surface of the cholesteric composition. The coating can thus be carried out with casein adhesives, polymer emulsions such as acrylic acid polymer, polyvinyl alcohol, cellulose esters such as cellulose acetate, and cellulose esters such as carboxymethylcellulose. Solutions, dispersions, suspensions and emulsions of these materials generally have sufficient viscosity to satisfy this criterion. The aqueous liquid form in which such materials are applied can contain from about 0.25 to 50% or more solids, but it is preferred that the lipophobic coating comprise from 0.5 to 10% solids. The following examples illustrate the process.

Example 1: A 10 mil black polyvinyl chloride sheet is silk-screened with a cholesteric composition to provide a one mil thickness of composition on the sheet in patterns of squares about one-sixteenth inch on a side with nine such squares per unit of the pattern. The cholesteric composition is a mixture of 25 parts of cholesteryl chloride having a MP of 94° to 95°C, 30 parts of cholesteryl oleyl carbonate showing a color play at 5° to 6°C, 39 parts of cholesteryl oleyl carbonate showing color play between 20° to 22°C, and 6 parts of Solvent Black 7 NJD. This material changes from a green color to a colorless condition at a temperature of 37°C.

The silk-screened sheet is then fed into a coating machine having a vacuum backing to hold the sheet to a cylinder as it is passed through an aperture 15 mils in height. A white casein glue is flowed onto the silk-screened sheet to provide an overall coating thickness of about 5 mils on the sheet. As the coated sheet is withdrawn from the aperture, it is heated on the surface by a hot air blast and by infrared radiation rapidly to set the casein. The sheet is then dried for 5 minutes.

The dried sheet bearing the cholesteric composition and the casein coating is sprayed with acrylic lacquer to provide a protective overcoating 2 mils thick. The sheet is then cut into strips so that each one contains a unit of 9 squares of the cholesteric composition.

These rectangular strips are then placed in a patient's mouth so that the cholesteric composition is subject to the oral temperature. The strip is withdrawn and if the squares on the strip show a green color, the patient's temperature is not above 37°C. If the squares are colorless, the patient's temperature is above 37°C.

Example 2: An 11" x 4" sheet of cellulose acetate butyrate is silk-screened with a cholesteric composition comprising 45 parts of cholesteryl oleyl carbonate, 45 parts of cholesteryl nonanoate, and 10 parts of cholesteryl benzoate (having a color play range of 26.5° to 30.5°C) in the form of an array of one-eighth inch squares.

The silk-screened sheet is then fed into a coating machine as described in Example 1 so that it is coated with white casein glue blackened by the addition of 20% carbon black. The coating is dried rapidly with heat and infrared radiation. The dried coating is then overcoated with acrylic lacquer as in Example 1.

The finished sheet has a color play temperature range of 26.5° to 30.5°C so that when a hand is placed on either surface of the sheet a vivid change of colors takes place. At the lower temperatures, the cholesteric composition squares are bright red and as the temperature increases they pass through orange, yellow, green, and blue to a deep violet. At higher temperatures the colors entirely disappear.

Example 3: A 10 mil black polyvinyl chloride sheet is sprayed through a mask with the cholesteric composition set forth in Example 1 to form a pattern of various geometric figures. The sheet is then coated with casein glue, dried, and overcoated with acrylic lacquer as in Example 1 to provide a large panel which loses its green color when heated to 37°C. The color does not return spontaneously, but the green color can be restored by briefly rubbing the acrylic surface.

Acrylic Resin

D.H. Baltzer; U.S. Patent 3,620,889; November 16, 1971; assigned to Vari-Light Corporation describes liquid crystal systems incorporating cholesteryl esters, which systems are relatively permanent and highly resistant to physical damage, the liquid crystal materials being incorporated in a plastic resin either in the form of a film or in the form of a casting.

While it has not been previously possible to incorporate liquid crystals in a plastic coating or other medium where the liquid crystals would be immobilized, it has been discovered that they can be incorporated in clear plastic resins, such as an acrylic resin, without inhibiting their ability to change color when heated. The resulting systems are protected against contamination and against marring and other abrasion which reduces their efficiency and useful life. The systems can be readily handled and reused, and consequently have opened numerous new fields of use for liquid crystals.

It also has been discovered that where liquid crystals are incorporated in a heat-curing resin and cured at elevated temperature to form a cast sheet, the resultant casting has the ability to turn from clear to opaque upon being subsequently cooled and to clear upon being reheated. Such systems lend themselves to a variety of uses. For example, they can be used as data storage elements on which data can be impressed using an infrared lamp as a heat source. By irradiating designated areas of the element, the designated areas will become clear so that upon scanning with a light source and a light detector, the clear areas will transmit light from the light source to the detector and hence activate the data retrieving mechanism. Such elements have the advantage of being erasable by using a thermoelectric cooling unit, it being only necessary to cool the element to the cloudy point to erase the data previously impressed on it, thereby rendering it reusable upon subsequent irradiation.

In the process, the liquid crystal compound or mixture is dissolved in a plastic resin. In a typical example, the liquid crystals may be mixed with an acrylic resin, such as Acryloid B-66 acrylic resin, which is a 40% solid solution in toluene. While different liquid crystal compounds will function best at slightly different concentrations depending upon their compatibility with the resin, which in turn depends upon molecular structure and weight, as well as other characteristics of the compound, it has been found that 1 to 3 parts by weight of the liquid crystal compound to 14 parts by weight of the resin usually produces excellent results. The resultant liquid crystal system is clear and may be applied as such to a test plate or other surface.

Upon application, the coating is allowed to dry at room temperature, thereby evaporating the solvent in the resin carrier. In order to insure complete solvent removal, the coated plate or the like may be heated in an oven for a short period of time at moderate temperature, i.e., 120° to 150°F, to drive out residual solvents.

Where two liquid crystal compounds are admixed, the resulting mixture is then dissolved with the resin in proportions previously stated. For example, highly successful systems have been formulated by mixing one-third cholesteryl nonanoate and two-thirds cholesteryl oleyl carbonate with a 40% solids solution of acrylic resin in a ratio of 1 part by weight of the combined liquid crystal compounds to 14 parts by weight of the resin.

While liquid crystal containing plastic coatings may be applied to an object which has been previously painted black where the natural color of the object is such that it is difficult to perceive the color changes, the process contemplates the addition of an insoluble black pigment to the coating mixture. Preferably, the insoluble pigment comprises a very fine particle size carbon black pigment which is admixed with the liquid crystals. The amount of carbon black added appears to be fairly critical in that the amount must be sufficient to provide a dark background and yet, if the quantity is excessive, the color change will be obscured. It has been found that using Cabot's Sterling R carbon black pigment, a concentration between 0.05 to 1% of carbon black by weight to the weight of the liquid crystals provides excellent visibility of the color phenomenon.

This is true both when the liquid crystals are applied directly to the object being tested or are incorporated in a plastic coating or casting. It is essential, however, that the pigment be insoluble when incorporated in the liquid crystals and/or in the liquid crystal, plastic resin system.

APPLICATIONS FOR CHOLESTERIC CRYSTALS

TEMPERATURE SENSING DEVICES

Skin Patch with Encapsulated Crystals

E. Flam; U.S. Patent 3,661,142; May 9, 1972; assigned to Johnson & Johnson describes a temperature sensing patch which comprises a flexible backing web having a pressure sensitive adhesive coated on one side for holding the flexible backing web in tight direct contact with the underlying skin. On the other or exposed side of the web is disposed a plurality of discrete temperature sensitive, color responsive indicators, each of which comprises a layer of encapsulated liquid crystals (sometimes abbreviated ELC's). Each of the indicators displays a color change spectrum corresponding to temperature changes within a different predetermined temperature range. Adjacent indicators on the patch preferably cover overlapping temperature ranges whereby, in effect, a patch is produced displaying a continuous color change over an extended temperature range.

In a preferred case the backing web is circular and the indicators on the surface are disposed in spaced pie-shaped sectors about the center. Other configurations may also be employed, e.g., square, rectilinear and the like. The web is substantially black in color and the pie-shaped sectors are separated or spaced by a border of contrasting color, preferably white.

Measuring Skin Temperature

A process described by *N.N. Goldberg and J.L. Fergason; U.S. Patent 3,533,399; Oct. 13, 1970; assigned to Westinghouse Electric Corporation* relates to a method of measuring skin temperature of living animals, including humans, and for demonstrating skin temperature patterns.

Skin temperature measurement is of great potential value in biologic study, and yet, because of the physical difficulties involved, has yielded data of limited usefulness and questionable reliability. Thermocouple and thermistor devices in current use, even when highly developed, have inherent drawbacks. They cannot be fastened or applied to the readily compressible irregularities in the surface of the skin without alteration in the surface with effects on the underlying vascular bed which cannot easily be quantitated. The dermatoglyphs, the normal markings and creases of the skin, are inaccessible to the most miniaturized of thermistors, and the least compression leads to cross radiation at the base of the depression with readings higher than in the natural state. The greatest difficulty, however, lies in the complexity of skin temperature patterns, particularly where the integument is reflected within limited areas over structures successively vascular and nonvascular as, for instance, on the dorsum of the human hand. Differences of 2° to 3°C may occur within an area of a few mm^2 and fluctuations may occur unevenly in these patterns from moment to moment. The placement of the thermistor probes over these patterns can only be regarded as arbitrary averaging, and even elaborate multiprobe devices can give only a crude picture of the actual temperature state.

Much effort has been directed to the problem of viewing objects by the radiation emitted from them which falls within the infrared portion of the electromagnetic spectrum. Objects emit infrared radiation, which is invisible to the naked eye, having an intensity which is a function of the temperature and the emissivity of the object. Therefore, the problem has been to convert a heat image which is invisible to the eye into an image which can be seen. Various schemes have been proposed for this purpose. One disadvantage which exists in most of the commonly known devices is that the member on which the heat image is imposed, for conversion to a visible image, is not sufficiently sensitive to distinguish between objects which are emitting radiation in intensities which are not greatly different.

Another common drawback is that in order to obtain a visible image a complex device is necessary to convert the heat pattern. The complexity frequently resides in the electrical components necessary for the device to function, often including a tube in which an electron beam is generated, resulting in a bulky and costly device which is unsuitable for many applications. This process provides thermally sensitive materials which are capable, by reason of their unique properties of converting a heat pattern imposed thereon into a visible pattern by exposure of the thermally sensitive materials to visible light.

As shown in Figure 2.1a, the process comprises a film of water-soluble plastic material **11** containing a pigment which absorbs transmitted light, that is, which has a color which is complementary to the color of the light incident to the film. The film of plastic and pigment is preferably applied to the skin **10** in liquid form and allowed to dry, thus forming a coating which conforms to the surface of the skin. To this film is applied another film **12** of liquid crystalline thermosensitive material, that is, a cholesteric material or combination of cholesteric materials, such as those described in U.S. Patent 3,114,836. The cholesteric materials are usually applied in organic solvents, sometimes in conjunction with film-forming organic materials. It is desirable that the films be of very thin form so that they are rapidly responsive to small temperature changes in the skin.

With further reference to Figure 2.1a, the temperature of the skin is quickly adopted by the thin films. The cholesteric material or materials in the film or layer **12** exhibit a color which is directly dependent on the temperature of the film and thus on the temperature of the skin. A cholesteric material, or preferably a combination of such materials, is selected which exhibits color changes in the temperature range of the skin of a human being, usually from 32° to 35°C. For animals and other living matter the temperatures may vary from below 5° to 40°C and higher. When a film of liquid crystals suitable in the range is applied, a brilliant array of colors appears within a few seconds. The coolest areas are red, and as the temperature increases the colors progress through yellow and green to blue and violet. Beyond the scale in either direction the field is black.

These changes occur with negligible time lag and are reversible almost indefinitely. Heat from skin is absorbed by the thin plastic film and by the cholesteric material, causing a thermal image to be formed on the layer **12** of the latter. Due to the variations in temperature of the skin from area to area, a temperature pattern is formed on the film of liquid crystals. Both the plastic film and the liquid crystalline film are so thin that there is practically no barrier to the transmission of the temperature pattern of the skin to the liquid crystals. An illuminating means **15**, such as a lamp, directs visible radiation through a filter **16** which transmits the desired portion of the spectrum onto the exposed face of the film **12** of thermally sensitive liquid crystalline material. The latter displays a visible image of light scattered by the liquid crystalline film which corresponds to the temperature pattern developed thereon corresponding to the temperature pattern of the skin and can be observed by an eye or camera **17**.

FIGURE 2.1: MEASUREMENT OF SKIN TEMPERATURE

a.

b.

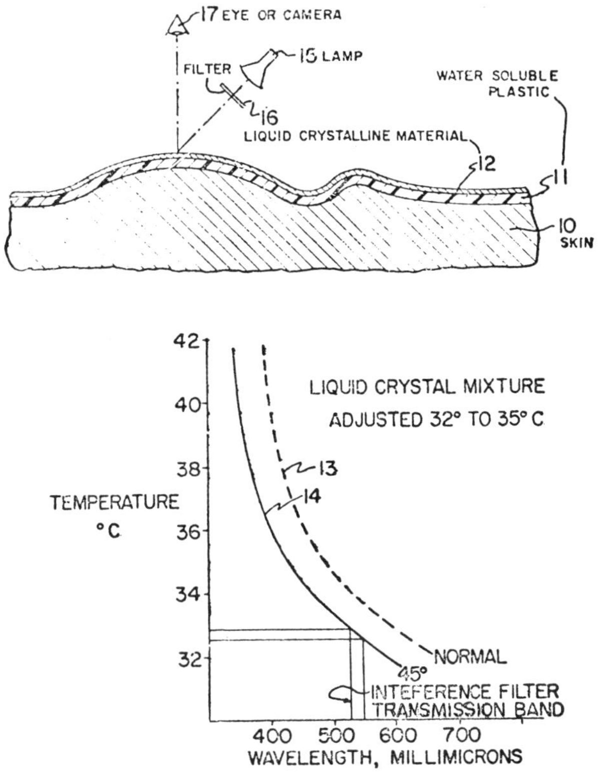

(a) Cross-Sectional View of Skin with Film of Liquid Crystalline Material Applied Thereto Irradiated with Visible Light

(b) Graph of Temperature Versus Wavelength of Light for Normal and 45° Angle of Incidence

Source: N.N. Goldberg and J.L. Fergason; U.S. Patent 3,533,399; October 13, 1970

For any particular cholesteric material or combination of materials, the colors for specific temperatures can be determined. Then when this material or combination of materials is incorporated in the film or layer **12**, the temperature of any area can be determined by the color of that area. This color is visible to the naked eye and can be recorded photographically if a permanent record is desired. Minute changes, as small as 0.1°C, in skin temperatures are readily discerned by significant changes in color of the particular area under observation.

Figure 2.1b shows typical temperature wavelength curves **13** and **14** (for normal and 45° angle of light incidence, respectively) for a typical combination of cholesteric materials It is possible, by varying the relative proportions of cholesteric materials in any combination, to produce compositions having temperature response ranges over any desired scale. As illustrated in Figure 2.1b, a mixture of cholesteric substances having a temperature response range from 32° to 35°C has been selected. The wavelength of the scattered light from this mixture varies with the temperature of the mixture and also with the angle of the beam of the light incident thereon. For any definite angle of incidence, the wavelength (and thus the color) of the scattered light is a measure of the temperature of the mixture of cholesteric materials and thus of the skin with which they are in contact. The following examples illustrate the process.

Example 1: A series of cholesteryl ester combinations was produced, each combination having a wavelength-temperature curve in the visible spectrum at temperatures close to body temperatures of living animals. The following compositions, consisting essentially of 5 to 45% cholesteryl nonanoate, 43 to 80% cholesteryl oleate, the balance selected from cholesteryl crotonate and dihydrocholesteryl benzoate, have been found useful in measuring the skin temperatures and recording skin temperature patterns (percentages being by weight).

Temperature Range, °C	Percent			
	Cholesteryl			Dihydro-cholesteryl
	Nonanoate	Oleate	Crotonate	Benzoate
24-26	10	70	20	-
41-44	45	40	-	15
39-42	25	60	-	15
25-29	30	48	7	15
18-24	20	55	10	15
22-25	35	43	7	15
30-34	40	40	5	15

A 10% solution of each of these compositions in chloroform was applied with a dropper to the surfaces of a series of Mylar polyethylene terephthalate films coated on the reverse side with carbon black or gold black and allowed to evaporate. The resulting films of cholesteric materials exhibit visible color patterns in white light and the colors can be calibrated with temperature as described in U.S. Patent 3,114,836. In this way the temperature ranges listed above were determined for each composition.

Each of these compositions in solution in chloroform is applied over a plastic film, such as polyvinyl alcohol admixed with carbon black, applied to a living body, and temperatures of the treated areas were determinable under various conditions.

Example 2: Another series of cholesteric compositions was prepared and calibrated as in Example 1. These compositions, consisting essentially of 30 to 44% oleyl cholesteryl carbonate, 46 to 60% cholesteryl nonanoate, the balance being cholesteryl benzoate, also exhibited visible temperature patterns in white light near body temperature, as shown in the following table.

Temperature Range, °C	Percent		
	Oleyl Cholesteryl Carbonate	Cholesteryl Nonanoate	Cholesteryl Benzoate
37-40	30	60	10
36-39	32	58	10
35-38	34	56	10
34-37	36	54	10
33-36	38	52	10
32-35	40	50	10
31-34	42	48	10
30-33	44	46	10

Each of these compositions was applied to parts of human bodies, hands, feet, back, upper arm and chest, over a 20 micron black polyvinyl alcohol film, and observed visually as well as photographically. Such effects as vasoconstriction in the extremities such as the hands, on smoking, for example, were readily observed. Effects of diseases were clearly evidenced by these compositions. This series illustrates how a combination of cholesteric materials can be modified slightly in composition to provide a group of compositions which exhibit temperature responses at slightly different temperature ranges.

Temperature-Limit Detection

According to a process described by *J.L. Fergason and N.N. Goldberg; U.S. Patent 3,594,126; July 20, 1971; assigned to Westinghouse Electric Corporation* temperature-limit detection is obtained by fast-cooling a cholesteric, liquid-crystal composition from an above-limit temperature to a temperature below the glass-transition temperature (T_g), which is the temperature at which the material will change to a mobile substance from a glassy substance. The liquid-crystal composition is compounded so that the glass-transition temperature matches the desired limit temperature, and the above-mentioned first above-limit temperature is preferably sufficiently higher than the limit temperature that the material, before the fast-cooling, has a characteristic color different from that which it will assume when the limit temperature is exceeded.

The fast-cooled material retains its high-temperature color in its glassy state, unless and until the glass-transition temperature or limit temperature is exceeded, at which time it changes to a different distinctive color. This color change is effectively irreversible and serves as an indication, many hours or days afterward, that the limit temperature was exceeded. To be suitable for use in accordance with the process, a cholesteric liquid-crystal material must be sufficiently sluggish in its color-changing action that, when fast-cooled, it retains its high-temperature color, and it must also be sufficiently stable, when exposed for a long period of time to a temperature below the glass-transition temperature, to retain its high-temperature color.

One such material, useful for indicating whether frozen food has been permitted to achieve a temperature as high as 10°C, consists of about 45% oleyl cholesteryl carbonate, 45% cholesteryl monanoate, and 10% dicholesteryl sebacate, the percentages being by weight. The glass-transition temperature is manifested as a change in slope when any of the primary thermodynamic properties of a material (such as volume, heat control, index of refraction) are plotted against temperature. This change in slope falls in the same temperature range as that in which the mechanical softening point occurs. At T_g the long chain molecules become free to rotate about their valence bonds. However, the molecules don't become entirely free from mutual restraint; otherwise melting would occur.

Clinical Thermometers

F. Davis; U.S. Patent 3,576,761; April 27, 1971; assigned to Liquid Crystal Industries describes thermometric compositions, elements, and articles which will record temperature changes, either permanently or for some shorter time. The compositions comprise a mixture of one or more cholesteric substances and one or more cholesteryl halides, together with an amount of oil-soluble dye sufficient to obstruct reversion of the mixture to the cholesteric state after the mixture has been subjected to a temperature at which it does not exhibit color in the cholesteric state.

Cholesteric substances used according to the process can be chosen from a wide range of compounds exhibiting the cholesteric phase. Derivatives of cyclopentanophenanthrene are desirably used. There are a number of factors to be considered in selecting such derivatives: all of the ring systems should be in the transconfiguration, the 3-substituent (on the A ring) should be in the β-configuration, and there should be no more than two axial methyl groups. Unsaturation at the 5,6 carbon atom bond can have an effect on the melting point, but otherwise has little effect on the formation of the cholesteric phase. Thus, derivatives of such cyclopentanophenanthrenes as cholesterol, campesterol, ergosterol, β-sitosterol, stigmasterol, and like materials can be used.

It is preferred to utilize alkyl and aryl derivatives of the cyclopentanophenanthrene materials, particularly those derivatives which are esters of alkanoic or aralkanoic acids, or mixed alkyl esters of the cyclopentanophenanthrene material and carbonic acid. The alkanoic acids used can contain from 1 to 24 or more carbon atoms in the molecule, and can be saturated or unsaturated and straight or branched chain. It is preferred to utilize esters comprising higher fatty acids containing from 9 to 22 carbons or lower saturated or unsaturated phenalkanoic acids having 1 to 3 carbons, and mixed carbonate esters comprising alkanols having from 1 to 22 carbons and cholesterol are also among the preferred cholesteric substances.

It will be appreciated by those skilled in the art that a pure cholesteric substance may have only a narrow color play range. However, where this color change does not occur at the temperature of interest, several stratagems permit coverage of a broad range of temperatures from 0°C, and even down to -40°C, up to and above 250°C. One method of varying the color play temperature range is to prepare a substance at a desired purity level, more impurities usually lowering the temperature range.

One convenient method of carrying out this adjustment is to admix a plurality of chemically distinct cholesteric substances having different color play temperature ranges until the desired temperature range is obtained. Another method of adjusting the color play range is to prepare the substance in a highly purified form and to admix enough of a less refined aliquot or aliquots of the substance with the purer material until the desired change of color play range is obtained. For instance, in this latter aspect, a 99.99% pure oleyl cholesteryl carbonate can be prepared and admixed with less refined material. All parts, proportions, percentages and ratios herein are by weight unless otherwise stated.

The following tabulation will exemplify a few of the many color play regions obtained with the cholesteric substance or substances:

Substance(s):	Color play region (° C.)
80% cholesteryl oleyl carbonate 20% cholesteryl acetate	0–4
95% cholesteryl oleyl carbonate 5% cholesteryl acetate	16–18
65% cholesteryl oleyl carbonate 25% cholesteryl nonanoate 10% cholesteryl benzoate	17–23
45% cholesteryl oleyl carbonate 45% cholesteryl nonanoate 10% cholesteryl benzoate	26.5–30.5
Cholesteryl cinnamate	180–260
80% cholesteryl nonanoate 20% cholesteryl propionate	45–65
20% cholesteryl butyrate 80% cholesteryl nonanoate	55–75
20% cholesteryl methyl carbonate 80% cholesteryl nonanoate	22–47
90% cholesteryl cinnamate 10% cholesteryl nonanoate	140–250
Cholesteryl benzoate	146–178
Cholesteryl butyl carbonate	45–below 0
20% cholesteryl dihydrocinnamate 80% cholesteryl nonanoate	64–67

As noted above, a desired melting range can be obtained by varying the purity of cholesteric substances. It is usually found that increasing the purity raises the temperature of the color play region and a narrowing of the range is also frequently obtained. It will of course be appreciated that the presence of excessive quantities of impurities will ultimately entirely prevent obtaining of the cholesteric phase, especially if the impurities themselves are not cholesteric substances. The cholesteric substance(s) can also comprise up to 5% or so of miscible materials such as fatty acid triglycerides to lower the range. It is most desirable to protect the cholesteric substance from the milieu to obviate the imbibition of impurities by the cholesteric substances and thereby to maintain the desired color play temperature.

As an illustration, cholesteryl oleyl carbonate is prepared as described in *Detection of Liquid Crystals,* AD 620,940, U.S. Department of Commerce (August 1965). A portion of the cholesteryl derivative is purified by solvent extraction and washed with methanol. The purified cholesteryl material is found to have a color play temperature of 21° to 22°C. Admixing 80 parts of this material with 20 parts of an unpurified material provides a color play temperature of 15° to 16°C.

The compositions also contain a cholesteryl halide. Although cholesteryl fluoride can be prepared, the desired halides for use herein are cholesteryl chloride, cholesteryl bromide, cholesteryl iodide, and mixtures of these halides. The preferred halide is cholesteryl chloride.

The cholesteryl halide serves to provide a single color over a broad range of temperatures in which the cholesteric substance or substances are in the cholesteric phase. Thus, the compositions show a single color below their transition to the condition wherein they do not scatter visible light, i.e., the condition in which they become colorless. The color below the transition point can be selected according to the amount of cholesteryl halide used. As the quantity of halide is increased from about 15% of the composition up to about 40%, the color usually varies from deep violet to deep red. The quantity of halide used will also vary according to the particular cholesteric substances utilized. It is generally preferred to use about 20 to 30% of the cholesteryl halide in the thermometric composition.

These halides are conveniently prepared by refluxing the cholesterol with an excess (twice or more, stoichiometrically) of a thionyl halide for 48 to 72 hours and distilling the mixture thereafter to remove unreacted material. Generally, the purity of the halides is sufficient to permit the desired change of phase from the cholesteric. It is preferred that the halides be at least 90% pure. Such halides usually have a tendency to raise the color play temperature range of the cholesteric substance(s).

The reversion of the cholesteric substance in the presence of the cholesteryl halide from a noncholesteric phase subsequent to exposure to the second temperature to the color of the cholesteric phase is prevented by the use of an oil-soluble dye, preferably a nitrogenous dye. The oil solubility is determined by considering the solubility of the dye in oleic or stearic acid and like long chain fatty acids. The desirable dyes for use herein are the disazo and azine dyes.

Dyes of this type and the processes for their production are described in *Colour Index,* pages 3409 to 3421, Volume 2. Broadly speaking, these dyes are prepared by the reaction of aniline or its derivatives. The nigrosines and indulenes are among the preferred dyes. Typical preferred dyes include C.I. Solvent Black 7 (C.I. 50415B), Solvent Black 7 NJD, C.I. Solvent Blue 7 (C.I. 50400), C.I. Solvent Black 3 (C.I. 26150), etc. These are all nitrogen based dyes having a plurality of benzenoid rings.

The thermometric compositions so formulated will exhibit the typical colorful scattering of white light which is found to be typical of the cholesteric state. Yet, when the transition temperature is reached so that the substance passes into a noncholesteric phase and the temperature is subsequently returned to the original temperature, the composition will not exhibit such cholesteric properties as the scattering of light. For many uses it is preferred to use thin layers of cholesteric materials, preferably layers on the order of from about 10 microns to about 5 mils in thickness.

Since a mil is one-thousandth of an inch, this thickness is approximately equal to from 10 microns to about 0.12 to 0.15 mm. Such thin layers can be used, for example, with infrared sensing devices to provide an image which will remain as long as the unchanged portions are not subjected to a temperature higher than the transition temperature. If such a temperature cannot easily be maintained under ambient conditions, a color or monochromatic photograph or other reproduction can be made of the image on the thermometric element.

Such elements have especial utility in recording the highest temperature to which an object has been exposed or which exists in the object. Thus, the thermometric element can be utilized in maximum temperature thermometers, especially in clinical thermometers. The range for the transition temperature would be chosen to be between the normal oral or other temperature to be measured and the highest temperature to which sensitivity is desired.

For instance, a clinical thermometer can be prepared from a thermometric element which passes from the color play range at 99.5°F. Such a thermometer could be inserted in the patient's mouth for a short period of time, removed and read, or alternatively, taken to a central area, and there read at will to determine whether the patient has a significantly elevated temperature. It can readily be appreciated that such thermometric elements also provide a means of keeping a permanent or semipermanent record of a patient's progress.

It will further be appreciated from the present description that a plurality of thermometric elements can be deposited on the same substrate with various transition temperatures. For example, thermometric compositions having transition temperatures of 98.6°, 99.5°, 100°, 101° and 103° could be placed on the same substrate. After removal from a patient's mouth, if all of the elements still exhibit the color of the cholesteric phase, the patient does not have an abnormally elevated temperature. On the other hand, if one, two or more elements have gone to a noncholesteric phase (as evidenced by the absence of any color), the degree of pyrexia can readily be determined.

Figure 2.2a shows a simple thermometric article comprising a substrate or base **11** which is nonabsorbent with respect to the thermometric composition. On substrate **11** is coated an area **12** of a cholesteric composition according to this process. The cholesteric composition is overcoated with a thin film of material **13** to prevent contamination of the composition by dust, fluids, vapors, and the like. While the cholesteric composition is shown as a rectangular area in Figure 2.2a, it will be appreciated that the configuration of the area can be any plane geometric figure.

FIGURE 2.2: CLINICAL THERMOMETERS

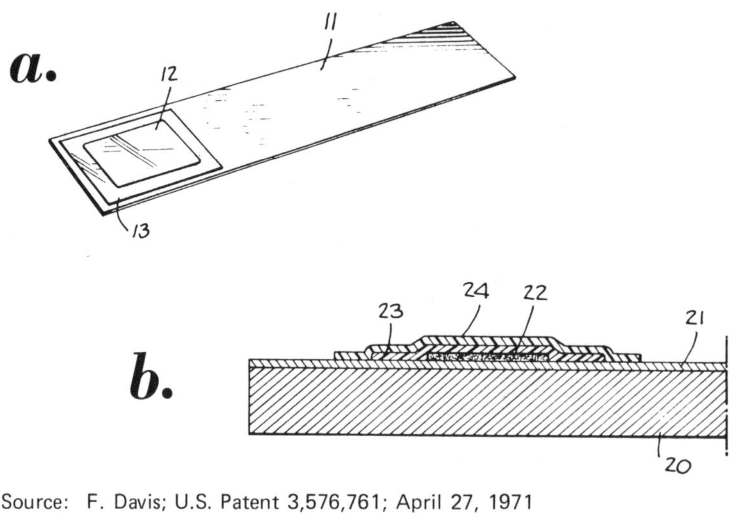

Source: F. Davis; U.S. Patent 3,576,761; April 27, 1971

It may be desirable to coat various different thermometric compositions onto a substrate in the form of numerals or letters which can either directly be read to obtain the temperature or which can be read by means of a code. In mass clinical use, for example, it may be desirable that the thermometric articles indicate temperature in terms of a set of code letters so that the patients will not be alarmed by knowledge that their temperatures are abnormal. Figure 2.2b shows a second form of the process which provides relatively inexpensive disposable clinical thermometers. Here, the base or substrate **20** is a piece of black paperboard stock which is coated with a thin layer of polyethylene **21** to protect the paper from the thermometric composition **22** coated thereon. **22** is then overcoated with a layer of immiscible material **23** such as casein glue, polyvinyl alcohol, etc. Where the thermometric article is to be utilized in an aqueous or humid environment it may be desirable to provide a second overcoating **24** of water-insoluble material such as polyethylene and methacrylate.

Example 1: A thermometric composition is prepared by mixing 30 parts of cholesteryl oleyl carbonate showing a color play at 5° to 6°C, 39 parts of cholesteryl oleyl carbonate showing a color play at 20° to 22°C, 25 parts of cholesteryl chloride having a melting point of 94° to 95°C, and 6 parts of Solvent Black 7 NJD nigrosine dye. At room temperature this mixture displays a green color in thin films.

When the temperature of a thin film of this material is heated to 37°C, the color disappears and does not reappear after cooling below 37°C, even two weeks after exposure to the higher temperature. Bulk quantities of the mixture behave in the same manner. Agitating the composition after it has been heated above 37°C and returned to room temperature (about 23°C), restores the original color.

Example 2: A 10-mil (0.25 mm) thick sheet of black polyvinyl chloride is coated by a silk-screen technique with a pattern of squares of the composition prepared in Example 1. The thickness of the composition coated on the sheeting is about 2 mils. The sheeting and the thermosensing composition are then overcoated with casein white liquid glue and dried rapidly. After curing to dry the glue fully, a protective coating of impregnated acrylic polymer is sprayed onto the glue overcoat.

The sheeting is then cut into thin strips such as shown in Figure 2.2a. These strips containing the area of thermosensing material are used to determine the oral temperatures of patients. Those with temperatures higher than the normal 37°C cause the composition to become colorless and to remain in that condition. The colorless thermometers can be reused by merely rubbing the surface of the acrylic overcoat to restore the original color.

Further compositions according to the process appear below, where all amounts are in parts: ChCl is cholesteryl chloride melting at 94° to 95°C; High ChOIC is cholesteryl oleyl carbonate showing a color play at 20° to 22°C; Low ChOIC is cholesteryl oleyl carbonate showing a color play at 5° to 6°C; ChNo is cholesteryl nonanoate; and the temperature is that at which the thermosensing compositions became colorless and remain in that condition.

Example	ChCl	High ChOIC	Low ChOIC	ChNo	Dye	Temperature, °C
3	27	73	-	-	6*	42
4	27	58	-	15	6*	45.5
5	27	50	23	-	6*	38
6	27	59	14	-	6*	35
7	25	75	-	-	6**	48

*Solvent Black 7 NJD as in Example 1
**Spirit Nigrosine SSB

It will be appreciated that the compositions of Examples 3 through 7 could be used to prepare thermometers as set forth in Example 2.

OPTICAL FILTERS

Notch Filters—Opposite Rotary Sense

J.E. Adams; L.B. Leder and W.E.L. Haas; U.S. Patent 3,679,290; July 25, 1972; assigned to Xerox Corporation describe an optical filter system capable of transmitting light at substantially all wavelengths of incident radiation while simultaneously reflecting radiation at a single wavelength band or plurality of wavelength bands within the incident radiation. The system employs liquid crystal films having optically negative properties but which are opposite in intrinsic screw sense.

The process makes use of the unique optical properties of liquid crystalline substances which possess optically negative characteristics. Birefringence, also referred to as double refraction, is an optical phenomenon characteristic of some solid crystals and most liquid crystal substances. When a beam of unpolarized light strikes a birefringent substance it is split into two polarized components whose transverse vibrations are at right angles to each other.

The two components are refracted at different angles through the substance and emerge as beams of polarized light. By the term liquid crystalline substances which have optically negative characteristics, as used herein, is meant those for which the extraordinary index of refraction η_E is smaller than the ordinary index of refraction η_O. Cholesteric liquid crystal substances exhibit this property. For a detailed description of this phenomenon see *Optical Crystallography,* Wahlstrom, 4th Edition, Wiley and Sons, Inc., New York.

The molecules in cholesteric liquid crystals are arranged in very thin layers with the long axes of the molecules parallel to each other and to the plane of the layers within each layer. Because of this configuration of the molecules the direction of the long axes of the molecules in each layer is displaced slightly from the corresponding direction in adjacent layers. This displacement is cumulative over successive layers so that the overall displacement traces out a helical path. A comprehensive description of the structure of cholesteric liquid crystals is given in *Molecular Structure and the Properties of Liquid Crystals,* G.W. Gray, Academic Press 1962.

Cholesteric liquid crystals have the property that when the propagation direction of plane polarized or unpolarized light is along the helical axis thereof, i.e., when the light enters in a direction perpendicular to the long axes of the molecules, this light is essentially unaffected in transmission through thin films of such liquid crystals except for a wavelength band centered about some wavelength λ_0 where $\lambda_0 = 2$ np with n representing the index of refraction of the liquid crystal substance and p the pitch or repetition distance of the helical structure.

The bandwidth $\Delta\lambda$ of this wavelength band centered about λ_0 will typically be of the order of about $\lambda_0/14$. For light of a wavelength λ_0 the cholesteric liquid crystal, under these conditions exhibits selective reflection of the light such that approximately 50% of the light is reflected and approximately 50% is transmitted, (assuming negligible absorption which is usually the case) with both the reflected and transmitted beams being approximately circularly polarized.

For light having wavelengths around λ_0 but not at λ_0 the same effect is present but not completely. The transmitted light is not circularly polarized but is instead elliptically polarized. The cholesteric liquid crystals which exhibit this property of selective reflection of light in a region centered around some wavelength λ_0 are said to be in the Grandjean or disturbed texture. If λ_0 is in the visible spectrum the liquid crystalline film appears to have the color corresponding to λ_0 and if λ_0 is outside the visible spectrum the film appears colorless.

Further depending upon the intrinsic rotatory sense of the helical nature of the substance, i.e., whether it is right-handed or left-handed, the light that is transmitted in the region about λ_0 is either right-hand circularly polarized light (RHCPL) or left-hand circularly polarized light (LHCPL). The transmitted light is circularly polarized with the same sense of polarization as that intrinsic to the helical nature of the substance. Thus, a cholesteric liquid crystal having an intrinsic helical structure which is left-handed in sense will transmit LHCPL and one having a helical structure which is right-handed in sense will transmit RHCPL.

These cholesteric liquid crystal substances will be indentified in order to conform with popular convention, by the kind of light which is reflected at λ_0. When a film is said to be right-handed it is meant that it reflects RHCPL and when a film is said to be left-handed it is meant that it reflects LHCPL.

A right-handed cholesteric liquid crystal substance transmits LHCPL essentially completely at λ_0 whereas the same substance reflects almost completely RHCPL. Conversely a left-handed film is almost transparent to RHCPL at λ_0 and reflects LHCPL. Since plane polarized or unpolarized light contains equal amounts of RHCPL and LHCPL, a cholesteric liquid crystal film is approximately 50% transmitting at λ_0 for these sources when the liquid crystal is in its Grandjean texture. Figure 2.3a illustrates the intensity of the output for a liquid crystal substance having a λ_0 of 4,000 A when white light is directed upon it.

A further unique optical property of optically negative liquid crystal films is that contrary to the normal situation when light is reflected, such as by a mirror, where the sense of the circular polarization of the reflected light is reversed, this same phenomenon does not occur with light reflected by these liquid crystal films. The sense of the circular polarization of light reflected from these liquid crystal substances is not reversed but rather remains the same as it was before it came into contact with the liquid crystal substance. For example, if RHCPL having a wavelength λ_0 is directed at a right-hand film having λ_0 = 2 np it is substantially completely reflected and, after reflection, remains RHCPL. If the same light were to be directed on a mirror the reflected light would be LHCPL. The process utilizes these unique properties of optically negative liquid crystal substances to provide an optical filter system.

According to the process the two optically negative liquid crystal films having the same λ_0 value but which are opposite in intrinsic rotatory sense are arranged in combination and incident light is directed upon the device, preferably in a manner such that the direction of light propagation is along the helical axes of the liquid crystal substances. The wavelengths of the incident light which are far from λ_0, i.e., those which are outside the wavelength band centered around λ_0 are essentially completely transmitted by the liquid crystal films. Consider, however, what occurs within the band of wavelengths centered around λ_0. For purposes of discussion it will be assumed that the incident light initially contacts the left-hand liquid crystal film and subsequently the right-hand film; however, it is to be recognized that the same effect would be obtained it the position of the films relative to each other were to be reversed.

The LHCPL component of the incident light around λ_0 strikes the left-hand liquid crystal film and is essentially reflected. The RHCPL component of the incident light around λ_0 is essentially completely transmitted by the left-hand film. Upon coming in contact with the right-hand liquid crystal film the RHCPL component is essentially completely reflected with the reflected light remaining RHCPL and being directed back upon the left-hand film. Since the reflected light is still RHCPL, it is essentially completely transmitted by the left-hand film. Thus, the combination of liquid crystal films having the same λ_0 values but being opposite in intrinsic rotatory sense effectively transmits all of the incident light with the exception of the wavelength band centered around λ_0. This result is illustrated in Figure 2.3b.

Hereinafter any pair of liquid crystal films, each having an opposite intrinsic rotatory sense to that of the other, but which together will essentially completely transmit all incident radiation with the exception of a wavelength band centered about some wavelength λ_0

will be referred to as a complementary matched pair of liquid crystal films.

The effect achieved by such a combination of liquid crystal films may be utilized in various ways. For example, where a broad band continuous source which also contains a narrow band source, e.g., a xenon lamp, is used as a light source the narrow band radiation may be removed from the source without removing any of the broad band radiation. Moreover, any number of optical filters consisting of combinations similar to that described can be stacked in series to remove any number of different bands of radiation from a light source. Very important practical advantages of such filters are that they may be provided in relatively large sizes and are relatively inexpensive since they are not difficult to make and the materials are themselves not expensive.

The technique may also be utilized to provide enhanced image brightening. It is known in the art that the pitch of a cholesteric liquid crystalline film may be quite sensitive to stimuli such as temperature, pressure, shear magnetic fields, electric fields, chemical vapors, ultraviolet light and others. Thus, when the pitch of the cholesteric liquid crystal substance is changed in response to some stimulus the color of the reflected light will change. These cholesteric liquid crystal films can therefore be used as monitoring techniques to detect the presence of such stimuli or to indicate changes in the stimuli themselves.

The optical filter combinations can be utilized to intensify the light reflected by the system. It can be seen that by arranging a complementary matched pair of cholesteric liquid crystal films in the path of incident light, the reflected light will be more intense by a factor of up to two than in the situation when only a single film is utilized for this purpose. Thus, such monitoring techniques can be made substantially more sensitive through the use of the optical filter system of the process.

The process will be more fully appreciated and understood from the following detailed description of various preferred forms thereof particularly when read in conjunction with the accompanying drawings. Figure 2.3a is a graphical illustration of the intensity of the output obtained when incident white light is directed on a single cholesteric liquid crystal film having a certain λ_0. Figure 2.3b is a diagrammatic illustration describing the effect which occurs when incident light is directed upon a typical optical filter of the process. Figures 2.3c and 2.3d are schematic side cross-sectional views of typical optical filters constructed according to the process. Figures 2.3e and 2.3f are graphical illustrations of the intensity of the output obtained when incident white light is directed upon the optical filters shown in Figures 2.3c and 2.3d respectively.

Figure 2.3g is a schematic side cross-sectional view of an example of the process suitable for detecting various foreign stimuli or changes therein. Figure 2.3h is a graphical illustration showing λ_0 vs composition for compositions containing various amounts of two cholesteric liquid crystal substances having opposite intrinsic screw sense.

Referring to Figure 2.3c there is shown an optical filter, generally designated **10**, having cholesteric liquid crystal films **11** and **11′** with protective members **12** and **12′** respectively and separated by optional spacing member **14**. The cholesteric liquid crystal films are selected to have the same λ_0 value but to be opposite in intrinsic rotatory sense. In this example, for purposes of illustration the films each have λ_0 value of about 5,000 A with one being right-handed and the other left-handed. It should be recognized that the position of the two films with respect to each other may be varied without affecting the operation of the device, i.e., either one may be positioned in front of the other.

Any suitable cholesteric liquid crystalline substance, mixtures thereof or compositions comprising cholesteric liquid crystals or compositions having liquid crystalline characteristics may be employed in the optical filter systems. Compositions containing cholesteric liquid crystals and nematic liquid crystalline substances may also be utilized as the liquid crystal films of the optical filter system; and it has been found that such compositions may contain up to 98% by weight of the nematic component yet continue to function in accordance with the process. Any suitable nematic liquid crystalline substance may be employed in this manner.

FIGURE 2.3: LIQUID CRYSTAL OPTICAL FILTER SYSTEM

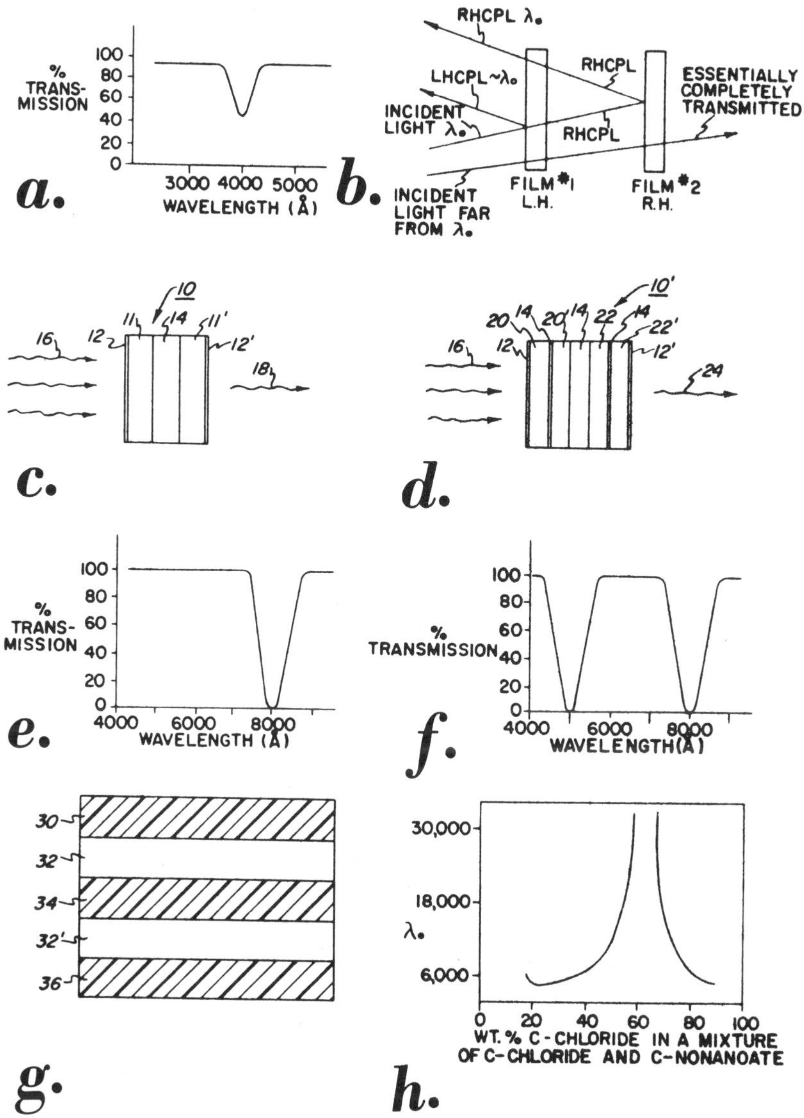

Source: J.E. Adams, L.B. Leder and W.E.L. Haas; U.S. Patent 3,679,290; July 25, 1972

The liquid crystal films **11** and **11′** are typically tacky, soft, viscous, glassy or liquid and thus are preferably covered with protective outer layers **12** and **12′** respectively to protect the films from foreign matter such as dust, insects or the like. The purpose of the protective outer layers is to keep the liquid crystal films in place and free from any contamination. Thus, the protective outer layers may be of any suitable material, flexible or rigid, which is optically transparent to the incident light radiation and which is nonreactive with the liquid crystalline films or other components of the optical filter.

Typical suitable materials for this purpose include glass, clear plastics such as, for example, Tedlar, Mylar, polyethylene, polypropylene or the like, and any other materials having the required characteristics. It is further preferred to utilize materials which have an index of refraction about the same as that of the liquid crystal films to minimize loss of light. The protective outer layers would typically be of a thickness of from about 0.25 to about 10 mils.

Spacer member **14** is an optional component of the optical filter system. The spacer member may be any suitable material which is optically isotropic and transparent to the incident light radiation. Typical suitable materials include glass, fused silica and any other materials having the required characteristics. It is further preferred to utilize materials which have an index refraction about the same as that of the liquid crystal films to minimize reflection loss of light.

For optimum results the optical filter **10** is preferably arranged in the path of a light beam from a light source in a manner such that the incident radiation, in this case visible light, represented by arrows **16** reaches the filter at normal incidence with the liquid crystal films being preferably disposed so that the helical axes of the liquid crystal films are in the direction of the light propagation. The characteristics of the emergent beam **18** are graphically illustrated in Figure 2.3e. It is seen that the optical filter essentially completely transmits all wavelengths of incident radiation except for those in the band centered around λ_0 (in this case λ_0 = about 5,000 A) for the liquid crystal films, whereas the wavelengths centered about λ_0 are substantially completely reflected and removed from the emergent beam.

Of course it should be understood that where the incident radiation is not at normal incidence to the liquid crystal films so that the helical axes of the liquid crystal substances are not exactly along the direction of the light propagation, the optical filter will continue to function according to the process, however with some deterioration in the bandwidth of the reflected wavelength band. Details relating to the compositions of the liquid crystal films and the preparation of the optical filter are given in Example 1.

Figure 2.3d illustrates another example of the process. There is seen an optical filter having two sets of complementary matched liquid crystal films stacked in series. Thus, liquid crystal films **20** and **20′** have λ_0 values of about 5,000 A and are right-handed and left-handed respectively; and liquid crystal films **22** and **22′** have λ_0 values of about 8,000 A and are left-handed and right-handed respectively. The characteristics of the emergent beam **24** are shown in Figure 2.3f. A detailed description of the compositions employed for the liquid crystal films and the preparation of the optical filter is set forth in Example 2. Either pair of complementary matched liquid crystal films can be placed closest to the incident light directed upon the filter. Alternatively the individual liquid crystal films can be located in any random order and continue to give the same results.

Figure 2.3g illustrates an example of the process where the image brightening characteristics of the process are utilized. There is seen a clear cover sheet **30** arranged above a pair of liquid crystal films **32** and **32′** having the same λ_0 but being of opposite rotary sense and separated by clear spacer **34**. The bottom member **36** of the device can be any optically transparent material and preferably is a sheet of black Mylar. Such a device can be used for thermography, printed circuit hot spot detection and the like. The Mylar surface of the device is placed against the surface of interest and any shift in the color of the reflected light can be observed. By arranging two liquid crystal films in this manner the intensity of the reflected light is increased by approximately a factor of two. Details concerning the preparation of the device are found in Example 3.

Figure 2.3h graphically illustrates the λ_0 values which can be obtained from a mixture of two cholesteric liquid crystal substances. It has been found that the pitch of two component mixtures of certain cholesteric liquid crystals is a strong function of composition. Over a wide range of materials the pitch of a mixture can be accurately represented by a weighted average of the ingredients. Further, if components with opposite intrinsic screw sense, i.e., right-handed and left-handed, are mixed there will exist one composition corresponding to no net rotation or infinite pitch. A detailed discussion of this phenomenon is found in *Liquid Crystals and Ordered Fluids,* J.E. Adams, W. Haas and J.J. Wysocki, Page 463, Plenum Press, 1970.

In this exemplary case the compositions are made from cholesteryl nonanoate, a left-handed fatty ester and cholesteryl chloride, a right-handed liquid crystal substance. The percentage composition of the two components of the mixture are plotted against λ_0 where $\lambda_0 = 2$ np. It can be seen that at a particular composition the left-handed and the right-handed components essentially compensate and the result is an infinite pitch. Moving away from this point the sense of the composition is determined by the dominant component. Defining the percent of cholesteryl chloride in any mixture as A and that in the composition having infinite pitch as A* it was found that when $A<A^*$ the films are left-handed and for the cases where $A>A^*$ the films are right-handed. Thus, liquid crystal films having any desired λ_0 value may be prepared by resorting to this technique.

The following examples illustrate the process. The behavior of the optical filters in the examples is observed by measuring their transmission spectrum with a Cary Spectrometer. The reflection spectrum for each of the filters can be inferred from its transmission spectrum since there is only negligible absorption in each instance.

Example 1: A composition of 65% anisylidene-p-n-butylaniline (ABUTA) in cholesteryl oleyl carbonate (COC) is prepared. This composition is left-handed and has a λ_0 value of about 8,000 A. A composition of 86% cholesteryl chloride (CC) in cholesteryl nonanonate (CN) is prepared. This composition is right-handed and has a λ_0 value of about 8,000 A. A thin film of the ABUTA in COC composition is applied to a thin glass plate and a thin layer of Mylar sheet is placed over the free surface of the liquid crystal film.

A thin film of the CC in CN composition is then applied to the free surface of the thin glass plate and subsequently the free surface of the latter liquid crystal layer is covered with another thin glass plate. Incident light from a broad band incandescent source of visible light is directed upon the optical filter at normal incidence. The transmission spectrum obtained is shown in Figure 2.3e. It can be seen that the filter transmits substantially completely all of the incident radiation with the exception of the wavelength band centered about 8,000 A.

Example 2: Two complementary matched pairs of liquid crystal films are incorporated into an optical filter in the following manner. A composition of 57% cholesteryl formate (CF) in cholesteryl nonanoate (CN) is prepared. This composition is left-handed and has a λ_0 value of about 5,000 A. A composition of 90% cholesteryl chloride (CC) in cholesteryl oleyl carbonate (COC) is prepared. This composition is right-handed and has a λ_0 value of about 5,000 A. A thin film of the CF in CN composition is applied to a thin glass plate and a second thin glass plate placed over the free surface of the liquid crystal film. A thin film of the CC in COC composition is then applied to the free surface of the second glass plate and subsequently a thin layer of Mylar sheet is placed on the free surface of the latter liquid crystal layer.

Then to the free surface of the first glass plate is applied a thin layer of the 65% ABUTA in COC composition (described in Example 1). Over the liquid crystal layer is placed a third glass plate and to the free surface of the third glass plate there is applied a thin layer of the 86% CC in CN composition (described in Example 1). This latter liquid crystal layer is then covered with a thin layer of Mylar sheet. Incident light from a broad band incandescent source of visible light is directed upon the optical filter at normal incidence. The transmission spectrum obtained is shown in Figure 2.3f. It can be seen that the filter transmits substantially completely all of the incident radiation with the exception of two wavelength bands centered about 5,000 A, and 8,000 A respectively, these bands being substantially completely reflected.

Example 3: A detection device suitable to detect the presence of stimuli such as temperature, pressure, chemical vapors and others is constructed in the following manner. A composition of 10% CC in COC is prepared. This composition is left-handed and has a λ_0 value of about 5,900 A. A composition of 10% CC in cholesteryl bromide (CB) is prepared. This composition is right-handed and has a λ_0 value of about 5,900 A. A thin layer of the CC in COC composition is applied to one surface of a thin glass plate and a second glass plate is placed over the free surface of the liquid crystal layer.

To the free surface of the first glass plate there is applied a thin layer of the CC in CB composition and subsequently a thin film of Black Mylar sheet is placed over the free surface of the latter liquid crystal film. The device is used by placing the Black Mylar component against any surface of interest and viewing through the glass cover plate. By employing this structure for the detection device the light reflected is intensified by a factor of approximately two over a similar device which has only one liquid crystal film.

Notch Filters—Same Intrinsic Rotary Sense

J.E. Adams, Jr. and J.L. Dailey; U.S. Patent 3,711,181; January 16, 1973; assigned to Xerox Corporation describe an optical filter system capable of transmitting light at substantially all wavelengths of incident radiation while simultaneously rejecting radiation at a single wavelength band or plurality of wavelength bands within the incident radiation. The system employs optically negative liquid crystal films which are the same in intrinsic rotatory sense and means for converting circularly polarized light of one sense, e.g., right-handed, to circularly polarized light of the opposite sense, e.g., left-handed.

Cholesteric liquid crystals have the property that when the propagation direction of plane polarized or unpolarized light is along the helical axis thereof, i.e., when the light enters in a direction perpendicular to the long axes of the molecules, this light is essentially unaffected in transmission through thin films of such liquid crystals except for a wavelength band centered about some wavelength λ_0 where $\lambda_0 = 2$ np with n representing the index of refraction of the liquid crystal substance and p the pitch or repetition distance of the helical structure.

The bandwidth $\Delta\lambda$ of this wavelength band centered about λ_0 will typically be of the order of about $\lambda_0/14$. For light of a wavelength λ_0 the cholesteric liquid crystal, under these conditions exhibits selective reflection of the light such that approximately 50% of the light is reflected and approximately 50% is transmitted, assuming negligible absorption which is usually the case, with both the reflected and transmitted beams being approximately circularly polarized.

For light having wavelength around λ_0 but not at λ_0 the same effect is present but not completely. The transmitted light is not circularly polarized but is instead elliptically polarized. The cholesteric liquid crystals which exhibit this property of selective reflection of light in a region centered around some wavelength λ_0 are said to be in the Grandjean or disturbed texture. If λ_0 is in the visible spectrum, the liquid crystalline film appears to have the color corresponding to λ_0 for normal viewing and normal incidence, and if λ_0 is outside the visible spectrum, the film appears colorless.

Further depending upon the intrinsic rotatory sense of the helix, i.e., whether it is right-handed or left-handed, the light that is transmitted in the region about λ_0 is either right-hand circularly polarized light (RHCPL) or left-hand circularly polarized light (LHCPL). The transmitted light is circularly polarized with the same sense of polarization as that intrinsic to the helix. Thus, a cholesteric liquid crystal having an intrinsic helical structure which is left-handed in sense will transmit LHCPL and one having a helical structure which is right-handed in sense will transmit RHCPL.

Hereinafter, these cholesteric liquid crystal substances will be identified in order to conform with popular convention, by the kind of light which is reflected at λ_0. When a film is said to be right-handed it is meant that it reflects RHCPL and when a film is said to be left-handed, it is meant that it reflects LHCPL.

A right-handed cholesteric liquid crystal substance transmits LHCPL essentially completely at λ_0 whereas the same substance reflects almost completely RHCPL. Conversely, a left-handed film is almost transparent to RHCPL at λ_0 and reflects LHCPL. Since plane polarized or unpolarized light contains equal amounts of RHCPL and LHCPL, a cholesteric liquid crystal film is approximately 50% transmitting at λ_0 for these sources when the liquid crystal is in its Grandjean texture.

According to the process there is provided an optical filter system employing at least one pair of optically negative liquid crystalline films in conjunction with means for converting circularly polarized light of one sense to circularly polarized light of the opposite sense. Each pair of liquid crystal films comprises two individual films which have the same intrinsic rotatory sense and which are arranged in the path of an incident light beam in a manner such that they reflect substantially the same wavelength band within the incident radiation as will be explained in detail further below. By positioning each film of any film pair employed in a particular apparatus on either side of the means for converting the circularly polarized light from that of one sense to that of the opposite sense a highly effective and efficient optical filter is constructed.

Figure 2.4a is a schematic side cross-sectional view of a typical optical filter constructed according to the process. Figure 2.4b is a diagrammatic illustration describing the effect which occurs when incident light is directed upon the typical optical filter shown in Figure 2.4a.

Referring to Figure 2.4a, there is shown a typical optical filter, generally designated **10**, of the process comprising in this illustrative instance, optically negative liquid crystalline films **12** and **14**, with optional protective elements **16** and **18** respectively, positioned on either side of element **20** converting circularly polarized light of one sense to that of the opposite sense. Liquid crystalline films **12** and **14** are comprised of substances selected so as to have the same intrinsic rotatory sense, i.e., both are either right-handed or left-handed, and to reflect substantially the same wavelength band within the incident radiation.

Each film may be comprised of different liquid crystalline substances which have the same λ_0 value; or the identical liquid crystalline substance may be used for both films. For optimum results, the optical filter is preferably placed in the path of incident radiation represented by the arrows, so that the filter is normal to the incident radiation although it will be appreciated by those skilled in the art that the angle of incidence of the incident radiation upon the filter may be other than normal.

When the incident radiation is not normal to the filter and the helical axes of the liquid crystal substances are not exactly aligned along the direction of the light propagation, the filter system will continue to function in the manner described but with some deterioration of the bandwidth of the reflected wavelength band. Good results are obtained where the angle of incidence is in the range of from about 80° to 100°. Substantially all of the incident radiation will be transmitted by the filter with the exception of some wavelength band centered around some wavelength λ_0 which corresponds to the λ_0 value of liquid crystalline films.

Figure 2.4b shows in detail the proposed theoretical operating mechanism for the system of the process. The spatial and angular separations are for purposes of illustration only. The incident beam is panchromatic and normal to the surface of the filter.

FIGURE 2.4: OPTICAL NOTCH FILTER

a.

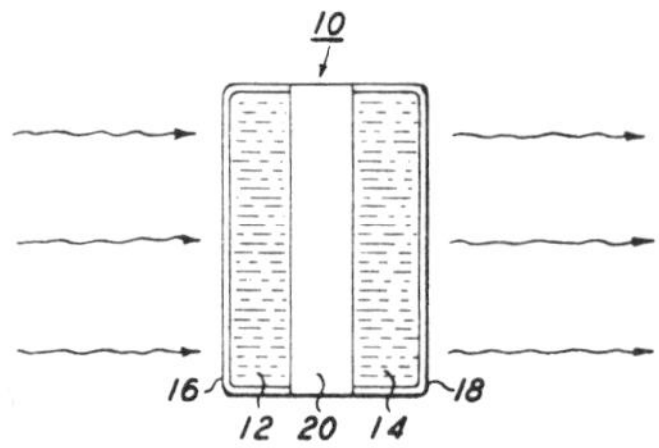

b.

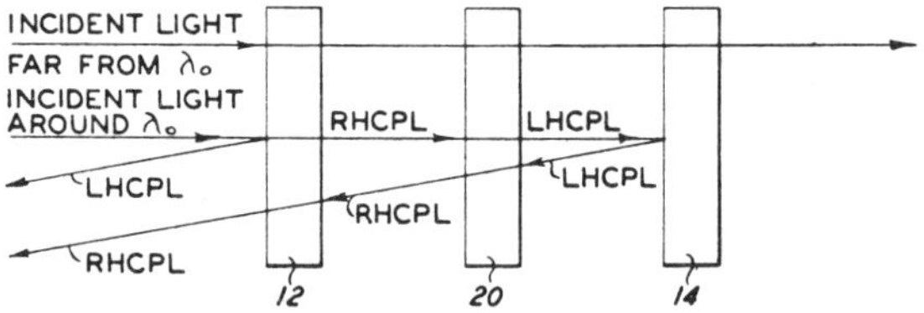

Source: J.E. Adams, Jr. and J.L. Dailey; U.S. Patent 3,711,181; January 16, 1973

In Figure 2.4b like numerals have been used to identify elements of the optical filter which are identical with those in Figure 2.4a. It will be assumed for purposes of illustration that liquid crystal films **12** and **14** are left-handed in intrinsic rotatory sense although, of course, it will be understood that they could both be right-handed. Referring now to Figure 2.4b it is seen that within the incident light which strikes the optical filter, the wavelengths of incident light which are far from λ_0 pass through the filter substantially completely unattenuated. Consider however what occurs with respect to those wavelengths of incident light which are around, and including, λ_0.

Within this wavelength region the LHCPL component of the incident light is substantially completely reflected by left-handed liquid crystal film **12** and the RHCPL component of the incident light is substantially completely transmitted by film **12**. When the RHCPL component contacts element **20** it passes through substantially completely unattenuated but now emerges from the element as LHCPL.

Subsequently, the now LHCPL strikes left-handed liquid crystalline film **14** and is substantially completely reflected with the reflected light remaining LHCPL. The light beam then is directed back on to element **20** which again reverses the sense of the polarization of the light while transmitting it therethrough substantially completely unattenuated, the light now becoming RHCPL again. The RHCPL is then substantially completely transmitted by liquid crystal film **12** thus resulting in the substantially complete reflection by the filter of the wavelength band centered around λ_0.

Any pair of optically negative liquid crystalline films having the same intrinsic rotatory sense and which, when each individual film of the pair is positioned on either side of a means for converting circularly polarized light of one sense to circularly polarized light of the opposite sense, will cooperate to reflect substantially completely a wavelength band within the incident radiation while substantially completely transmitting the remainder of the incident radiation will be referred to as a cooperative matched pair of optically negative liquid crystal films.

It can be seen that the optical filter system of the process offers a great number of advantages. For example, where the same optically negative liquid crystalline substance is used for both films of a cooperative matched pair the filter is not only automatically tuned, i.e., both films reflect substantially the same wavelength band, but the tuning of the filter is automatically stable to any change in the pitch of the liquid crystalline substance in response to environmental perturbations such as temperature, chemical vapors, etc. since the change will be identical in both films. Moreover an optical filter may be constructed so as to be relatively insensitive to the presence of a particular stimulus by selecting a liquid crystalline substance whose pitch is relatively stable to changes in the particular stimulus over a given range of environmental conditions in which the filter is used. On the other hand, there are certain applications where it is desirable to use different liquid crystal substances having different pitch vs stimulus coefficients for the films of a cooperative matched pair.

For example where the filter device is to be used as a detector to indicate changes in the stimulus present, e.g., temperature or chemical vapors, the filter can be tuned for a particular desired ambient level and when this stimulus level shifts the two films will respond differently thus becoming mismatched and drifting off null. This may be convieniently done since the pitch of certain optically negative liquid crystalline substances is extremely sensitive to particular stimuli whereas the pitch of others is less sensitive to the same stimuli.

The following examples illustrate the process. All parts and percentages recited are by weight unless otherwise specified. The behavior of the optical filters described in the examples is observed by measuring the transmission spectra of the filters using a Cary Spectrometer. The reflection spectrum of each of the filters is inferred from its transmission spectrum since there is only negligible absorption in each case.

Example 1: A first right-handed optically negative liquid crystalline composition containing about 50% cholesteryl iodide and about 50% cholesteryl chloride and having a λ_0 value of about 5,900 A is prepared. A second right-handed optically negative liquid crystalline composition containing about 10% cholesteryl chloride and about 90% cholesteryl bromide and having a λ_0 value of about 5,900 A is also prepared. A thin film of the first liquid crystal composition is applied to one surface of a half wave plate available from Polaroid Corporation and a glass plate is placed over the free surface of the liquid crystal film.

A thin film of the second liquid crystal composition is then applied to the other surface of the half wave plate and a glass plate is placed against the free surface of the second liquid crystal film. This optical filter is placed in the path of a light beam emitted from a broad band incandescent source of visible radiation so that the radiation strikes the filter at normal incidence. The optical filter substantially completely transmits all of the incident radiation with the exception of a wavelength band centered about a wavelength of about 5,900 A which is reflected.

The procedure described in Example 1 is repeated with the exception that the liquid crystalline composition listed for each example is used for both liquid crystal films of the optical filter.

Example 2: A liquid crystalline composition containing about 25% cholesteryl chloride and about 75% cholesteryl nonanoate and having a λ_0 value of about 5,200 A.

Example 3: A liquid crystalline composition containing about 80% cholesteryl chloride and about 20% cholesteryl oleyl carbonate and having a λ_0 value of about 6,800 A.

Example 4: A liquid crystalline composition containing about 20% cholesteryl chloride and about 80% cholesteryl bromide and having a λ_0 value of about 5,900 A.

Example 5: A liquid crystalline composition containing about 20% anisylidene-p-n-butylaniline and about 80% cholesteryl oleyl carbonate and having a λ_0 value of about 4,000 A.

Band Pass Filter

J.E. Adams and W.E.L. Haas; U.S. Patent 3,669,525; June 13, 1972; assigned to Xerox Corporation describe an optical filter system capable of transmitting a single wavelength band or a plurality of wavelength bands of incident radiation while simultaneously rejecting substantially all other wavelengths of incident radiation The optical filter system utilizes liquid crystal films having optical negative properties.

One of the common types of optical filters is the band pass filter. A well known type of band pass filter is the so-called interference filter which, generally speaking, comprises alternating layers of a dielectric material having a relatively high index of refraction and a dielectric material having a relatively low index of refraction. By making the layers of the proper thickness, the reflections of certain bands of wavelengths from the boundaries between the materials are reinforced and thereby removed from the transmitted beam. The other wavelengths which are grouped together in a plurality of orders of widely spaced bands pass through the material. These prior art interference filters have tended to be relatively expensive devices because of the problems inherent in their manufacture.

This process involves arranging at least one film of a liquid crystalline substance having optically negative properties between a linear polarizer and a linear analyzer with the latter two elements having a predetermined relationship between their axes of polarization such that the cooperative action of the two elements is effective to prevent the transmission of incident plane polarized or unpolarized light through the linear analyzer.

The following examples illustrate the process. All parts and percentages recited are by weight unless otherwise specified. The behavior of the optical filters described in the examples is observed by measuring the transmission spectra of the filters using a Cary spectrometer. The reflection spectrum for each of the filters is inferred from its transmission spectrum since there is only negligible adsorption in each case.

Example 1: A composition containing 85% of anisylidene-p-n-butylaniline (ABUTA) in cholesteryl oleyl carbonate (COC) is prepared. This composition has a λ_0 value of about 2.0 microns. A thin layer of the liquid crystalline mixture is placed on a Polaroid sheet and subsequently another Polaroid sheet is placed over the liquid crystal layer so that the axes of polarization of the two Polaroid sheets have an angular relationship of about 90° to each other. This optical filter is then placed in the path of a light beam emitted from a broad band source of infrared radiation. The filter is positioned in a manner such that the incident radiation is normal to the filter. The optical filter substantially completely reflects all of the incident radiation with the exception of a wavelength band centered about a wavelength of about 2.0 microns which is transmitted.

Example 2: Compositions containing 90% cholesteryl chloride (CC) in cholesteryl oleyl carbonate (COC) and 40% cholesteryl chloride (CC) in cholesteryl nonanoate (CN) are prepared. A thin layer of the composition of CC in COC is placed on a Polaroid sheet and a thin glass plate is placed on the free surface of the liquid crystal layer. A thin layer of the composition of CC in CN is then applied to the free surface of the glass plate and subsequently a second Polaroid sheet is placed over the free surface of the latter liquid crystal layer.

The two Polaroid sheets are arranged in a manner such that their respective axes of polarization have an angular relationship of about 90° to each other. This optical filter is then placed in the path of a light beam emitted from a broad band incandescent source of visible radiation. The filter is positioned so that the incident radiation is normal thereto. The filter substantially completely reflects all of the incident radiation with the exception of two wavelength bands centered about wavelengths of about 5,000 A and about 8,000 A which are transmitted.

Tuning Method for Plural Filters

J.E. Adams and L.B. Leder; U.S. Patent 3,697,152; October 10, 1972; assigned to Xerox Corporation describe a method for tuning optical devices constructed with one or more liquid crystal films possessing optically negative characteristics which comprises rotating the liquid crystal films with respect to the beam of incident light.

Optical filters are well known and widely used devices some of which can be constructed to essentially completely transmit substantially all wavelengths of incident light radiation while simultaneously rejecting a band of undesired wavelengths. Conversely others can be constructed to essentially completely reject all wavelengths of incident light radiation while simultaneously transmitting only a band of desired wavelengths. The former type may be referred to as optical notch filters while the latter are commonly referred to as band pass filters.

One type of prior art band pass filters, called interference filters, typically comprises alternating layers of dielectric material having a relatively high index of refraction and a dielectric material having a relatively low refractive index. By making the layers of the proper thickness, the reflections of certain bands of wavelengths from the boundaries between the materials are reinforced and thereby removed from the transmitted beam. The other wavelengths which are grouped together in a plurality of orders of widely spaced bands pass through the material.

It is known in the art how to tune the band pass filter, that is, to shift the wavelength regions at which these filters reject certain wavelength bands or transmit desired wavelength bands, by tilting the filter so as to change the angular incidence of the light beam directed upon the filter. Generally speaking, however, these prior art optical filters are quite sensitive to small changes in the angular incidence of the light beam thus resulting in large changes in color for relatively small rotation of the filter. A further characteristic is that the bandwidth of the transmitted band of wavelengths deteriorates rapidly when the filter is rotated and, consequently, the filter cannot be rotated over a relatively large wavelength span thus limiting the effectiveness of this technique with respect to these prior art filters.

Recent advances in liquid crystal technology have brought about the discovery that liquid crystalline substances having optically negative characteristics, are extremely useful when utilized in optical filter systems. Detailed descriptions of an optical notch filter system and a band pass filter system respectively utilizing such liquid crystalline substances are given in U.S. Patent 3,679,290 and U.S. Patent 3,669,525. It would be desirable to have a method for tuning optical devices employing liquid crystal films having optically negative properties for such a method would make it possible to adapt these devices for various specific conditions thereby broadening their versatility and increasing their numerous advantageous characteristics.

The process described below involves rotating these optical devices themselves, or the liquid crystal films independently of the other members of the devices, with respect to the angular incidence of the beam of incident light. Furthermore, in those optical devices which have a plurality of liquid crystal films, the individual films may be rotated independently or they may be rotated in tandem to achieve various desired results.

In this discussion, the cholesteric liquid crystal substances will be identified in order to conform with popular convention, by the kind of light which is reflected by the particular substance in the region around λ_0. When a substance or a film is said to be right-handed it is meant that it reflects RHCPL and when it is said to be left-handed it is meant that it reflects LHCPL.

A right-handed cholesteric liquid crystal substance transmits LHCPL essentially completely at λ_0 whereas the same substance reflects almost completely RHCPL at λ_0. Conversely, a left-handed film is almost transparent to RHCPL at λ_0 and reflects LHCPL. Since plane polarized or unpolarized light contains equal amounts of RHCPL and LHCPL a cholesteric liquid crystal film is approximately 50% transmitting at λ_0 for these sources when the liquid crystal is in its Grandjean or disturbed texture. These properties make such cholesteric liquid crystal substances extremely valuable when utilized in optical devices.

The dispersive reflection of a disturbed cholesteric liquid crystal film is described by the equation:

$$\lambda=\lambda_0 \text{ COS } \tfrac{1}{2}\left[\text{SIN}^{-1}\left(\frac{\text{SIN } \phi i}{n}\right)+\text{SIN}^{-1}\left(\frac{\text{SIN } \phi s}{n}\right)\right]$$

A detailed discussion of this dispersive reflection phenomenon is given in J.L. Fergason, *Mol. Cryst.* 1, 293 (1966). Where ϕ_i equals the angle of incidence of the incident light (assuming ϕ_i for normal incidence equals zero); ϕ_s equals the angle of reflection of the reflected light; and λ_0 equals the reflected wavelength for the condition where ϕ_i equals ϕ_s equals zero. The geometry for this phenomenon is shown in Figure 2.5a.

It can be seen from the equation set out above that the reflected wavelength is related to the angles ϕ_i and ϕ_s and the optical properties of the film. The equation does not take into account the intensity of the incident light. The intensity of a given reflection is a function of the orientation distribution of scattering sites. It has been found that in the disturbed state the distribution of helical axes is predominantly perpendicular to the substrate. Specular and near specular dispersive scattering is large compared to reflections far from specular. Therefore, for normal incidence, i.e., where ϕ_i equals zero, the wavelengths around λ_0 are most strongly scattered. However, if the film is then tilted with respect to an incident beam of white light such that ϕ_i is not equal to zero any longer then the wavelength most effectively scattered is given by the equation:

$$\lambda=\lambda_0 \text{ COS}\left[\text{SIN}^{-1}\left(\frac{\text{SIN } \phi i}{n}\right)\right]$$

A detailed discussion of this phenomenon is given in "Optical Properties of Certain Cholesteric Liquid Crystal films", Adams, Haas and Wysocki, *Journal of Chemical Physics,* Volume 50, Number 6, 2458 to 2464 (1969). Therefore, rotating a liquid crystal film with respect to an incident light beam constitutes a tuning technique. In the case of a band pass filter which has one liquid crystal film and transmits only one band of wavelengths, the transmitted band can be shifted to another region of the light spectrum.

For multiband pass filters which have more than one liquid crystal film and transmit a plurality of bands of wavelengths corresponding to the number of films present, the individual transmitted bands may be shifted independently of each other by rotating one film while maintaining the others in their original position or they may be shifted in tandem by rotating the filter itself.

With respect to optical notch filters which utilize complementary matched pairs of liquid crystal films a variety of results may be achieved by resort to the process. By a complementary matched pair of liquid crystal films is meant two individual films arranged in a manner such that they both have the same λ_0 value with one being a left-handed film and one being a right-handed film. Of course, both films may have the same intrinsic λ_0 value and thus could both be arranged in the path of the incident light beam in a manner such that the light is directed on the films at normal incidence. Consider, however, a situation where the two films do not have the same intrinsic λ_0 value.

In such a case, the reflection spectra of the two individual films may be brought into coincidence by rotating one film while the other is held fixed. At this point the two films would then constitute a complementary matched pair. Of course, with respect to any complementary matched pair of films, whether the individual films which make up the pair are intrinsically matched or matched in the manner described above, the notch itself, i.e., the band of wavelengths which is substantially completely reflected by the filter, can be shifted by rotating the matched pair of films in tandem. Moreover, where this type of filter has a plurality of complementary matched pairs of liquid crystal films so that the filter substantially completely reflects a plurality of wavelength bands the respective reflected bands may be shifted independently or in tandem.

Figure 2.5a illustrates the light scattering characteristics of a liquid crystal film having optically negative properties. Figure 2.5b is a schematic side cross-sectional view of a typical optical band pass filter. Figure 2.5c is a schematic side cross-sectional view of a typical optical notch filter. Figure 2.5d is a graphical illustration showing the shift in the center wavelength of the transmitted band for a band pass filter for the change in the angular incidence of the incident light beam. Figure 2.5e is a graphical illustration showing the change in the reflection spectrum of a typical optical notch filter as one of the liquid crystal films is rotated with respect to the incident light beam.

Referring to Figure 2.5b there is seen an exemplary optical band pass filter **10** of the type described in U.S. Patent 3,669,525. The filter is shown for purposes of illustration as consisting of a liquid crystal film **12** having optical negative properties, and positioned between a linear polarizer **14** and a linear analyzer **16**, the latter two members having a predetermined angular relationship between their axes of polarization such that the cooperative action of these two members is effective to prevent the transmission of incident unpolarized light through the linear analyzer.

Preferably, to obtain optimum results, the optical filter is arranged in the path of the incident light beam, represented by arrows **18**, in a manner such that the incident light beam is normal to the liquid crystal film. Thus, the emergent beam **20** contains only a wavelength band centered about some wavelength λ_0, the remainder of the wavelengths of radiation within the incident light beam having been extinguished by the cooperative effect of the linear polarizer and the linear analyzer.

Of course, λ_0 in any instance is determined by the particular liquid crystal substance, mixture of liquid crystal substances or composition comprising a liquid crystal substance employed for the liquid crystal film. In this exemplary instance the liquid crystal film is comprised of a composition of 23% cholesteryl chloride by weight in cholesteryl nonanoate. Details concerning preparation of the optical filter are given in Example 1.

Consider now what occurs when the liquid crystal film is rotated with respect to the incident light beam such that the angular incidence of the light beam on the film deviates from normal incidence. The results obtained are illustrated by Figure 2.5d. The vertical axis of the graphical illustration is a measure of λ_0 in Angstroms whereas the horizontal axis is expressed as the angular deviation, in degrees, from normal incidence of the incident light beam upon the liquid crystal film. Thus, it is seen that at normal incidence λ_0 for the liquid crystal film is about 5500 A.

When the film is rotated to an extent such that the angular incidence of the incident light beam upon the film deviates from normal incidence by 20° the λ_0 value for the film is shifted to about 5350 A. When rotating the liquid crystal film the λ_0 value thereof can be shifted to various other wavelengths as is clearly shown. Although the method has been described with respect to rotating only the liquid crystal film, it should be apparent that for this particular example of band pass filter the same results would be obtained by rotating the filter itself since the filter has only one liquid crystal film and further the cooperative action of the linear polarizer and linear analyzer is not greatly affected by modest changes in the angular incidence of the incident light beam.

Referring to Figure 2.5c there is seen an exemplary optical notch filter **30** of the type described in U.S. Patent 3,679,290. The filter **30** is shown for purposes of illustration as having two components. One component has a thin liquid crystal film **32** comprising a composition of 24.5% by weight of cholesteryl chloride (CC) in cholesteryl decanoate (CD) positioned between a thin glass plate **34** and thin Mylar sheet **36**.

FIGURE 2.5: TUNING METHOD FOR PLURAL FILTERS

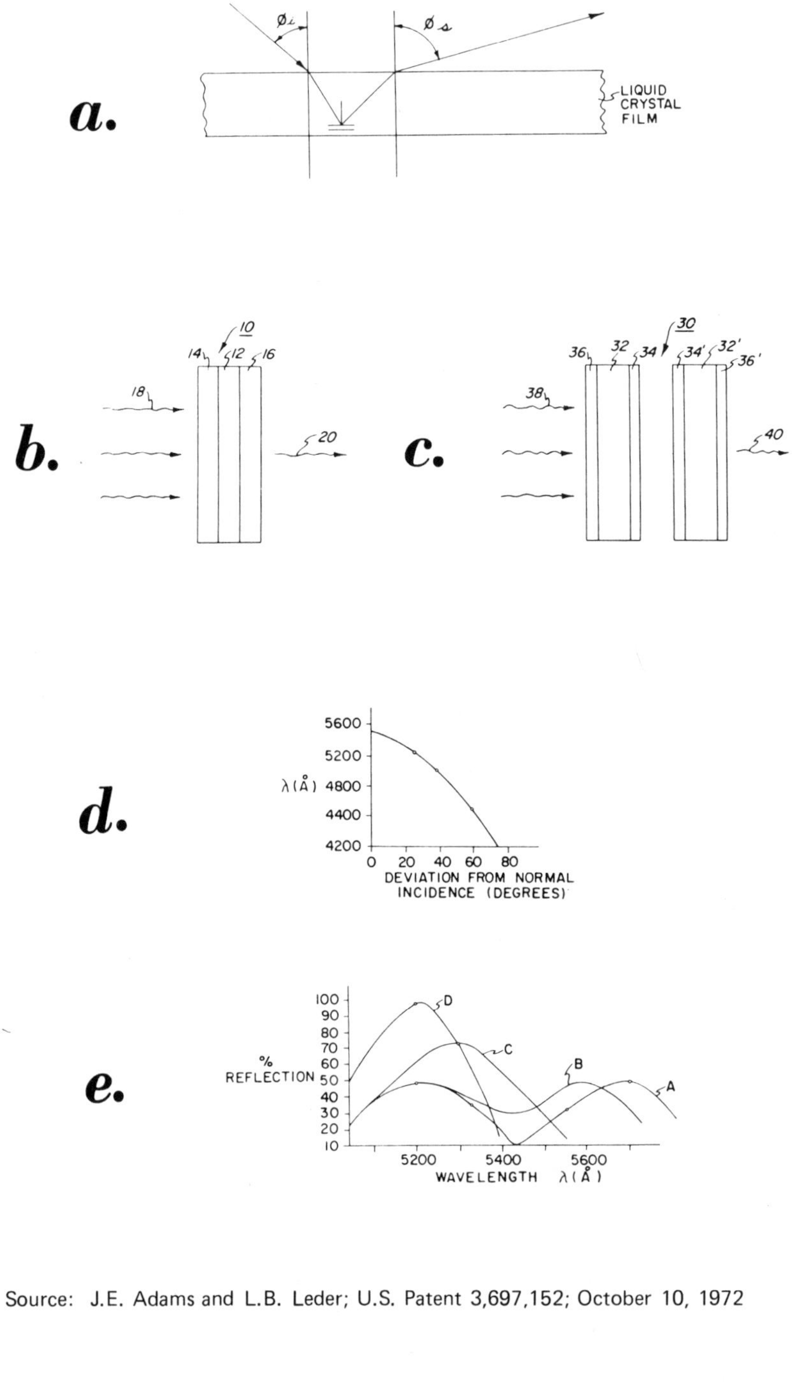

Source: J.E. Adams and L.B. Leder; U.S. Patent 3,697,152; October 10, 1972

This liquid crystal composition is left-handed and has a λ_0 of about 5200 A. The other component has a thin liquid crystal film **32'** comprising a composition of 20% by weight CC in cholesteryl bromide (CB) positioned between a thin glass plate **34'** and a thin Mylar sheet **36'**. The liquid crystal composition of the latter component is right-handed and has a λ_0 of about 5700 A. While it is clearly apparent that the two liquid crystal films could be selected to have the same intrinsic λ_0 values, i.e., for incident light directed upon them at normal incidence they would reflect the same wavelength band, for purposes of illustrating the process in this exemplary instance they have been chosen to have different intrinsic λ_0 values. Details concerning the preparation of the filter are shown in Example 2.

The reflection spectra of the optical filter of Figure 2.5c for various conditions are shown in Figure 2.5e where the percent reflection of the filter is plotted against wavelength (A). Curve **A** shows the condition of the reflected beam when both films **32** and **32'** are positioned in a manner such that the incident light beam, represented by arrows **38** strikes both films at normal incidence. There are seen two discrete wavelength bands, which are partially reflected by the filter, one centered around a wavelength of about 5200 A and corresponding to liquid crystal film **32** with the other centered around a wavelength of about 5700 A and corresponding to liquid crystal film **32'**.

When liquid crystal film **32'** is rotated by 15° and film **32** is held fixed the reflection spectrum of the filter changes. The spectrum for this situation is illustrated by Curve **B**. Rotating film **32'** by 30° while keeping film **32** normal to the light beam produces the reflection spectrum shown in Curve **C**. Curve **D** illustrates the result which occurs when film **32'** is rotated by 42°. It can be seen that the filter now substantially completely reflects light in the wavelength band centered around 5200 A since the individual reflection spectra of films **32** and **32'** have now been brought into coincidence, i.e., both films now reflect the incident light most strongly in the region around 5200 A.

Thus, the filter now constitutes a typical optical notch filter since it transmits substantially all of the incident light while simultaneously rejecting substantially completely the wavelengths of light in the narrow band centered around 5200 A. At this condition the emergent beam **40** for optical filter **30** contains all the wavelengths of radiation of the incident light except for the reflected band centered about 5200 A.

Films **32** and **32'** have been selected with varying intrinsic λ_0 values in order to illustrate the process. Of course, it is apparent that they could be selected so as to have the same intrinsic λ_0 values. In any event, when the reflection spectra of the two films are coincident, the band of wavelengths which is reflected by the pair can itself be shifted by rotating the pair in tandem.

Generally speaking, the liquid crystal films of the optical filters described can be rotated to an extent such that the incident light beam deviates from normal incidence by up to 90°. It should be recognized, however, that as any liquid crystal film is rotated in this manner there is a corresponding gradual deterioration in the bandwidth of the transmitted or reflected wavelength band. For example, at a deviation of 60° from normal incidence the bandwidth is about two times greater than that obtained when the incident light is at normal incidence.

Of course, the transmitted or reflected band will disappear completely when the deviation from normal incidence reaches 90°. Moreover, from the foregoing, it can be seen that λ_0 for any liquid crystal film suitable for use in these optical filters is greatest when the incident light strikes the film at normal incidence. Thus, as any liquid crystal film is rotated in accordance with the method the λ_0 value of the film becomes progressively smaller. The shift in λ_0, in Angstroms, for any film which can be achieved through the use of the process is dependent on the particular liquid crystal film itself. Broadly speaking, the shift in λ_0 for any film can be any number of Angstroms up to λ_0.

The following examples illustrate the process. All parts are given by weight unless otherwise specified. The behavior of the optical filters described in the examples is observed by measuring the transmission spectra of the filters using a Cary Spectrometer. The reflection spectra of the filters are inferred from their transmission spectra since there is only negligible absorption in each case.

Example 1: A composition of 23% cholesteryl chloride (CC) in cholesteryl nonanoate is prepared. The composition has a λ_0 value of about 5500 A. A thin layer of the liquid crystal composition is applied to a Polaroid sheet and subsequently another Polaroid sheet is placed over the free surface of the liquid crystal layer. The Polaroid sheets are arranged so that their respective axes of polarization have an angular relationship of about 90° to each other. The optical filter is positioned in the path of a light beam emitted by a broad band incandescent source of visible light so that the incident radiation is at normal incidence to the filter. The filter is then rotated over varying angles and the transmitted wavelength band measured for each position. The results obtained are shown in Figure 2.5d.

Example 2: Compositions of 24.5% of cholesteryl chloride (CC) in cholesteryl decanoate (CD) and 20% CC in cholesteryl bromide (CB) respectively are prepared. The former is a left-handed composition with a λ_0 of about 5200 A while the latter composition is right-handed and has a λ_0 of about 5700 A. A thin layer of one composition is applied to a thin glass plate and a thin Mylar sheet then applied over the free surface of the liquid crystal layer. A thin layer of the second liquid crystal composition is similarly arranged between a thin glass plate and thin Mylar sheet.

The two components of the optical filter are then arranged in the path of a light beam emitted from a broad band incandescent source of visible light with the incident light being normal to the filter. The two components of the filter are positioned so that the thin glass plate members of each component are adjacent to each other and further so that they may be rotated independently of each other. The component having the 20% CC in CB composition is rotated over varying angles and the transmitted wavelength band measured for each position. The results obtained are shown in Figure 2.5e.

Bandwidth Narrowing

A process described by *J.E. Adams and W.E.L. Haas; U.S. Patent 3,720,456; March 13, 1973; assigned to Xerox Corporation* provides a method for narrowing the bandwidth of an optical signal. The process comprises directing a linearly polarized optical signal through an optically negative liquid crystal film and a linear analyzer arranged in tandem. The optically negative liquid crystal film is chosen to have characteristics such that some of the light is extinguished whereas the remainder escapes extinction and is transmitted.

According to a preferred form of the process, where the incident optical signal is randomly polarized (or unpolarized) a linear polarizer element is arranged before the liquid crystal film to convert the optical signal to linearly polarized light. The pitch and the thickness of the optically negative liquid crystal film are selected so that some portion of the incident light is extinguished and the remainder is transmitted, the relative amounts of each being determined by the extent to which it is desired to narrow the bandwidth. According to the method, the bandwidth of the optical signal can be narrowed to substantially any desired width.

Induced Optical Activity

According to a process described by *F.D. Saeva and J.J. Wysocki; U.S. Patent 3,780,307; December 18, 1973; assigned to Xerox Corporation* the electronic transitions of extrinsically optically inactive materials become circularly dichroic when these materials are introduced into an optically negative liquid crystalline environment thereby providing unusual and highly advantageous properties in such compositions. The circularly dichroic optical properties induced in the normally extrinsically optically inactive materials can be utilized for imaging, optical filters and other related applications. The following examples illustrate the process.

Example 1: A composition made up of 2% of 4-methoxy-1-phenylazonaphthalene and 98% of a 60%/40% mixture of cholesteryl chloride/cholesteryl nonanoate is prepared. The composition is placed in an electric cell comprising two ⅛ x 1 inch optically transparent conductive tin oxide-coated quartz plates (made by subliming an optically transparent conductive tin oxide layer onto the quartz plates) separated by a 1 mil thick Mylar spacer. The cell is then placed in a Cary 60 Spectropolarimeter in the circular dichroism mode. A large negative induced circularly dichroic band (cotton effect) is observed at 390mμ for the azo compound incorporated in the optically active cholesteric liquid crystalline mesophase.

An electric field of about 200 volt/mil is placed across the transparent electrodes and the induced optical activity is no longer observed. The electric field causes the cholesteric mesophase to be converted to the nematic mesophase. The electric field is removed from the electrodes and the induced circularly dichroic band reappears.

Example 2: An optically negative liquid crystalline composition is formed by mixing together about 60% p-methoxybenzylidene-p-n-butylaniline (MBBA) and about 40% cholesteryl chloride. At 25°C the composition has a pitch band centered at about 1.8μ and an absorption band centered at about 325mμ. A film of the composition is placed between parallel glass supports separated by a 1 mil thick Mylar spacer. This combination is placed in a polarizing microscope and the absorption band is observed to have induced circularly dichroic activity. At 21°C the optical activity in the induced band is about 112° at 489mμ.

The temperature of the liquid crystal film is then reduced to about -5°C. The pitch band is observed to move to about 750mμ. The absorption band is observed to remain at about 325mμ, however, the optical activity becomes about 185°. Thus a thermal change is converted into a change in the activity of the absorption band and a change in location in the pitch band. The liquid crystal film can be arranged between a linear polarizer and a linear analyzer to provide a device which is capable of functioning as an image converter.

Example 3: An optically negative liquid crystalline composition is formed by mixing together about 1% MBBA and about 99% of a 33.9% cholesteryl chloride-66.1% cholesteryl nonanoate mixture. A film of the composition is placed between quartz plates separated by a 1 mil Mylar spacer. The combination is placed in a Cary 60 Spectropolarimeter in the circularly dichroic mode.

At 25°C the pitch band is observed at about 600mμ and preferentially transmits right-hand circularly polarized light. The induced circularly dichroic absorption band, located at about 325mμ preferentially transmits left-hand circularly polarized light. The device can be used to generate circularly polarized light at both 600 and 325mμ, with each respective transmitted band being circularly polarized in the opposite sense from that of the other.

DISPLAY AND IMAGING SYSTEMS

Electrically Controlled Imaging Systems

J.J. Wysocki, J.E. Adams, J.H. Becker, R.W. Madrid and W.E. Haas; U.S. Patent 3,718,380; February 27, 1973; assigned to Xerox Corporation describe a system for transforming an optically negative liquid crystalline substance to an optically positive liquid crystalline mesophase by an applied electrical field, and an imaging system where the electrical field-induced transition images a liquid crystalline member.

Figure 2.6a is a cross-sectional view of a liquid crystalline imaging member. Figure 2.6b is an isometric view of an example of a liquid crystalline imaging member where the desired image is defined by the shape of the spacing member. Figure 2.6c is an isometric view of a liquid crystalline imaging member where the desired image is defined by the shape of at least one of the electrodes. Figure 2.6d is an isometric view of an example where a typical liquid crystalline imaging member is viewed between polarizers.

FIGURE 2.6: IMAGING SYSTEMS

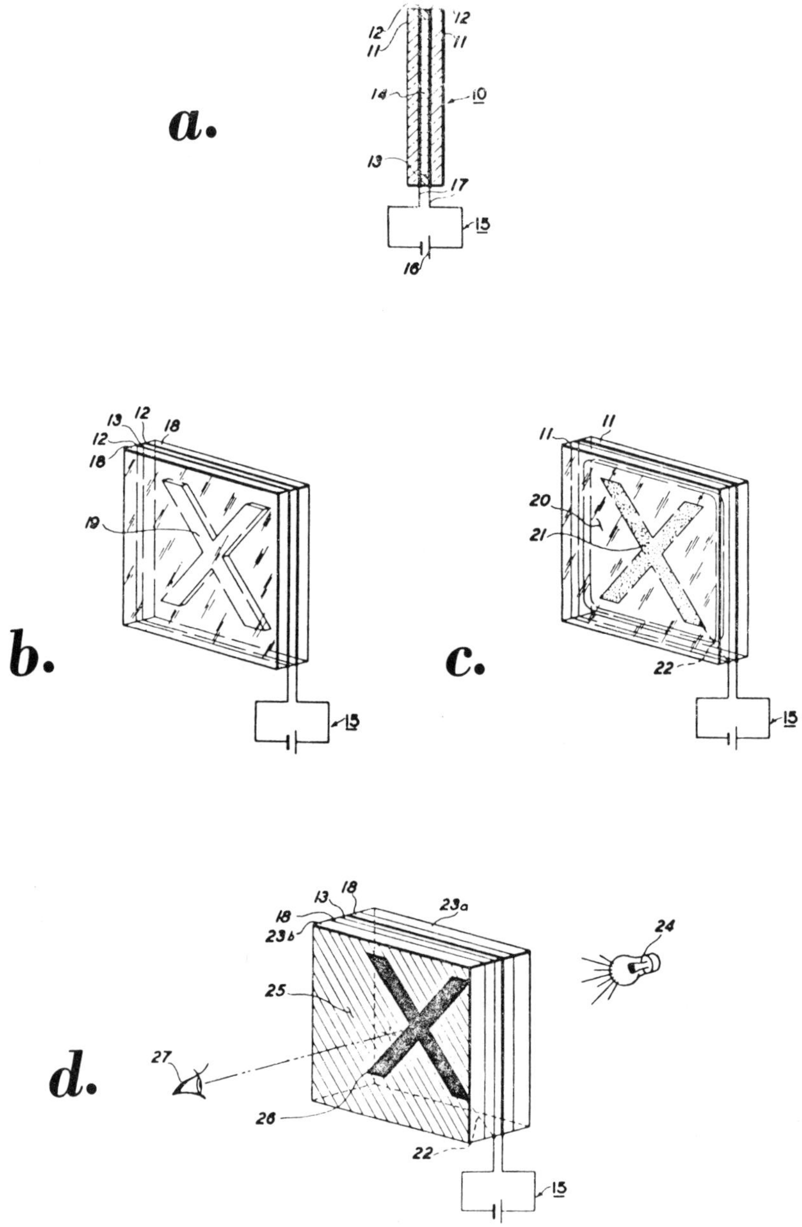

(continued)

FIGURE 2.6: (continued)

e.

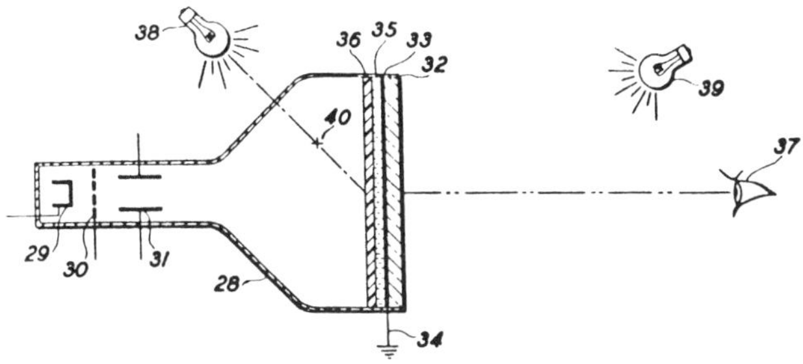

f.

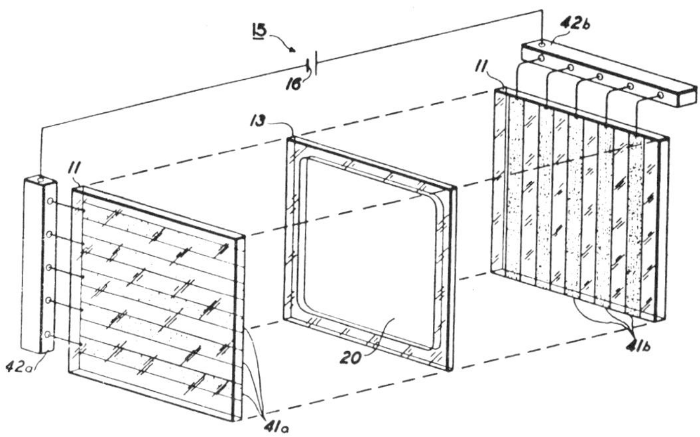

g.

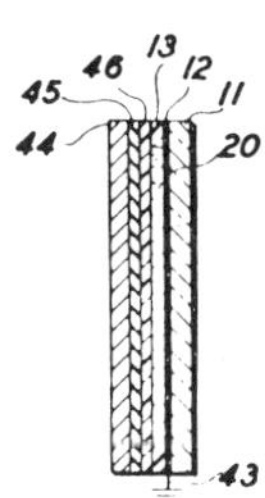

(continued)

FIGURE 2.6: (continued)

h.

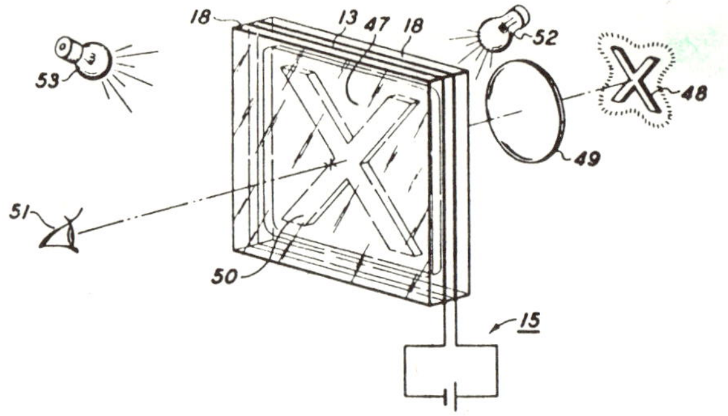

i.

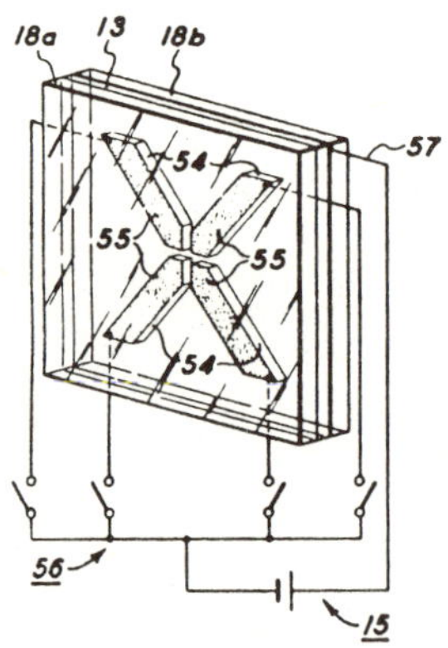

j.

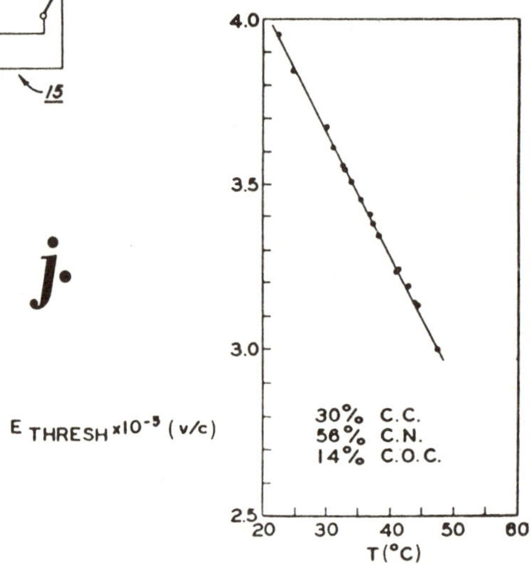

Source: J.J. Wysocki, J.E. Adams, J.H. Becker, R.W. Madrid and W.E. Haas; U.S. Patent 3,718,380; February 27, 1973

Figure 2.6e illustrates an imaging system where a liquid crystalline imaging member is imaged by an electron beam address system. Figure 2.6f is an exploded isometric view of an imaging system where a liquid crystalline imaging member is imaged by an X-Y address system. Figure 2.6g is a cross-sectional view of an imaging system where a liquid crystalline imaging member is imaged by an electrostatic latent image on an electrostatic latent image support surface. Figure 2.6h is an isometric view of a liquid crystalline imaging member imaged by a thermal image projection address system. Figure 2.6i is an isometric view of a uniplanar, multiple cell, liquid crystalline imaging member. Figure 2.6j is a threshold field vs temperature plot for the electrical field induced cholesteric-nematic phase transition of a specific cholesteric liquid crystal mixture.

In Figure 2.6a a typical liquid crystalline imaging member is shown in cross-section where a pair of transparent plates **11** having substantially transparent conductive coating **12** upon the contact surface, comprise a parallel pair of substantially transparent electrodes. An imaging member where both electrodes are transparent is preferred where the imaging member is to be viewed using transmitted light; however, a liquid crystalline imaging member may also be viewed using reflected light thereby requiring only a single transparent electrode while the other may be opaque. The transparent electrodes are separated by spacing member **13** which contains voids which form one or more shallow cups which contain the liquid crystalline film or layer which comprises the active element of the imaging member. A field is created between the electrodes by an external circuit **15** which typically comprises a source of potential **16** which is connected across the two electrodes through leads **17**. The potential source may be either DC, AC or a combination therof.

It has been discovered that when cholesteric liquid crystals or a mixture of cholesteric liquid crystalline substances is used in an electrode sandwich such as that described in Figure 2.6a, that high electrical fields across the liquid crystal film cause an electrical field-induced phase transition to occur where the optically negative cholesteric liquid crystalline substance transforms into an optically positive liquid crystalline state. This transition is believed to be the result of the cholesteric liquid crystal transforming into the nematic liquid crystalline mesophase structure.

Cholesteric liquid crystals in the cholesteric state are typically translucent, for example, like a milky white, opalescent layer, when first placed in the unbiased electrode sandwich. When the high electric field is placed across the liquid crystal film, the field-induced phase change is observable because the liquid crystal film becomes transparent in areas where the field is present. When viewed between polarizers with transmitted light, for example as described in conjunction with Figure 2.6d below, the areas in which the field-induced phase change has taken place appear dark, while the unchanged, translucent, light scattering and birefringent, cholesteric areas still retain their light appearance.

Any other means suitable for enhancing the contrast of the imaged areas may also be used in place of the polarizers. Hence, it is seen that either field or nonfield areas in a liquid crystalline imaging sandwich may be used to create the desired image, with or without the addition of means for image enhancement. The optical negative-positive, cholesteric-nematic phase transition has also been confirmed by experiments using a polarizing microscope (see examples below).

In the liquid crystal imaging members described in Figure 2.6a the electrodes may be of any suitable transparent conductive material. Typical suitable transparent, conductive electrodes include glass or plastic substrates having substantially transparent and continuously conductive coatings of conductors such as tin, indium oxide, aluminum, chromium, tin oxide, or any other suitable conductor. These substantially transparent conductive coatings are typically evaporated onto the more insulating, transparent substrate. NESA glass, a tin oxide coated glass, is a commercially available example of a typical transparent, conductive electrode material.

The spacer, **13** of Figure 2.6a, which separates the transparent electrodes and contains the liquid crystal film between the electrodes, is typically chemically inert, transparent, not birefringent, substantially insulating and has appropriate dielectric characteristics. Materials suitable for use as insulating spacers include cellulose acetate, cellulose triacetate, cellulose acetate butyrate, polyurethane elastomers, polyethylene, polypropylene, polyesters, polystyrene and polycarbonates.

The liquid crystal imaging film **14** may comprise any suitable cholesteric liquid crystal or mixture of cholesteric liquid crystals. Typical cholesteric liquid crystals include derivatives from reactions of cholesterol and inorganic acids, for example, cholesteryl chloride, cholesteryl bromide, cholesteryl iodide, cholesteryl fluoride, cholesteryl nitrate; esters derived from reactions of cholesterol and carboxylic acids; for example cholesteryl crotonate; cholesteryl nonanoate; cholesteryl hexanoate; cholesteryl formate; cholesteryl docosonoate and cholesteryl chloroformate.

Mixtures of liquid crystals can be prepared in organic solvents such as chloroform, petroleum ether, methylethyl ketone and others, which are typically subsequently evaporated from the mixture thereby leaving the liquid crystalline mixture. Alternatively, the individual liquid crystals of the liquid crystalline mixture can be combined directly by heating the mixed components above the isotropic transition temperature.

In Figure 2.6b the example of the liquid crystal imaging member described in Figure 2.6a is shown with the desired image defined by the shape of the void areas in the spacer gasket **13**. As before, transparent electrodes **18** are separated by the spacer, but the entire desired image area **19** comprises the liquid crystal film or layer. In this example the entire inner faces of the transparent electrodes comprise substantially transparent conductive coating **12** and the conductive coatings are electrically connected to external circuit **15**.

In operation there is an electrical field across the entire area of the spacer **13**, however the image caused by the electrical field-induced phase transition in the liquid crystal film, occurs only in the image area where the liquid crystal film is present. Again here, depending upon whether the desired image is to be viewed by transmitted or reflected light, both or only one of the electrodes, respectively, may be transparent.

In Figure 2.6c another preferred example of the liquid crystal imaging member described in Figure 2.6a is shown. In Figure 2.6c the desired image is defined by the shape of an electrode, and therefore by the shape of the corresponding electrical field. The imaging member here comprises transparent plates **11** separated by the spacer gasket having void area **20** filled with liquid crystals, and comprising substantially the entire area of the spacer layer **13**. The desired image is defined by the shape of the substantially transparent conductive coating shown at **21**, which is affixed to the inner surface of one or both of the transparent support plates, and is affixed only in the desired image configuration.

The example illustrated in Figure 2.6c shows only one of the two electrodes in image configuration, however it will be understood by those skilled in the art that both electrodes could easily be made in a matched pair to define the same desired image. When the single image electrode configuration is used, the second electrode will comprise transparent plate **11** with substantially transparent conductive coating **12** upon the entire area of the inner surface of the transparent electrode. It is noted that a very thin, or substantially invisible conductor **22** is necessary in this example to electrically connect the electrode in the desired image configuration to the external circuit which is similarly connected to the conductive coating of the opposite electrode.

In operation this system will produce an electric field only in areas where there are parallel electrodes, i.e., between the electrode in the desired image configuration, and the opposite electrode, whether or not the second electrode is also in the desired image configuration. Again here, one of the electrodes may be opaque if it is desired to observe the imaged member by reflected light rather than transmitted light.

In Figure 2.6d a liquid crystal imaging member comprising a pair of substantially transparent electrodes **18** sandwiching a spacer **13** containing a liquid crystal film is shown being observed between polarizers **23**. As described earlier, cholesteric liquid crystals are typically translucent, and when placed in the electrical field characteristic of the advantageous system of the process, the optical negative-positive, cholesteric-nematic phase change occurs thereby rendering the transformed portion of the liquid crystalline film substantially transparent to transmitted light.

At the same time, those areas remaining in the cholesteric state remain light scattering and translucent. When such a liquid crystalline imaging sandwich is observed between polarizers, light from source **24** is planar polarized while passing through polarizer **23a**, scattered by the translucent cholesteric liquid crystalline substance in nonimage areas **25** (same as **20** in Figure 2.6c), and is transmitted by the field-induced nematic state areas **26**.

A viewer **27** then sees the planar-polarized light which passes through polarizer **23b** which originated from the source and was scattered and passed through the nonimage portion of spacer **13**. Although the light was polarized by polarizer **23a** in a plane crossed with the plane of polarizer **23b**, the scattering effect of the cholesteric liquid crystal between the substantially transparent electrodes scattered sufficient amounts of the originally planar-polarized light to allow some of it to pass through polarizer **23b**. However, in the image areas **26**, the effect of polarizers **23**, when these polarizers have their respective planes of polarization crossed, is to cut off the light transmitted through polarizer **23a** and transformed image area so that the image area appears dark, as illustrated.

In Figure 2.6e another example is illustrated where an electron beam address system is provided for the generation of an imagewise field across a liquid crystalline imaging member. In Figure 2.6e the electron beam address system is within vacuum tube **28**, and the address system itself comprises electron gun **29**, accelerator **30** and deflector **31** which are provided with electrical leads through the vacuum tube so that suitable electrical circuitry may be connected therewith to operate the electron beam imaging system. The liquid crystalline imaging member in conjunction with the electron beam address system comprises substantially transparent substrate **32** upon which substantially transparent conductive coating **33** is affixed. The substantially transparent coating is also grounded at **34**.

The liquid crystalline imaging layer **35** may be coated directly onto the substantially transparent conductive layer, provided the liquid crystalline material has a sufficiently high viscosity to support itself on a surface even when the surface is oriented vertically or horizontally with the liquid crystalline imaging layer on the lower surface of such horizontal imaging member. Alternatively, transparent insulating layer **36** may be placed over the liquid crystalline imaging layer to protect and support the liquid crystalline material.

The impingement of electrons directed from the electron gun and impinging upon layer **36** or the liquid crystalline imaging layer creates a momentary field when taken in conjunction with the grounded transparent conductive layer. This momentary field across the liquid crystalline layer causes the optical negative-positive, cholesteric-nematic phase transition, thereby creating an image visible to an observer **37**.

The electron beam address system may be used in conjunction with an electroded liquid crystalline imaging member where the sum of the fields created by the electrode system and the electron beam address system is sufficient to create a total field of field strength greater than the threshold field value of the particular liquid crystalline imaging member.

The electron beam address system does not itself illuminate the image which it creates. However, external sources of light **38** and **39** are illustrated to show that the liquid crystalline imaging member comprising the face of the vacuum tube may be observed by light transmitted, as from light source **38**, or reflected, as from light source **39**. Alternatively, transmitted light may originate from a source placed inside the vacuum tube at a location indicated, for example, at point **40**. However, the in-tube source of illumination should be so placed so as to not interfere with the electron beam which creates the image on the face of the tube.

The substantially insulating overlayer **36** may be selected from materials which may exhibit secondary emission effects when bombarded by the initial electron beam, and such secondary emissions may be the imaging emissions, or, alternatively, the secondary emission system may be suitable as a method for erasing the previous image on the liquid crystalline imaging member.

In Figure 2.6f an X-Y address system suitable for imaging a liquid crystalline imaging member is illustrated in exploded isometric view. The liquid crystalline imaging film is placed in void area **20** within the transparent and substantially insulating spacer gasket **13**. The liquid crystalline film and the spacer are sandwiched between a pair of substantially transparent electrodes comprising transparent support plates **11** upon which strips of substantially transparent, conductive material **41** is coated. The substantially transparent electrodes are oriented so that conductive strips **41b** and conductive strips **41a** on the respective electrodes cross each other in an X-Y matrix or grid.

Each conductive strip in each set of parallel strips **41a** and **41b**, is electrically connected to a circuit system **42** which is suitable for selective or sequential operation. Through selection systems **42** and the external circuit **15** including source of potential **16**, an electric field suitable for creating the optical negative-positive cholesteric-nematic phase transition can be created across selected points or a selected sequence of points in the illustrated imaging system. It will be understood that substantially transparent conductive strips **41** may vary in width from a very fine, wire-like structure to any desired strip width. In addition, one support plate may be opaque where the imaging system is to be observed from one side, using only reflected light.

In Figure 2.6g the advantageous liquid crystalline imaging member is shown in a system where the member is imaged by an electrostatic latent image on an electrostatic latent image support surface such as a photoconductive layer. In Figure 2.6g typically transparent, support substrate **11** is shown with typically substantially transparent conductive coating **12** which is grounded at **43**. Adjacent the substantially transparent conductive coating spacer gasket **13** contains liquid crystalline imaging layer **20** between the substantially transparent grounded electrode and an electrostatic latent image support surface.

The electrostatic latent image support surface may be any substantially insulating surface which is capable of supporting an electrostatic latent image. The electrostatic latent image supported by the surface, when taken in conjunction with the grounded substantially transparent conductive layer creates an imagewise electrical field across the liquid crystalline imaging layer, and the field causes the optical negative-positive, cholesteric-nematic phase transition.

A typical electrostatic latent image support surface is a photoconductive plate, for example as typically used in xerography. As illustrated, conductive supporting layer **44** supports a layer of photoconductive material **45**, typically comprising selenium, mixtures of sulfur and selenium, phthalocyanine, or any other photoconductive insulating layer. Between photoconductive insulating layer **45** and liquid crystalline imaging layer **20** there may be an insulating protective layer **46** which may retain the liquid crystalline substance in some examples, and at the same time layer **46** protects both the liquid crystalline layer and the photoconductive insulating layer.

An electrostatic latent image may be created on the photoconductive layer by a variety of means, for example, by typical xerographic process steps as described in U.S. Patent 2,297,691, by corona charging, or by breakdown techniques described, for example in U.S. Patent 2,982,647, and U.S. Patent 2,825,814 and 2,937,943, or by charge patterns conforming to selected shaped electrodes or combinations of electrodes and created by the TESI discharge technique described in U.S. Patents 3,023,731 and 2,919,967.

The electrostatic latent image can be created on the photoconductive insulating layer before the photoconductive layer is brought into proximity with the liquid crystalline imaging layer. However, in certain cases it is possible to create the electrostatic latent image on the photoconductive insulating layer while the photoconductor is in close proximity with the liquid crystalline imaging layer. The resulting image is then viewed through substantially transparent layers **11** and **12**.

In Figure 2.6h a liquid crystalline imaging member is shown imaged by a thermal image projection address system. Here the liquid crystalline imaging member comprises substantially transparent electrodes **18** separated by spacing gasket **13** which encloses a cholesteric liquid crystal with space **47** comprising substantially all of the area within the spacer gasket.

In this example a source of a thermal image **48**, here a heat source in the desired image configuration, is shown in position with conventional means **49** for focusing and projecting a thermal or optical image, and the thermal image **50** appears in the liquid crystalline film in the areas where the liquid crystalline substance is heated into the temperature range required for the optical negative-positive, cholesteric-nematic phase transition, while at the same time, the liquid crystalline imaging member is biased by external circuit **15** so that the field across the cholesteric liquid crystalline film is sufficient to cause the phase transition when the imaged area of the film reaches the transition threshold temperature.

An observer **51** sees the transformed image area become transparent to transmitted light from source **52**, or go blank in reflected light from source **53**, as the phase change occurs in response to the thermal image. If the imaging member is observed while between polarizers, as described in Figure 2.6d, the imaged area of the liquid crystal imaging member would appear dark to the observer while the nonimaged areas **47** would remain substantially unchanged as a result of light scattering.

Figure 2.6i illustrates a uniplanar, multiple cell, liquid crystalline imaging member suitable for use with the electrical field-induced phase transition imaging system of the process. In the imaging member of Figure 2.6i substantially transparent electrodes **18** are separated by the typical spacer gasket which contains voids **54** which contain the liquid crystal film. Corresponding areas **55** are the shaped, substantially transparent, conductive coating on the inner surface of electrode **18a**. Each one of these cells **55** is capable of being selectively imaged either individually, or jointly, as desired, through the use of switching system **56** in external circuit **15** which is also connected to the substantially transparent conductive layer on the inner surface of substantially transparent electrode **18b**, by electrical lead **57**.

It will be appreciated that uniplanar, multiple cell, imaging members such as the one in Figure 2.6i may be designed so that various combinations of the desired image cells may be imaged to create any desired figure or character in any language or number system. It will also be appreciated, that either a character shaped spacer as described in Figure 2.6b, or a character-shaped electrode system as described in Figure 2.6c, in addition to the combination shaped spacer and electrode system here described in Figure 2.6i, can be used in such a uniplanar, multiple image system.

The following examples further define the process with respect to transforming a cholesteric liquid crystalline substance to the nematic liquid crystalline mesophase by an applied electrical field. The parts and percentages are by weight unless otherwise indicated.

Example 1: The electrical field-induced optical negative-positive, cholesteric-nematic mesophase transition is observed in the following liquid crystal imaging member: a mixture of liquid crystal materials of about 30% cholesteryl chloride, about 56% cholesteryl nonanoate, and about 14% cholesteryl oleyl carbonate is prepared. A layer of the cholesteric liquid crystal mixture is placed between and in contact with a pair of substantially transparent electrodes of NESA glass. The transparent electrodes are separated by about one-fourth mil and the electrodes are connected in circuit with an electrical generator and a source of white light is placed behind the entire imaging member.

The liquid crystal mixture normally appears as a light scattering, white, opalescent, translucent sheet. An electrical field of field strength of about 10^5 volts/cm is then applied across the liquid crystal layer. In the presence of this field, the imaged portion of the liquid crystal layer assumes a colorless, transparent appearance.

The same imaging member is viewed with a linear analyzer while plane-polarized light is normally incident on the imaging member. By viewing with the linear analyzer, the optical negative-positive, cholesteric-nematic transition is observed because it is impossible to extinguish the field of view with the linear analyzer when the cholesteric liquid crystal layer is viewed with plane-polarized light incident before the phase transition takes place.

After the phase transition takes place, the normally incident, plane-polarized light emerges plane-polarized through the electrical field-induced image areas, and may be extinguished when viewed with the linear analyzer. However, the transformed liquid crystal material is not optically isotropic. Plane polarized light obliquely incident on the imaging member does not emerge plane-polarized, indicating that the film is acting as a birefringent material with an optic axis perpendicular to the surface of the liquid crystal layer, i.e., parallel to the applied electrical field.

Example 2: The electrical field-induced mesophase transition is confirmed by observations of a liquid crystal imaging member using a Leitz Ortholux polarizing microscope with a 50X objective lens and convergent light. The cholesteric liquid crystal mixture disclosed in Example 1 is placed between a pair of substantially transparent electrodes. The lower electrode is a NESA glass plate, as described in Example 1.

The upper electrode is a glass microslide about 0.2 mm in thickness, upon which a substantially transparent (optical density of about 0.3) conductive coating of chromium is evaporated. The conductive coatings of the transparent electrodes are spaced apart by a spacing gasket about one-fourth mil in thickness, comprising Mylar, a polyethylene terephthalate polyester. The liquid crystal mixture contacts both transparent electrodes and is contained within the spacing gasket.

This imaging member or cell is then observed through the Leitz Ortholux polarizing microscope, and the uniaxial character of the film and the sign of the birefringence are determined. Cholesteric liquid crystals have negative birefringence characteristics (see Gray, G.W., *Molecular Structure and the Properties of Liquid Crystals,* Academic Press, New York, 1962, page 40).

Without applying a field to the imaging member, a gypsum first-order-red plate is inserted and a color figure results which is characteristic of optically negative uniaxial materials, (as reported in Wahlstrom, E.E., *Optical Crystallography,* John Wiley & Sons, New York, 1966, 3rd edition, page 202); here the first and third quadrants of the conoscopic patterns are yellow while the second and fourth are blue. Then a field of about 10^5 volts/cm is applied. With the applied field, the conoscopic color pattern quickly inverts to a pattern characteristic of optically positive uniaxial material; that is, the first and third quadrants become blue while the second and fourth become yellow.

Nematic and smectic liquid crystalline mesophase structures exhibit positive birefringence characteristics (see Gray, supra). It is therefore clear that the electrical field-induced transition is either a cholesteric-nematic or a cholesteric-smectic phase transition. It is believed that the phase transition of the process is a cholesteric-nematic phase transition. Such a phase transition induced by magnetic fields has previously been reported in Sackmann, Meiboom & Snyder, *J. Am. Chem. Soc.,* 89, 5981 (1967).

Example 3: Using the cholesteric liquid crystal imaging member described in Example 2, the temperature dependence of the threshold of the phase change inducing field is measured. The experimental data for threshold field necessary for phase transition versus temperature are shown in Figure 2.6j where the field is expressed in (volts/cm) x (10^{-5}), and the temperature is expressed in degrees Centigrade. It is observed that the threshold field decreases rapidly with increasing temperature. The isotropic temperature of the mixture of about 30:56:14 cholesteryl chloride:nonanoate:oleyl carbonate, respectively, is about 57°C. This isotropic temperature is the point at which the liquid crystal mixture changes to the completely liquid physical state.

The threshold field versus temperature relationship of Figure 2.6j defines the operating range for the electrical field-induced cholesteric-nematic phase transition imaging member of the process having a cholesteric liquid crystal imaging layer comprising the mixture of about 30% CC:56% CN:14% COC. The preferred operating ranges for this mixture are: temperatures in the range of about 20° to about 57°C and electrical field strengths in the range of about 2.7×10^5 to about 4.0×10^5 volts/cm of liquid crystalline material.

Additional studies involving electrically controlled imaging systems are described by *J.J. Wysocki, J.E. Adams, J.H. Becker, R.W. Madrid and W.E.L. Haas; U.S. Patents 3,711,713; January 16, 1973; 3,707,322; December 26, 1972; and 3,652,148; March 28, 1972; all assigned to Xerox Corporation.*

Matrix-Addressed Display

A process described by *W.B. Hatfield and D.L. White; U.S. Patent 3,837,730; September 24, 1974; assigned to Bell Telephone Laboratories, Incorporated* relates to a multicell matrix-addressed display in which each cell comprises a thin film of a host material in which a pleochroic guest material is dissolved.

The discovery in recent years of a number of new electrooptic effects in liquid crystal materials has aroused considerable interest in these materials for various display applications of practical importance. Among the potentially attractive features of liquid crystal display devices are their simplicity, low cost, and low power operation. Moreover, such devices have the potential for exhibiting relatively high brightness and good contrast ratios under ambient light conditions.

For applications requiring the display of more than say four or five characters, it has been recognized that matrix addressing of a liquid crystal display is required to keep the number of individual input leads to the display at a manageable level. For example, a display composed of an array of 50 x 100 spots has 5,000 distinct addressable points. Clearly, a matrix addressing scheme (i.e., an x-y coincidence method) in which the number of input leads would be 150 is generally preferred over an individual spot-by-spot addressing technique which would require 5,000 separate external wiring connections.

Various types of matrix-addressed liquid crystal devices are known in the display art. See, for example, J.J. Wysocki, et al, *Proceedings of the Society Information Display,* 13, 114 (1972) and *Electronics,* 45, Number 22, 53 (172). However, these and other known display devices suffer from one or more of the following disadvantages: they are relatively dim in appearance, they must be operated in a so-called projection mode, they require a relatively high voltage for operation, or they exhibit such low storage time that the size of the matrix array and therefore its characteristic display capabilities are limited.

This process provides a high-capacity matrix-addressed liquid crystal display characterized by low voltage operation and advantageous brightness and contrast ratio properties. The method includes a plurality of liquid crystal cells arranged in a matrix array of rows and columns. Each cell comprises a cholesteric-type liquid crystal material characterized by positive dielectric anisotropy, with a pleochroic dye dissolved in the liquid crystal material.

Quiescently, in the absence of an applied voltage, the proposed cell is relatively dark. The application to the cell of an energizing voltage above a well-defined transition voltage causes the cell to become transparent. When the voltage applied to a transparent cell is reduced to a point just below the transition voltage (to a so-called holding voltage), the transparent condition of the cell persists as a metastable state for a considerable time. Eventually, the metastable state decays to the dark state.

A row-by-row matrix-addressed display composed of the proposed liquid crystal cells, each operated at either its energizing or its holding voltage, exhibits advantageous brightness and contrast-ratio characteristics and can be constructed to form relatively large arrays. Moreover, the transition voltage of each cell can be established at a value that is compatible with complementary MOS integrated circuits. A display system including such circuits requires only microwatts of power.

The specific illustrative liquid crystal cell or device shown in Figure 2.7 includes two parallel glass plates **12** and **14** having a thin dye liquid crystal film **16** sandwiched between them. Thin conductive coatings **13** and **15** of, for example, In_2O_3-SnO_2 on the respective facing surfaces of the plates provide transparent electrodes by means of which a uniform electric field can be established through the depicted film. Spacers **18** and **20** made, for example, of a glass frit or a polymer such as Mylar or Teflon maintain the thickness of the film at any desired value in the range of approximately 5 to 30 microns measured in the z direction.

FIGURE 2.7: LIQUID CRYSTAL CELL

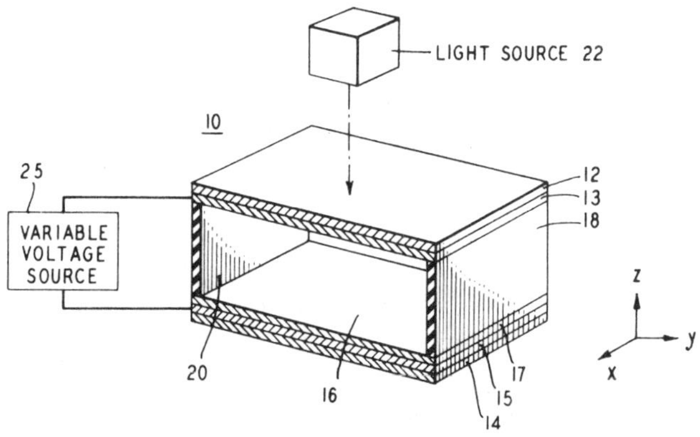

Source: W.B. Hatfield and D.L. White; U.S. Patent 3,837,730; September 24, 1974

Light source **22** shown in Figure 2.7 comprises, for example, a conventional light bulb that provides ambient illumination for the device **10**. The source **25** is connected to apply a variable voltage between coatings **13** and **15** of the cell **10**. This voltage may be either DC or AC in nature. For AC operation, frequencies up to about 20 kHz may be used, with frequencies of 50 to 1,000 Hz being preferred. Typical magnitudes for the voltage applied by the source **25** to the coatings will be specified later below.

The device may be operated in the so-called transmission mode (in which case element **17** is omitted). In that operating mode the light source is disposed on one side of the device and an observer, i.e., a human being, would position himself on the other side of the device to view light transmitted through the transparent plate **14**. Alternatively, the top surface of the layer **15** may be coated with a highly reflecting material **17**, which is, for example, silver. Or a diffuse or specular reflecting mirror coating (not shown) may be placed on or spaced apart from the bottom surface of the plate **14**.

In any of these last-mentioned cases the observer would position himself to view light that is reflected from such a coating and then transmitted back through the upper plate **12**. In other words in that operating mode, the reflection mode, both the source **22** and the observer are located on the same side of the device.

Film **16** included in the device includes as one main component thereof a liquid crystal material that exhibits helical molecular ordering. This component may comprise, for example, a cholesteric liquid crystal material of the type described by F.J. Kahn in "Cholesteric Liquid Crystals for Optical Applications," *Applied Physics Letters,* March 1971, pages 231 to 233.

However, for reasons set forth below, it is generally advantageous to constitute the liquid crystal material in the film **16** as a mixture of a nematic liquid crystal and an optically active material such as, for example, a cholesteric material of the type described in the aforementioned Kahn article. (The term optically active is employed to characterize a material in which the plane of polarization of linearly polarized light is rotated as the light passes through the material.)

Other examples of optically active materials suitable for inclusion in the mixture are the active amyl ester of N-(p-cyanobenzylidene-p-aminocinnamic acid), hereinafter designated CBACAA, the active amyl ester of N-(p-carboxybenzylidene)-p-aminobenzonitrile and l-menthol. Advantageously, the molecules of the optically active material should be similar to the host liquid crystal (i.e., long and rod-like), as is the case with CBACAA.

Such a mixture exhibits a helical molecular structure whose pitch is a function of the relative concentrations of the nematic liquid crystal material and the optically active material. More specifically, if the densities of the mixed substances differ only slightly, the ratio of the pitch of the pure optically active material to that of the mixture will be equal to the concentration of the optically active component of the liquid crystal mixture.

Accordingly, by employing such a mixture, it is possible in effect to formulate a substance whose helical pitch assumes any desired value. This provides an important design flexibility that makes it possible to tailor the switching voltage and contrast ratio properties of the device **10** to meet the requirement of various specific applications. It is characteristic of such a mixture that it exhibits a different structural order property than do cholesterics or mixtures of cholesterics alone. In turn, this property is the basis for maximizing the anisotropy and absorption characteristics of the thin film.

One advantageous liquid crystal mixture adapted to be included in the device of Figure 2.7 comprises a nematic liquid crystal mixture having positive dielectric anisotropy mixed with an optically active material such as the aforementioned CBACAA. The nematic liquid crystal mixture comprises, for example, equal parts by weight of the N-(p-butoxy), N-(p-hexyloxy), and N-(p-octanoyloxy)benzylidene-p-aminobenzonitriles, or mixtures of 4-N-(4'-n-hexylbenzylidene)aminobenzonitrile and 4-N-(4'-n-propylbenzylidene)aminobenzonitrile, or mixtures of 4'-cyanophenyl-4-n-butylbenzoate and 4'-cyanophenyl-4-n-heptylbenzoate, as reported by A. Boller, H. Scherrer, M. Schodt and P. Wild, *Proceedings of the I.E.E.E.,* 60, 1002 (1972).

Mixtures of the nematic liquid crystal and the optically active material in the range of about 5 to 15% by weight of the optically active material impart to the mixture a helical molecular ordering with a pitch value of about 1.0 to 3 microns. Both of the above materials included in the liquid crystal mixture include rod-like molecules and exhibit a positive dielectric anisotropy. In such a material the electric susceptibility parallel to the longitudinal axis of a molecule thereof is greater than the electric susceptibility perpendicular to the longitudinal molecular axis.

When the helical pitch of the mixture approximates the wavelength of visible light directed at the thin film, the light propagates through the film in right- and left-handed elliptically polarized modes. When the pitch exceeds about 10 wavelengths, the mixture appears to the incident light very much like a nonhelical nematic-type structure which propagates linearly polarized modes. In accordance with the principles of the process, this second mentioned propagation mode is to be avoided on a steady-state basis in the absence of an above-transition-value electric field in the film. Hence, as a practical matter, the pitch of the mixture is usually designed to be less than about 10 wavelengths of the incident visible light.

The pitch of the helical configuration of the mixture should not be made much less than a wavelength of the incident visible light. If it is made much less, the voltage required across the film **16** to establish an electric field therein sufficient to unwind the helix may be too large for some applications of practical importance. Accordingly, the pitch is usually designed not to be less than about one wavelength. A second main component included in the thin film is a pleochroic material such as, for example, a soluble nonionic dye. One such suitable dye is 2-(4-N,N-diethylaminoazophenyl)-5-nitrothiazole mixed in the liquid crystal mixture in a concentration of about ¼ to 1% by weight of the dye.

It is characteristic of such a dye that it absorbs light preferentially: its optical absorption spectrum is a strong function of the polarization of the incident light with respect to the orientation of the dye molecules. Specifically, when the electric vector representative of light polarization is parallel to the longitudinal axis of such a dye molecule, the molecule is relatively absorptive of that portion of the incident light whose wavelengths fall within a prescribed range. As a result of this absorption, the dye molecules dispersed in the film exhibit a characteristic color. The device is then said to be in its dark state. On the other hand, when the electric vector of the incident light is perpendicular to the longitudinal axis of such a dye molecule, the molecule appears relatively transparent. Accordingly, under that set of circumstances, the film or some preselected localized portion thereof assumes a relatively clear or bright state.

An important factor for achieving high contrast ratio and brightness in a device of the type described herein is that the transition moment vector of the pleochroic guest has an orientation with respect to the crystallographic axis of the liquid crystalline host that is most nearly parallel thereto. Ideally the transition moment axis of the dye molecule should be most nearly parallel to its long molecular axis. The order of the dye relative to that of the liquid crystal optic axis depends on the order of the liquid crystal host molecules and the nature of the dye molecules.

By selection of appropriate dye molecules, the ordering of the dye molecules may be greater than that of the host molecules. Dyes which exemplify this characteristic are 2-(4-N,N-diethylaminoazophenyl)-5-nitrothiazole and, in addition, the nematic liquid crystalline dye N-(p-methoxybenzylidene)-1-amino-4-(4'-cyanophenylazo)naphthalene. Classes of elongated dyes which also meet this requirement are azoic dyes, azo dyes, quinone dyes, merocyanine dyes, disperse dyes and liquid crystalline dyes.

When an electric field is applied to a cholesteric liquid crystal material that is characterized by positive dielectric anisotropy, the pitch of the material increases and eventually, when the electric field reaches a certain critical or transition field, E_c, the pitch becomes essentially infinite; i.e., the material appears to be nematic. The critical field for such a material is defined by

$$E_c = 1/q \quad \sqrt{K/\Delta\epsilon}$$

where q = pitch in cm, K = elastic constant, and $\Delta\epsilon$ = dielectric anisotropy. If the pitch is made relatively long, say only several pitches equal the thickness of the cell, and the dielectric anisotropy is large, as it is in the liquid crystal materials specified herein, the corresponding critical voltage can be controlled to be quite small.

When the voltage applied by the source **25** of Figure 2.7 to the coatings **13** and **15** reaches a point at which the corresponding field in the film reaches the so-called critical or transition value, the helical molecular ordering is in effect unwound to establish a homeotropic molecular ordering. For film thicknesses and helical pitches having values in the ranges discussed above, AC voltages less than 10 to about 30 volts rms are effective to accomplish such molecular reordering.

It is advantageous to establish initially in the film a helical ordering in which the helical axes in the immediate vicinity of the electrodes **13** and **15** lie in a plane parallel to the x-y plane. This arrangement is achieved by applying to the coatings **13** and **15** a surfactant, for example N,N-dimethyl-N-octadecyl-3-aminopropyltrimethoxysilyl chloride, which causes the surface molecules of the film **16** to align with their longitudinal axes perpendicular to the coatings. Even with such an imposed surface alignment, however, major bulk portions of the film are characterized by a helical ordering whose axes are on the average at an angle to the coatings. Hence, the absorption phenomenon will still occur in the film to a considerable extent in the absence of an applied electric field.

When an electric field exceeding the transition value is applied to such a device, the bulk helical ordering is unwound and as a result the film is relatively transmissive. When the applied field is entirely removed from such an arrangement (or at least reduced below the transition value to or below a relatively low value of field designated the spontaneous value), the film almost immediately exhibits both absorption and scattering of incident light. This is a stable no-field (or low-field) effect which causes the film to present a milky or cloudy appearance. Switching of the device between these clear and cloudy states may be done in a relatively high speed manner. As indicated, these states correspond with the establishment in the film of above-transition and zero (or low) fields, respectively.

The process is based on the discovery that when the voltage applied across the film of Figure 2.7 is first lowered below the critical or transition voltage V_c to a value above the spontaneous voltage V_s, the visible state of the cell **10** remains substantially unchanged. In time cholesteric domains generated by line dislocations are observed growing from some nucleation sites in the liquid crystal mixture. If the applied voltage is reduced to a point just under the critical voltage (say to a value about 95% of the critical voltage), the number of such domains is relatively small and their rate of growth is relatively slow.

In that case the liquid crystal mixture remains in its clear metastable state for a considerable length of time, for example, minutes. If the applied voltage is reduced to a point considerably less than the critical voltage (say to a value about 50% of the critical voltage but still above the spontaneous voltage), the number and rate of growth of such domains is increased. In this latter case the mixture remains in the metastable clear state for a shorter interval, for example, seconds or fractions of a second.

Thus, a cell of the type depicted can be controlled to exhibit plural distinct states. Above V_c there is only one state, namely the clear nematic state. Below V_s there is only one state, namely the turbid cholesteric condition. Between V_c and V_s there are two states, the stable colored state and the metastable clear state which in time switches to the colored or turbid condition as a result of the aforementioned domain-growth phenomenon.

Accordingly, by applying to the cell a so-called holding voltage having a value between V_c and V_s, the liquid crystal mixture is either maintained in its colored-turbid state (if the previously applied voltage was below V_c) or maintained temporarily in its clear state (if the previously applied voltage was above V_c). Switching from the colored state to the clear state is achieved by temporarily raising the applied voltage above V_c. To go almost immediately from the clear to the colored state is accomplished by lowering the applied voltage below V_s.

Significantly, the aforementioned phenomena are polarity independent. Hence, either a positive or a negative voltage having a value intermediate V_c and V_s is effective to maintain a previously energized or clear cell in its metastable clear state. Similarly, either a positive or a negative holding voltage is effective to maintain a previously colored state in its turbid condition. In accordance with one aspect of the process, plural cells each of the type shown and described above are combined to form a matrix-addressed liquid crystal display.

High Frequency Color Display System

J.E. Adams, J.H. Becker and W.E.L. Haas; U.S. Patent 3,836,230; September 17, 1974; assigned to Xerox Corporation describe a liquid crystal color display system which utilizes a plurality of films or layers of optically negative liquid crystalline materials as passive light discriminating screens. The display system may be operated in a two color mode or a three color mode and has the capability of producing various color hues. The system lends itself to operation at very high frequency.

Figure 2.8a is a partially schematic, cross-sectional view of one form of the color display system. Figure 2.8b is a partially schematic, cross-sectional view of a display screen. Figure 2.8c is a partially schematic, cross-sectional view of another form of a display screen. Figure 2.8d is a partially schematic, cross-sectional view of another form of a display screen used according to the process. Figure 2.8e is a partially schematic, cross-sectional view of an example of a display screen including means to shift the light reflection band of one liquid crystal film.

FIGURE 2.8: HIGH SPEED COLOR DISPLAY SYSTEM

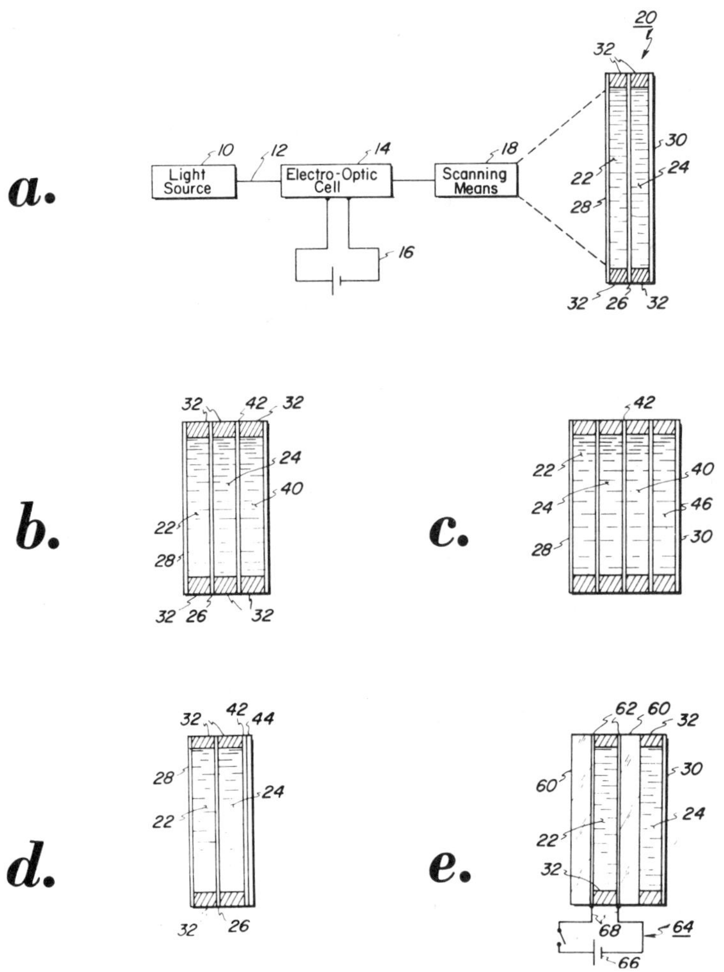

Source: J.E. Adams, J.H. Becker and W.E.L. Haas; U.S. Patent 3,836,230; Sept. 17, 1974

In Figure 2.8a a light source **10** directs a beam of light **12** through electrooptic cell **14**, which is connected to a source of electrical potential **16**, and in which the state of polarization of the light beam is varied in accordance with a predetermined input signal. From the electrooptic cell, the light beam is directed into a scanning means, or light beam deflector, **18**, which causes the light beam to be scanned or deflected in a predetermined pattern across a display screen, generally designated **20**, comprising optically negative liquid crystal films **22** and **24** arranged on either side of optional support member **26** and enclosed by optional protective layers **28** and **30** and gaskets **32**.

The liquid crystal films are comprised of optically negative liquid crystalline materials which have opposite intrinsic rotary sense, i.e., one of the materials is right-handed and the other is left-handed, and different light reflection bands. It should be recognized that the incident light may strike the display screen at other than normal incidence. Preferably the angle of incidence will be from about 0° to about 30°. For optimum performance of the display system of the process the incident light will be directed upon the display screen at normal incidence.

In operation of the advantageous color display system, a substantially completely linearly polarized light beam, which contains at least those wavelengths of radiation which correspond to, or are at least within, the light reflection bands of the optically negative liquid crystal films is directed through electrooptic cell **14** which is capable of adjusting the state of polarization of the light in accordance with some predetermined sequence of signals such that the light exiting from the cell is right-hand circularly polarized (RHCPL), left-hand circularly polarized (LHCPL), or some combination thereof. (Linearly polarized light would correspond to equal amounts of each sense.) This pattern is then projected or displayed upon the display screen by scanning means **18** and the image formed on the screen may be viewed in reflection or in transmission.

According to the form of the process illustrated in Figure 2.8a, the option of two colors, or some adjustable hue which is a combination thereof, is available at each point on the display screen. For example, consider that optically negative liquid crystal film **22** is right-handed and exhibits a light reflection band having a center wavelength λ_1 and film **24** is left-handed and exhibits a light reflection band having a center wavelength λ_2. Of course, it will be understood that the position of the films in Figure 2.8a is merely illustrative; the same effect would be obtained if the position of the films relative to each other were to be reversed, i.e., the left-handed film arranged in front of the right-handed film.

When the electrooptic cell is adjusted to produce only RHCPL, wavelengths of light at and around λ_1 which are present in the incident beam will be reflected by film **22** and none of the incident light will be reflected by film **24** since the incident light does not contain any LHCPL. When the cell is adjusted to produce only LHCPL, wavelengths of light at and around λ_2 which are present in the incident beam will be reflected by film **24** and none of the incident light will be reflected by film **22** since the incident light does not contain RHCPL.

The electrooptic cell may also be adjusted to produce a mixture of RHCPL and LHCPL (which could correspond to linearly polarized light or elliptically polarized light) in which case, both liquid crystal films will reflect wavelengths of light at and around λ_1 and λ_2, respectively, which are present in the incident light beam and, consequently, the reflected light will represent some combination of the wavelengths at and around λ_1 and λ_2. Of course, it will be appreciated that the color hue of the light reflected by the display screen is adjustable where the electrooptic cell produces a combination of RHCPL and LHCPL. A range of color hues can be reflected, for example, by the technique of time averaging address.

For example, in a two color display system such as is illustrated in Figure 2.8a wherein, for purposes of illustration, λ_1 is in the red region of the visible spectrum and λ_2 is in the blue region, if a red to blue intensity ratio of 4:1 is desired for a total dwell time of about 20 milliseconds; optically negative liquid crystal film **22** may be made to reflect for about

16 milliseconds and liquid crystal film **24** made to reflect for about 4 milliseconds, and so forth.

The image created on the display screen **20** may be of a variety of configurations depending upon the manner in which the state of polarization of the light beam is changed and how the light beam is deflected. An image can be produced, for example, by a raster scan of the screen displaying, for example, a dynamic image such as a conventional line television image. The advantageous color display system of the process is capable of being operated at very high speed, for example, at television frame rates since there are available electrooptic cells which can operated at frequencies in the vicinity of about 10^{10} cps, as will be discussed in detail below herein. It is again noted that the display system may be viewed in both the transmission and reflection modes.

As stated above, the light beam **12** which is directed upon electrooptic cell **14** typically should be substantially completely linearly polarized and contain at least those wavelengths of radiation which correspond to, or are at least within, the light reflection bands of optically negative liquid crystal films **22** and **24**. Hence, in Figure 2.8a, light source **10** may be any source which is capable of providing linearly polarized light, either intrinsically or in combination with other elements.

Typical suitable light sources include multiple wavelength lasers which intrinsically emit substantially completely linearly polarized light; light emitting luminescent films and gas discharge arcs, e.g., xenon arcs, which should be used with some means for removing all but linearly polarized light with a preferred direction such as a linear polarizer; and broad band sources such as incandescent light sources which are employed with band pass filters such as, for example, those described in U.S. Patent 3,669,525. It should also be noted that the type of light source used in the system controls the degree of color separation which may be obtained since each optically negative liquid crystal film reflects a band of radiation typically having a bandwidth of about 200 A or more.

Hence, it can be seen that the quality of the color separation which may be attained according to the color display system is dependent to some extent upon the location of the center wavelengths of the light reflection bands of the optically negative liquid crystal films and to a larger extent, upon the quality of the light provided by the light source. Accordingly, it is preferred to use a laser which emits two appropriate colors as the light source because of the simplicity of the element and the quality of the light it provides.

The electrooptic cell may comprise any material which is capable of producing RHCPL, LHCPL or mixtures thereof in response to an electric field established by voltage from potential source **16** which may be AC, DC or combination thereof. The state of polarization of the light produced by the electrooptic cell depends upon the voltage applied. The applied electric field changes the index of refraction of the material and, consequently, changes the sense of the light output. Hence, the state of polarization of the light output may be varied in accordance with a predetermined pattern.

Typical suitable materials which may be used in the electrooptic cell include, for example, potassium tantalate-niobate, calcium pyroniobate, dihydrogen phosphate, ammonium dihydrogen phosphate, potassium dihydrogen phosphate, mixtures thereof and others. Electrooptic cells suitable for use in the color display system are commercially available. As noted above, since typical electrooptic cells are capable of operating at frequencies up to about 10^{10} cps, the color display system is capable of operating at very high speeds.

Scanning means, or light beam deflecting means, **18** may be any device capable of distributing the light beam across the display screen. Scanning means **18** may comprise an electrooptic device, an acoustooptic device, a high speed mirror system or a hybrid system such as, for example, an acoustooptic device in one direction combined with a high speed mirror system in another direction.

Any suitable liquid crystalline substance or composition having optically negative properties may be used to provide liquid crystal films or layers **22** and **24**. It is again noted that the films typically have different light reflection bands and are of the opposite handedness, i.e., one is right-handed and the other is left-handed. The liquid crystal films may be of any thickness and are preferably from about 0.5 to about 50μ. For optimum results the liquid crystal films will have a thickness of from about 0.5 to about 20 microns.

It should be noted that the effects obtained in accordance with the process do not depend critically on film thickness except to the extent that the film thickness should preferably be greater than λ_0 in order to produce full intensity in the reflected light. When the films are relatively thin, i.e., when they have a thickness less than wavelength λ_0, they will typically not selectively reflect light around λ_0 as efficiently as has been previously described. When the films are of relatively great thickness, then the molecules will usually not align properly and the observed behavior will not be consistent with the optimum results attainable with the advantageous color display system. For liquid crystal films which have a λ_0 in the visible region, film thickness of from about 3 to about 10 microns will provide optimum results.

The display screen shown in Figure 2.8e is similar to that shown in Figure 2.8a except that protective layer **28** and support member **26** are replaced with substantially transparent, isotropic electrodes which in this case comprise substantially transparent isotropic support plates **60** having substantially transparent conductive coating **62** upon the contact surface. A field can be created across the electrodes by an external circuit **64** which typically comprises a source of potential **66** which is connected across the two electrodes through leads **68**. The circuit may also contain suitable switching means. The potential source may be either AC or DC.

The electrodes may be of any suitable transparent and isotropic conductive material. Typical suitable transparent isotropic conductive electrodes include glass substrates having substantially transparent and continuously conductive coatings of conductors such as tin, indium oxide, aluminum, chromium, tin oxide, or any other suitable conductor. These substantially transparent conductive coatings are typically evaporated onto the more insulating transparent substrate. NESA glass, a tin oxide coated glass is a commercially available example of a typical transparent, conductive electrode material.

For ease of illustration only optically negative liquid crystal film **22** in Figure 2.8e has been shown as being sandwiched between a pair of electrodes. Alternatively and preferably, optically negative liquid crystal film **24** may also be arranged between a pair of electrodes in a similar manner. It should be noted here that the electrodes need not be all substantially transparent in all the examples where they are used. For example, if liquid crystal film **24** in Figure 2.8e were sandwiched between a pair of electrodes and the display screen were viewed in reflection the rear electrode could be opaque.

Support member **26** may be any suitable material which is optically isotropic and transparent to the incident wavelengths of radiation. Typical suitable materials include fused silica, glass and any other materials having the requisite optical and physical properties. It is preferred to utilize materials which have an index of refraction about the same as that of the liquid crystal films to minimize reflection loss of light. The liquid crystal films **22** and **24** need not be adjacent opposite surfaces of the same support member as is shown in Figure 2.8a; in another form the liquid crystal films may be arranged independently of each other and spatially separated. The support member typically has a thickness in the range of from less than 1 to about 100 microns or more.

The liquid crystal films are typically tacky, soft, viscous, glassy or liquid and therefore it is preferred to have the films overcoated or encapsulated, with thin protective layers **28** and **30**, respectively. The purpose of the protective layers is to keep the liquid crystal films confined and free of any contamination. Hence, the protective layers may be of any suitable material, flexible or rigid, which is isotropic, optically transparent to the incident radiation and which is nonreactive with the liquid crystal films or any other components of the device. It is preferred to use materials which have an index of refraction about the same as that of the liquid crystal films to minimize loss of light. Protective layers **28** and **30** typically have thickness of from about 0.25 to about 10 mils. It should be noted here that where the display screen is viewed in reflection, layer **30** may comprise a dark colored material so as to improve contrast.

Figure 2.8b illustrates a preferred form of a display screen suitable for use in the color display system. The display screen shown in Figure 2.8b is similar to display screen **20** in Figure 2.8a with the exception that the former includes an additional liquid crystal film and an optional filter. Thus, in the form illustrated in Figure 2.8b, elements which are common to the display screen shown in Figure 2.8a are identified by the same numeral.

Referring to Figure 2.8b, there is seen a display screen which includes three optically negative liquid crystal films **22, 24** and **40**, respectively. Liquid crystal films **24** and **40** are separated by optional filter **42**. The incorporation of the liquid crystal film **40** in the display screen increases the range of color hues which can be observed at any point on the screen. Liquid crystal film **40** may comprise any of the common materials and may be right-handed or left-handed.

Moreover, liquid crystal film **40** may have a λ_0 value which is less than, greater than, or intermediate between, the λ_0 values of films **22** and **24**. According to a preferred example liquid crystal films **22, 24** and **40** would have λ_0 values in the red, blue and green regions of the visible spectrum, respectively. The light source in a color display system employing the display screen illustrated in Figure 2.8b would, of course, emit wavelengths of light which are reflected by the liquid crystal film **40**.

According to this example, it would be possible to mix green light with red and blue light in the light reflected at any point on the display screen. It is to be noted, however, that this result will only occur when the light impinging on the display screen has at least some component which has the appropriate state of polarization.

For example, if liquid crystal film **40** is right-handed, it will only reflect RHCPL in the region of its reflection band and, consequently, wavelengths of light in the green would only be mixed in with the total light reflected at any point on the display screen when the light reaching the screen is completely RHCPL or contains at least some component which is RHCPL. The converse is true in the instance where liquid crystal film **40** is left-handed. As was discussed above, electrooptic cell **14** shown in Figure 2.8a is capable of providing RHCPL, LHCPL or combinations thereof in response to varying applied voltages.

It is possible to use filters in the display screen to improve device performance. These filters may be used to selectively attenuate various wavelengths of light or to attenuate all wavelengths of light equally. In the example illustrated in Figure 2.8b optional filter **42** is incorporated in the display screen to reduce the amount of light contributed by liquid crystal film **40** to the total light reflected at any point on the display screen.

According to the preferred case, where liquid crystal films **22, 24** and **40** have λ_0 values in the red, blue and green regions respectively, filter **42** would preferably be one which substantially attenuates red and blue light and, to a lesser degree, green light. It should be noted that the filter may be a neutral density filter, i.e., one which attenuates all wavelengths of light equally. The neutral density filter may be a developed silver halide emulsion, partially metallized glass or a screen or other pattern comprised of opaque areas and light transmitting windows.

Figure 2.8c illustrates another preferred form of a display screen suitable for use in the color display system. The display screen shown in Figure 2.8c is similar to the display screen shown in Figure 2.8b with the exception that the former includes an additional liquid crystal film **46**. This liquid crystal film is selected so as to have substantially the same light reflection band as that of liquid crystal film **40** but to have the opposite handedness. For example, where film **40** is right-handed, then film **46** will be a left-handed film. It will be recognized that, according to this example, wavelengths of light reflected by liquid crystal films **40** and **46** will always be present in the light reflected from any point on the display screen when the incident light beam includes these appropriate wavelengths. It should be noted that this result is obtained irrespective of the particular state of polarization imparted to the incident light by the electrooptic cell at any point. Providing liquid crystal films with substantially the same light reflection band and opposite handedness may be conveniently accomplished by forming mixtures of liquid crystal substances. It has been found that the pitch of two component mixtures of certain cholesteric liquid crystalline materials is a strong function of composition. Over a wide range of materials, the pitch (and, consequently, the light reflection band) of a mixture can be accurately represented by a weighted average of the ingredients.

Further, if components with opposite intrinsic screw sense are mixed, there will exist one composition corresponding to no net rotation or infinite pitch. On either side of this composition, the other compositions will either be right-handed or left-handed. Thus, liquid crystal films having any desired light reflection band and handedness may be prepared by resort to this technique. A detailed discussion of this phenomenon is found in *Liquid Crystals and Ordered Fluids,* J.E. Adams, W. Haas and J.J. Wysocki, page 463, Plenum Press, 1970.

Figure 2.8d illustrates another form of a display screen suitable for use in the color display system. In Figure 2.8d there is provided a reflecting surface **44** to reflect incident light. In this manner any wavelengths of light emitted by the light source may be incorporated in the light reflected at any point on the display screen. According to a preferred example where the light source emits wavelengths in the red, blue and green regions and optically negative liquid crystal films **22** and **24** have λ_0 values in the red and blue respectively, optional filter **42** preferably is one which substantially completely absorbs red and blue light and is substantially completely transparent to green light. The following example further defines the liquid crystal color display system. All parts and percentages are by weight unless otherwise indicated.

Example: A display screen is constructed by providing approximately 25 micron thick films of a 9:1 mixture of cholesteryl chloride:cholesteryl nonanoate (a right-handed composition having a λ_0 of about 6000 A) and a 1:3 mixture of cholesteryl chloride and cholesteryl nonanoate (a left-handed composition having a λ_0 value of about 5000 A) on opposite sides of a microscope slide. The free surfaces of the liquid crystal films are covered with microscope slides.

Light from a 25 watt concentrated zirconium arc lamp connected to a DC 50 watt power supply is directed upon the display screen through an electrooptic cell made up of a Glan Thompson polarizer and a quarter wave plate. The electrooptic cell is mechanically rotated to change the state of polarization of the incident light. At one orientation of the electrooptic cell the display screen reflects red light. The electrooptic cell is rotated 90° and the display screen reflects green light.

Series of Alphanumeric Images

J.J. Wysocki and R.W. Madrid; U.S. Patent 3,622,224; November 23, 1971; assigned to Xerox Corporation describe an imaging system having a multiplanar series of alphanumeric images in conjunction with an electrooptic imaging cell suitable for displaying one or more of the images.

Figure 2.9a is a cross-sectional, partially schematic view of an imaging member or cell. Figure 2.9b is an isometric view of an imaging cell where the desired image is defined by the shape of the spacing member. Figure 2.9c is a partially schematic, exploded isometric view of a multiplanar, multiple cell, imaging device where the desired images are defined by the shape of the respective spacing members.

FIGURE 2.9: ALPHANUMERIC ELECTROOPTIC IMAGING DEVICE

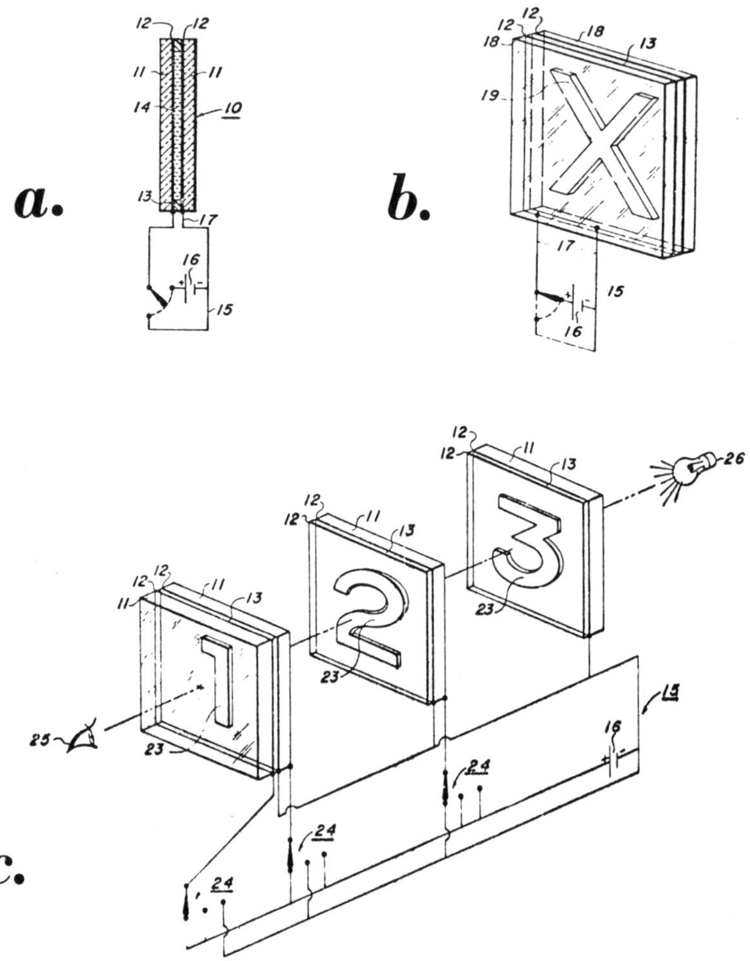

Source: J.J. Wysocki and R.W. Madrid; U.S. Patent 3,622,224; November 23, 1971

In Figure 2.9a a typical imaging cell **10** is shown in cross section where a pair of transparent plates **11** having substantially transparent conductive coating **12** upon the contact surface, comprise a parallel pair of substantially transparent electrodes. An imaging member wherein both electrodes are transparent is preferred where the imaging member is to be viewed using transmitted light; however, such imaging members may be viewed using reflected light thereby requiring only a single transparent electrode with the other opaque.

The transparent electrodes are separated by spacing member **13** which contains voids which form one or more shallow cups or tanks which contain a suitable film or layer of electrooptic material which comprises the active element of the imaging member. In operation a field is created between the electrodes by an external circuit **15** which typically comprises a source of potential **16** which is connected across the two electrodes through leads **17**. The potential source may be either AC, DC, or a combination thereof.

The preferred system for use in the imaging cell of the process is one in which cholesteric liquid crystals or a mixture of cholesteric liquid crystalline substances is used in an electrode sandwich such as that described in Figure 2.9a, and electrical fields across the liquid crystal film cause an electrical field-induced phase transition to occur wherein the optically negative cholesteric liquid crystalline substance transforms into an optically positive liquid crystalline state.

This transition is believed to be the result of the cholesteric liquid crystal transforming into the nematic liquid crystalline mesophase structure. Cholesteric liquid crystals in the cholesteric state are typically translucent, for example, like a milky white, opalescent layer, when first placed in the unbiased electrode sandwich. When the electric field is placed across the liquid crystal film, the field-induced phase transition is observable because the liquid crystal film becomes transparent in areas where the field is present. Such transparent areas allow transmitted light to pass through them substantially unaffected.

The basic electrooptic imaging cell described in Figure 2.9a is further described in Figure 2.9b where the desired image configuration is defined by the shape of the void areas in spacer gasket **13**. In this case, the desired image area may be thought of as the shape of these void areas, the shape of the spacer gasket, or more correctly the shape of the electrooptic material within the voids defined by the shape of the spacer gasket.

Transparent electrodes **18** are separated by the spacer, but the entire desired image area **19** comprises the electrooptic imaging material. In this example the entire faces of the transparent electrodes comprise substantially transparent conductive coating **12** and the conductive coatings are electrically connected to external circuit **15**. In operation, there is an electrical field across the entire area of the spacer, however the image caused by the response of the electrooptic material occurs only in the area **19** where the electrooptic imaging material is present.

In Figure 2.9c a preferred example of the multiplanar electrooptic imaging system is shown in a partially schematic, exploded isometric view. It will be understood that the alphanumeric units illustrated in Figure 2.9c as being spaced apart, are typically stacked in contact one behind the other. The individual imaging cells of the multiplanar system illustrated in Figure 2.9c are similar to the cell illustrated in Figure 2.9b. Substantially transparent supporting layers **11** have substantially transparent, electrically conductive layer **12** thereon, and the individual imaging cells comprise two such substantially transparent electrodes having a spacing gasket separating the electrodes.

The desired alphanumeric images in this example comprise the voids **23** in spacing members **13**, which are typically filled with electrooptic imaging material, and which correspond to void area **19** shown in Figure 2.9b. External circuit **15** having the source of electrical potential **16** therein, and also illustrated with a selective switching system **24** is used to selectively actuate the desired cell or cells of the multiplanar imaging system so that the desired image appears to the eye of the viewer **25** when transmitted light from source **26** passes through the substantially transparent imaging system. Here substantially transparent is intended to mean any degree of transparency sufficient to allow any or all of the system's images to be viewed when the system is in operation.

When the device illustrated in Figure 2.9c is used with an electrooptic imaging material in areas **23**, and the material responds to the application of an electrical field across it by becoming transparent, i.e., as cholesteric liquid crystalline materials typically do, the device may be more advantageously used with image contrast enhancing devices such as polarizers or crossed polarizers. For the devices used with electrooptic imaging materials which respond to the electrical field across them by becoming translucent and light scattering, i.e., as nematic liquid crystalline substances typically do, it may often be just as advantageous to use the system without the image contrast enhancing devices.

Although the device illustrated in Figure 2.9c may be constructed by simply stacking a series of electrooptic imaging cells like the one illustrated in Figure 2.9b, one behind the other, an advantageous way of conserving materials and making the desired imaging system as small as possible is to use transparent support members **11** having substantially transparent conductive coatings **12** on both sides of the support member, as illustrated in Figure 2.9c.

Variable Color Display Using Induced Deformational Stresses

E.N. Sharpless; U.S. Patent 3,768,886; October 30, 1973; assigned to Liquid Crystal Industries, Inc. describes an optical display device capable of producing an infinite variety of color variations and patterns. The process involves introducing a liquid crystalline material into a display device of special construction. Desirably the liquid crystalline material is selected which has a characteristic of variable light scattering and attendant transmittance at room temperatures or at least at those environmental temperatures under which the device is employed.

For example, a liquid crystalline material can be selected, which is capable of such variable transmittance at operating temperatures in the region of a projection bulb or other illuminating light source, for example one utilized in an advertising sign or other display arrangement. The selected liquid crystalline material preferably exhibits a variable scattering or transmiting characteristic when subjected to mechanical deformation, such as occasioned by shear or flow stresses.

Accordingly, the display device or liquid crystal cell is provided firstly with a light transmitting wall to permit viewing of the contained liquid crystalline material. Secondly, the liquid crystalline cell is associated with means for inducing mechanical stresses within the contained liquid crystalline material. This can be accomplished in a variety of ways: for example, the liquid crystalline cell can be constructed with means permitting the displacement of one wall structure thereof relative to another. Circulating means can be associated with the cell or display device for inducing flow and attendant shear stresses within the liquid crystalline material.

The display device is capable of a large number of applications, for example as an aesthetic novelty, decorative wall and table top panels, back drops for stages and other illuminated areas, and numerous analogous applications. For this purpose, background or environmental illumination is sufficient for viewing the infinite variety of color patterns resulting from a stress manipulation of the display device.

Figure 2.10a is an isometric view, partially broken away, of one form of a liquid crystal cell or display device. Figure 2.10b is a cross-sectional view of the device of Figure 2.10a and taken along reference line **1A–1A**. Figure 2.10c is an isometric view of a similar display device provided in conjunction with rotatable polarizing means. Figure 2.10d is an isometric view of another form of display device and incorporating modified means for inducing deformational stresses within the liquid crystal. Figure 2.10e is a cross-sectional view of the device as shown in Figure 2.10d and taken along reference line **IV–IV.**

Figure 2.10a shows a display device **10** in the form of a liquid crystalline support **12.** In this arrangement, the support is in the form of a flat container having apposed wall structures **14, 16** of any suitable size and shape. In the arrangement shown, the wall structures are substantially coextensive although this is not an essential requirement. In point of fact, one of the wall structures can be significantly smaller than the other wall structure, as long as one wall structure is joined about its periphery to the other wall structure.

Likewise, the wall structures need not be of square or rectangular configuration as illustrated but can be of some other configuration for example circular as illustrated in Figure 2.10d. It is contemplated however, that any geometrical or nongeometrical, symmetrical or nonsymmetrical shape can be employed for either or both of the wall structures. As noted previously, the wall structures need not be coterminous. Further, the wall structures need not be planar as shown in Figure 2.10b but instead one or both sides can be dished or otherwise configured as indicated in Figure 2.10e.

FIGURE 2.10: VARIABLE COLOR DISPLAY DEVICES

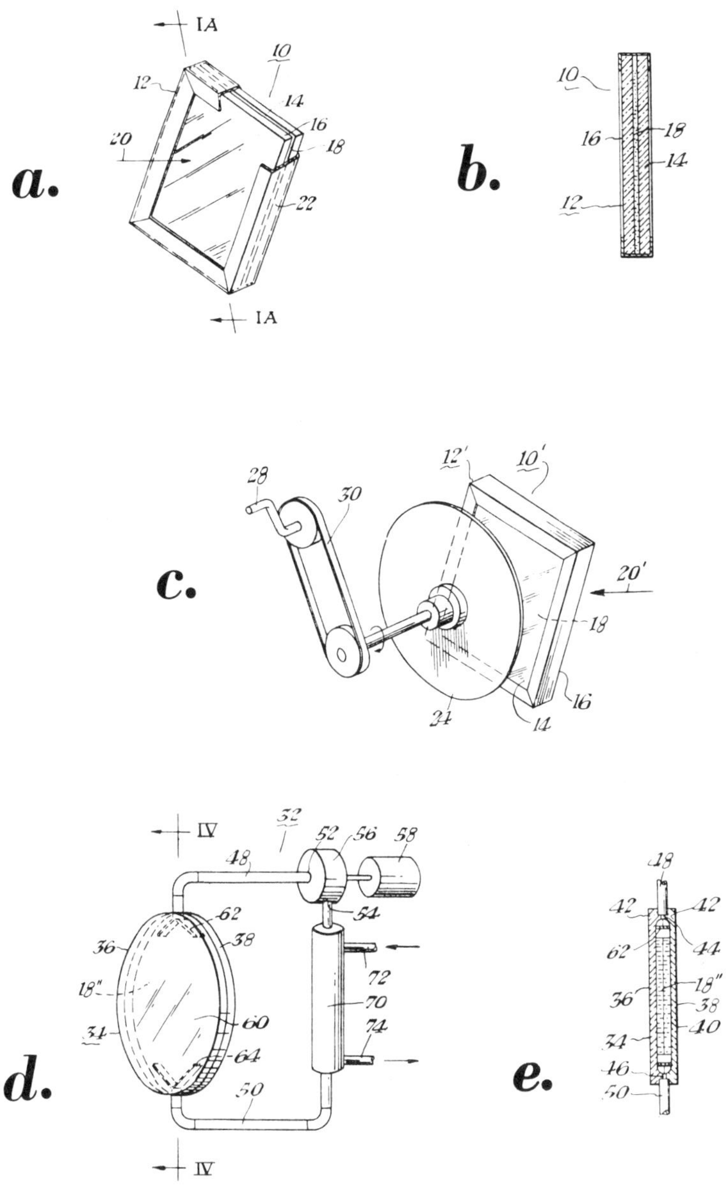

Source: E.N. Sharpless; U.S. Patent 3,768,886; October 30, 1973

Depending on the manner in which the wall structures **14, 16** are joined, the resilience and hence the thickness of either or both of the wall structures may or may not be critical. Such criticality, whether encountered depends on the manner in which deformational stresses are to be applied to a liquid crystalline material **18** confined between the wall structures. In the arrangement of the display device as illustrated in Figure 2.10a, at least one of the wall structures is sufficiently thin or is made of a suitably plastic material as to lend a resilient character to the wall structure.

Thus, the wall structure, such as the wall structure **16** can be bent or otherwise deformed toward the wall structure **14** when a force is applied more or less transversely thereto as denoted by arrow **20.** By thus bending one of the wall structures relative to the other, the liquid crystalline material, which is supported, in this example, between the wall structures in film-like form, is caused to flow generally away from the region of applied force (the arrow) to other regions of the volume confined within the liquid crystal support **12.** The application of the force **20** and the resultant flow of the liquid crystal develops shear and other deformational stresses within the liquid crystal. Such stresses modify the light scattering and attendant transmittance characteristics of the liquid crystal material and result in an endless variety of color changes and patterns.

In order to observe these aesthetic color changes, at least one of the wall structures is light transmitting, and preferably both of the wall structures are light transmitting to permit the display device **10** to be observed from either side thereof. Desirably, both of the wall structures are made transparent for optimum visual characteristics.

As noted previously, at least one of the wall structures is joined about its periphery to a surface of the other wall structure. In the Figure 2.10a arrangement such joining means are further arranged to peripherally seal one wall structure to a surface of the other. In the display device, such joining and sealing means includes a pressure-sensitive tape **22,** which is compatible with the material of the wall structures and covers their coextensive peripheral edges.

The liquid crystal is thereby sealed in the context of film thicknesses within the space defined by the slightly separated wall structures and the peripheral tape. It will be understood, of course, that the separation between the wall structures can be different from that illustrated, depending upon the relative quantity of liquid crystal which is used. Generally, a relatively thin film of liquid crystal should be enclosed between the wall structures to conserve the liquid crystalline material.

In those cases where the joining and sealing tape is quite flexible and more or less loosely applied at the wall edges or is at least somewhat elastic, one or both of the wall structures can be made thicker and hence less resilient. In such cases, an eccentric application of the deformational force will cause one of the wall structures to become slightly canted or angulated or otherwise displaced relative to the other in order to induce deformational flows in the liquid crystalline material. Such deformational flows are, of course, aided by the elasticity and/or edge slackness of the joining and sealing tape.

The liquid crystalline material is selected from one or more of those materials which exhibit variation in light scattering and attendant transmittance characteristics under deformational stresses. Desirably, such variations are within the visible range at room temperatures or at whatever ambient temperature conditions prevailing in the area of utilization of the display device. As an example of the latter situation, the liquid crystalline material can be one of those which exhibit visual stress variation in the aforementioned characteristics at or near body temperature, and is therefore useful when the display device is held in the observer's hand. Larger display devices can of course be bathed with infrared radiation, if their liquid crystals are not of the room temperature–visual variety.

On the other hand, display devices for outdoor use, as for signs and other advertising situations, require liquid crystalline materials exhibiting stress indicia at correspondingly lower temperatures.

Specific examples of liquid crystalline compositions useful for the process appear below, wherein all amounts are in parts, ChCl is cholesteryl chloride melting at 94° to 95°C; High ChOIC is cholestery oleyl carbonate showing a color play at 20° to 22°C; Low ChOIC is cholesteryl oleyl carbonate showing a color play at 5° to 6°C; ChNo is cholesteryl nonanoate; and the temperature is that at which the compositions become colorless.

Example	ChCl	High ChOIC	Low ChOIO	ChNo	Temperature, °C
1	27	73	-	-	42
2	27	58	-	15	45.5
3	27	50	23	-	38
4	27	59	14	-	35
5	25	75	-	-	48

Other alkanoic esters of cholesterol or alkyl carbonate esters of cholesterol can be used in the foregoing examples to provide a broad variety of temperatures and temperature ranges for the liquid crystalline material **18.** Likewise, other cholesteric materials such as corresponding derivatives of β-sitosterol, stigmasterol, ergosterol, and the like can be substituted with comparable results.

In Figure 2.10c, means are illustrated for enhancing the aesthetic effects or visual display of the display device **10'**. Such means encompass the use of polarizing means including one or two polaroid panels or the like, at least one of which is rotatable relative to this display device. One form of such polarizing means includes rotatably mounted disc **24** supported on shaft **26** which is rotated in this example by a hand crank **28** through a suitable transmission denoted generally at **30.** Obviously, the shaft can be motor-driven instead. In this arrangement, the polarizing disc or panel is of polaroid material.

Instead of rotating the disc, the liquid crystal container **12'** obviously can be rotated instead. In any event, such relative rotation of the display device relative to the polarizing means **24** effectively varies the angle of impinging illumination and thereby effects a gross color change in the variable color patterns produced by the display device when subjected to the more or less transverse deformation forces as denoted by arrow **20'**.

In the arrangement of the process as noted in Figure 2.10c, then, a great variety of color patterns can be produced in the display device through the application of the transverse forces. At the same time or sequentially, gross color changes in the produced color patterns can be effected by relative rotation between polarizing disc **24** and the liquid crystal cell **12'**. The unusual and unexpected results thus produced are engendered by the unexpected cooperation between the polarizing disc and the inherent circularly polarizing characteristic of liquid crystalline material **18'**. This characteristic, or dichroism, is present in varying degrees in the liquid crystalline materials.

A second polaroid panel or disc (or equivalent polarizing means) desirably is disposed on the other side of the container. The cooperative effect of the polarizing panels and the dichroic character of the liquid crystal greatly and surprisingly enhances the vividness of the visual liquid crystalline color patterns and the changes wrought therein by stress variations. Ancillary to such cooperation is the significant fact that the intensity of the visual color display is not diminished, when the polarizing panels (at least one of which is desirably rotatable when two are used) are crossed, owing to the dichroic characteristic.

With two polarizing media, one medium desirably is incorporated within the structure of the liquid crystal container, as by forming one of the wall structures **14'** or **16'** of polaroid or other polarizing material. Desirably that wall structure (e.g., the wall **16'** in Figure 2.10c) removed from the external polarizing filter **24** (if used) is so constructed so that the liquid crystal is advantageously disposed between the polarizing means. If rotation of the polarizing is not desired, the external polarizing filter can be omitted and both wall structures can be fabricated from light polarizing material with a constant but advantageous enhancement of color patterns.

Another form of the color display apparatus **32** is illustrated in Figures 2.10d and 2.10e. The liquid crystal container **34** forming part of the display apparatus is of modified construction in that it does not depend on the functional displacement of one wall structure relative to another. Instead, the liquid crystal container can be formed from a relatively rigid housing, including in this case bipartite components **36, 38** provided with inwardly dished surfaces **40** as better shown in Figure 2.10e.

Any suitable structural material can be employed for fabricating the housing components although at least one of the components is fabricated from a light transmitting material. Accordingly the housing components can be fabricated from glass, plexiglass, or other material of transparent or translucent character.

A quantity of liquid crystalline material **18"** is sealed between the housing components as evident from Figure 2.10e. This is accomplished by peripherally joining the thickened edge portions **42** of the housing components by means of a suitable cement.

In the event that the housing components are fabricated from certain plastic materials such as polymethacrylic resin, the thickened portions of the housing components can be joined by heat-sealing or solvent welding.

In order to create a variety of varying color patterns within the liquid crystalline material thus enclosed with the cell , means are provided for inducing fluid flows and attendant deformational stresses within the body of the liquid crystalline material. As mentioned above with reference to Figure 2.10a, the resultant shear and other deformational stresses lead to the changing color patterns of the liquid crystal.

One arrangement for so inducing flow of the liquid crystal includes means for circulating a quantity of liquid crystalline material throughout the volume enclosed by the housing components. In furtherance of this purpose, the housing components define a pair of opposed flow apertures **44, 46** when assembled as aforesaid.

Alternatively, the flow apertures can be bored in the liquid crystalline container after assembly thereof. The apertures provide in this example egress and ingress of liquid crystal relative to the container.

To circulate the liquid crystalline material throughout the container, a pair of conduit sections **48, 50** are joined to the container as shown. Such joining can be effected by threading the ends of the conduit sections and tapping the apertures or by cementing closely fitting ends of the conduit sections into the apertures or by other suitable securance means.

The conduit sections are joined to the inlet **52** and outlet **54** respectively of a circulating pump **56**, which is rotated in this example by a suitable prime mover such as electric motor **58.** For smaller sizes of the display apparatus and their correspondingly smaller pumps, a manually operated crank can be provided.

Operation of the pump circulates the liquid crystal material through the conduit sections and of course at a correspondingly slower rate through the liquid crystal container. The circulation thus afforded by the pump can be faster or slower depending upon the particular sequence of changing color patterns which is desired.

The flow patterns desirably are spread out across the face **60** of the liquid crystal container by suitable flow baffle means. One form of such flow baffle means includes a pair of V-shaped baffle members **64, 62** mounted as shown near the outlet and inlet apertures of the cell. Other shapes of baffle members can be substituted for the particular baffle members depending upon the flow characteristics desired within the container.

Additional studies with variable color display devices based on mechanical deformation are described by *E.N. Sharpless and F. Davis; U.S. Patent 3,647,279; March 7, 1972; assigned to Liquid Crystal Industries, Inc.*

Uniform Electrical Holding Field

J.J. Wysocki, J.H. Becker and G.A. Dir; U.S. Patent 3,790,251; February 5, 1974; assigned to Xerox Corporation describe an imaging system including an imaging layer comprising cholesteric liquid crystalline material in an electrical field capable of having field strengths in the cholesteric nematic phase transition field strength range or the Grandjean to focal-conic texture transformation field strength range of the liquid crystalline imaging material. The imaging material is typically imaged by placing an imagewise phase-transition or texture-transformation inducing electrical field across a layer of imaging material.

The process controls such an imaging system by maintaining a substantially uniform holding field across the entire surface of the imaging material, the holding field typically being maintained across the imaging material after it has been imaged. The holding field may be applied in the form of various phenomena. The particular type of holding field applied has field strengths below the cholesteric-nematic phase transition or the Grandjean to focal-conic texture transformation field strength thresholds of the imaging material.

The imaging system, and especially its image memory and image relaxation characteristics are controlled by varying the strength of the holding field, and by applying field pulses in imagewise configuration to image the liquid crystalline imaging material, or by pulsing the field uniformly across the entire surface of the imaging material to field strengths above the phase-transition threshold or the texture transformation threshold, thereby enhancing the image already formed in the imaging material.

In Figure 1.1a a typical liquid crystalline imaging member **10**, sometimes referred to as an electroded imaging sandwich, is shown in partially schematic, cross-section wherein a pair of substantially transparent plates **11** having substantially transparent conductive coating **12** upon the contact surfaces thereof comprises a parallel pair of substantially transparent electrodes. An imaging member where both electrodes are transparent is preferred where the imaging member is to be viewed using transmitted light; however, a liquid crystalline imaging member may also be viewed using reflected light thereby requiring only a single transparent electrode while the other may be opaque.

The transparent electrode is separated by spacing member or gasket **13** which contains one or more voids which contain the liquid crystalline film or imaging layer which comprises the active element of the imaging member. A field is created between the electrodes, for example, by an external circuit **15** which typically comprises a source of potential **16** which is connected across the two electrodes through leads **17**. The potential source may be either DC, AC, or a combination thereof.

In such an imaging system it has been found that when cholesteric liquid crystals or a mixture comprising cholesteric liquid crystalline materials is used in such an electrode sandwich, the electrical fields across the liquid crystalline film cause the optical appearance of such liquid crystalline films to visibly change in the field-affected areas.

When high electrical fields are placed across the liquid crystalline imaging member, an electrical field-induced phase transition occurs wherein the optically negative cholesteric liquid crystalline material transforms into an optically positive liquid crystalline state. This transition is believed to be the result of the cholesteric liquid crystal material transforming into an aligned nematic liquid crystalline mesophase structure. This cholesteric-nematic phase transition imaging system is fully described in U.S. Patent 3,652,148.

The cholesteric-nematic phase transition is the primary liquid crystalline imaging mechanism which is used in the process. Cholesteric liquid crystalline materials in the cholesteric state are typically translucent, for example, like a milky white, opalescent layer. This is the way that such materials frequently appear when first placed in the unbiased electrode sandwich.

When a high magnitude electric field is placed across the liquid crystalline film, the field-induced phase transition is observable because the liquid crystalline film becomes transparent

FIGURE 2.11: UNIFORM ELECTRICAL HOLDING FIELD TO IMPROVE IMAGING

a.

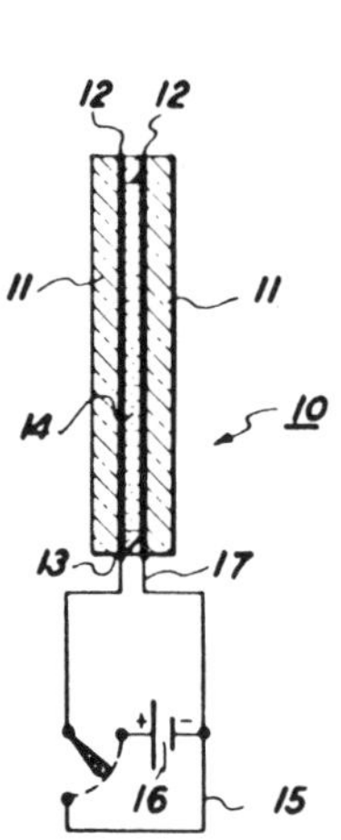

Cross-Sectional View of Liquid Crystal Imaging Member

b.

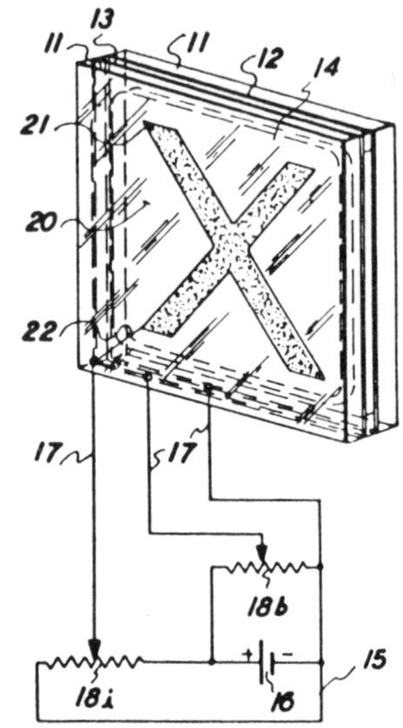

Schematic View of Specific Imaging Member

c.

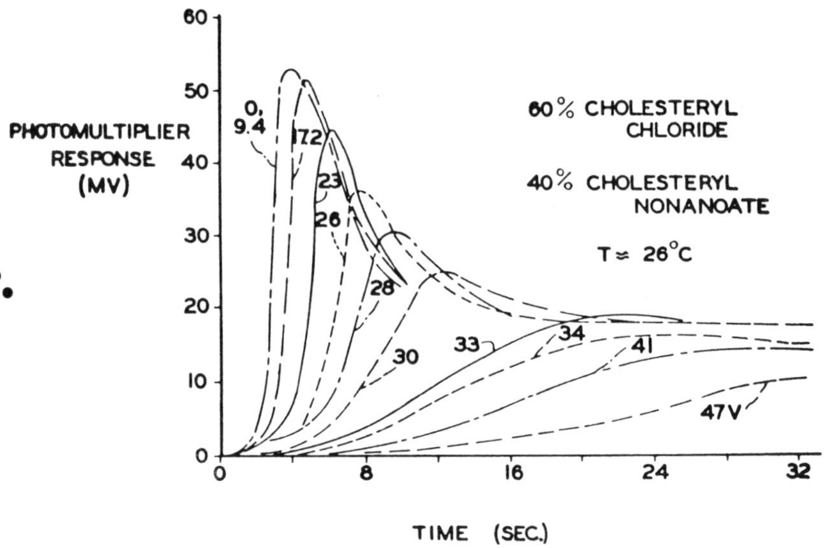

Performance Data

Source: J.J. Wysocki, J.H. Becker and G.A. Dir; U.S. Patent 3,790,251; February 5, 1974 in areas where the high electric field is present. When the imaging member is viewed between polarizers with transmitted light, the areas in which the field-induced phase change has taken place appear dark, while the unchanged, translucent, light scattering, optically active and birefringent, cholesteric areas still retain their white appearance. Although polarizers have just been described as a convenient means for enhancing the contrast of the imaged areas, other enhancement techniques may also be used in place of such polarizers. It is therefore clear that either field or nonfield areas in a liquid crystalline imaging member may be used as the desired image, with or without the addition of other means for image enhancement.

The same apparatus and materials used as an imaging system where the cholesteric-nematic phase transition mechanism is used may also be used to cause an electrical field-induced texture transition to occur wherein a cholesteric liquid crystalline material initially in its Grandjean or "disturbed" texture is transformed into its focal-conic or "undisturbed" texture. The voltages and field strengths typically used for imaging the liquid crystalline imaging members by the texture change mechanism typically have lower magnitudes than the voltages and field strengths used in the phase transition already described above. This electrical field-induced texture transition system is fully described in U.S. Patent 3,642,348.

The Grandjean texture is typically characterized by selective dispersion of incident light around a wavelength λ_0 (where $\lambda_0 = 2np$ where n equals the index of refraction of the liquid crystalline film and p equals the pitch of the liquid crystalline film) and optical activity for wavelengths of incident light away from λ_0. If λ_0 is in the visible spectrum, the liquid crystalline film appears to have the color corresponding to λ_0, and if λ_0 is outside the visible spectrum, the film appears colorless and nonscattering. The Grandjean texture of the cholesteric liquid crystalline material is sometimes referred to as the disturbed texture.

The focal-conic texture is also typically characterized by selective dispersion, but in addition this texture also exhibits diffuse scattering in the visible spectrum, whether λ_0 is in the visible spectrum or not. The appearance of the focal-conic texture state is typically milky white when λ_0 is outside the visible spectrum. The focal-conic texture of cholesteric liquid crystals is sometimes referred to as the undisturbed texture.

If the unbiased electrode sandwich containing an imaging composition comprising cholesteric liquid crystals predominantly in the Grandjean texture state is observed between polarizers, the imaging sandwich appears colored or black. When the electrical field is placed across the liquid crystalline film, with field strengths in the field-induced texture change range, the texture change to the predominantly focal-conic texture state is observable because the liquid crystalline film becomes white in the field-affected areas when the imaging sandwich is observed in transmitted or reflected light.

The texture change imaging system thereby produces white image (or field-affected) areas on a darker colored background. Like the phase-transition system, the texture change imaging system may also be used with or without the use of polarizers or other image enhancing devices.

In the liquid crystal imaging members of the process, the electrodes may be of any suitable transparent conductive material; however, opaque electrodes may be used, at least in part, in some embodiments. Typical suitable transparent, conductive electrodes include glass or plastic substrates having substantially transparent and continuously conductive coatings of conductors such as tin, indium oxide, aluminum, chromium, tin oxide, or any other suitable conductor. These substantially transparent conductive coatings are typically evaporated onto the more insulating, transparent substrate. NESA glass, a tin oxide coated glass, is an example of a typical transparent, conductive electrode material.

The spacer **13** in Figure 2.11a which separates the transparent electrodes and contains the liquid crystal film between the electrodes is typically chemically inert, transparent, substantially insulating and has appropriate dielectric characteristics. Materials suitable for use as such spacers include cellulose acetate, cellulose triacetate, cellulose acetate butyrate, and polyurethane elastomers.

Although the cholesteric-nematic phase transition mechanism and the Grandjean to focal-conic texture transition mechanism have both been described with reference to a preferred example where these transitions are caused to occur by the application of an electric field to a layer of cholesteric liquid crystalline material, there are other phenomena by which these transitions may be accomplished. Thus, it is possible to carry out the cholesteric-nematic phase transition and the Grandjean to focal-conic texture transition with electrical fields, magnetic fields, thermal fields, photochemical effects, static pressure effects, acoustic and ultrasonic fields, surface forces, and combinations of these phenomena.

The use of electrical fields to carry out these transitions has been described in U.S. Patent 3,642,348. The use of magnetic fields alone to effect the cholesteric-nematic phase transition has been described in the art. See Sackmann et al, *J. Am. Chem. Soc.* 89, 5981 (1967).

Thermal fields can be used with mixtures of right handed and left handed cholesteric materials or materials with an effective right handed or left handed nature in a chiral environment to cause the transitions. If the materials used in the mixture are suitably chosen, the composition can be converted to a nematic (compensated) state by addition or subtraction of thermal energy, i.e., a change in temperature.

At compensation the net chirality is zero. Thus, a thermal field can cause a cholesteric-nematic or nematic-cholesteric transition. Generally higher resolution can be obtained in transient thermal displays than with static thermal displays; however, additional memory and resolution enhancement can be obtained by appropriate biasing stimuli such as thermal, electrical or magnetic fields. Use of materials having a long relaxation time can also provide memory effects, typically of intermediate duration.

It has been described in the art that ultraviolet radiation can photodecompose suitable cholesteric liquid crystalline materials which causes a change in pitch (and therefore the color reflected by) of the resultant material. It has now been found that this phenomenon can be utilized to achieve a cholesteric-nematic phase transition. Furthermore, this procedure can provide a permanent or temporary display dependent upon the nature of the materials and whether recombination of the decomposition products is provided for. In an imaging device a mixture of right handed and left handed cholesteric liquid crystalline ingredients which include at least one photodegradable constituent such as cholesteryl iodide may be employed.

The action of the ultraviolet radiation will typically convert the photodegradable material into a different species thereby upsetting the initial ratio of right handed and left handed materials. By a suitable adjustment of the components of the mixture and the length of exposure, the resultant mixture is typically converted to the nematic state at a chosen temperature in the radiation-affected areas whereas the nonimaged or background areas are not converted in this manner.

An interesting feature of such an imaging scheme is that the character of the image can be temporarily (barring diffusion) modified by changing the sample temperature. Moreover, the mixture of materials having opposite handedness may be achieved not only by the initial combination of such materials but also by the imaging process itself, i.e., by conversion of part of the material of a given chirality to another of the opposite chirality by means of photochemical action.

It is known in the art that static pressure applied to individual cholesteric liquid crystalline compounds and also mixtures of such compounds usually reduces the pitch of the material. This effect may be utilized to obtain a nematic-cholesteric phase transition where a compensated composition is employed. It is to be expected that systems containing mixtures of right handed and left handed cholesteric liquid crystalline materials could be formulated wherein static pressure, rather than reducing the pitch of the mixture, would increase it and affect it enough to induce a nematic transition at a particular operating temperature. Static pressure would therefore typically change the effective rotary power of the compositions.

A further static pressure effect which would be useful to achieve the phase transition is based on the fact that the threshold electric field is a unique feature of any system. Thus, for a given applied voltage, a change in the thickness of the liquid crystalline imaging layer will change the strength of the applied field. When the layer is deformed by static pressure so that the layer thickness is reduced, the initial applied voltage which was initially not sufficient to cause the phase transition will correspond to a higher field and thus cause the desired cholesteric-nematic phase transition to occur.

This mode of operation has particular advantage for a controlled memory display since the same applied voltage may serve as both the transforming and the holding voltage. This mode of operation typically allows retention of pressure patterns on the liquid crystalline imaging layer long after the pressure itself is removed therefrom. The combination of pressure and voltage can thus be advantageously utilized to sharpen the threshold for the cholesteric-nematic transition and to alter the transformation and relaxation times for this transition.

The foregoing considerations relative to static pressure fields are also pertinent to dynamic pressure fields. Additionally, pressure may give rise to internal voltages (piezoelectric effects) which alter the transitions.

It is known in the art that surfaces may interact with cholesteric liquid crystalline materials principally in establishing the texture of the material. Additionally it is also well known that such surface interactions can be modified by surface treatments, for example, such as by rubbing the surface or by applying thereto surface coating of materials such as lecithin or the like. Generally speaking, imagewise changes in surface interactions to achieve the phase transition in imagewise fashion may typically be effected.

These various stimuli discussed above have been described with respect to their efficacy in facilitating the cholesteric-nematic phase transition. These stimuli, individually or in combination, typically influence the Grandjean-focal-conic texture change as well and therefore the stimuli have utility in texture change display systems as well as in phase transition display systems.

Figure 2.11b is a partially schematic, isometric view of a liquid crystalline member preferred for use in the process. This imaging member, like the one illustrated in Figure 2.11a, comprises substantially transparent plates **11** having substantially transparent conductive coatings, here separately designated **12, 20** and **21** upon the contact surfaces of the transparent plates.

Spacing member gasket **13** and liquid crystalline film or imaging layer **14** are identical to those described in conjunction with Figure 2.11a. In Figure 2.11b the substantially transparent conductive coating **12** which is behind the liquid crystalline imaging composition is a single, substantially continuous, electrically conductive, transparent coating.

However, transparent continuous conductive coatings **20** and **21** on the substantially transparent plate illustrated in front of the liquid crystalline imaging composition are in complementary imagewise configurations which are separated by an insulating space or material at the boundaries of the imagewise configuration electrode designated **21**. Because imagewise electrode **21** is not electrically connected to its complementary, coplanar background electrode **20**, an appropriately electrically insulated conductive lead **22** is provided to connect the imagewise electrode with the external circuitry **15**.

Although in Figure 2.11b the imagewise electrode is illustrated by a more dense dot pattern indicating its substantially transparent and conductive nature, the complementary background electrode **20** typically comprises exactly the same substantially transparent, electrically conductive material which comprises the imagewise electrode **21**, and the two areas are illustrated in Figure 2.11b only for the purpose of clarity in distinguishing the image and background areas of the coplanar, complementary imagewise and background electrodes.

In Figure 2.11b the external circuit **15** which is connected to the various electrodes in the imaging member by conductive electrical leads **17** is illustrated containing source of potential **16** and potentiometers **18i** and **18b** as suitable means for controlling the voltage and therefore the field strengths in the image and background areas, respectively, in the imaging member. The electrical circuit illustrated in Figure 2.11b is one suitable means for the application of the desired imaging and holding fields across the imaging composition.

In operation, an imaging composition which is typically in its predominantly Grandjean texture state is typically made to adopt the focal-conic texture by adjusting potentiometers **18i** and **18b** to establish between electrodes **12** and **20** and electrodes **12** and **21** electric fields within the Grandjean to focal-conic texture transformation field strength range of the imaging composition thus causing the liquid crystalline film to appear white or light scattering. Potentiometer **18i** is then adjusted to establish between electrodes **12** and **21** an electric field within the cholesteric-nematic phase transition field strength range of the imaging composition and the field is maintained until the composition between electrodes **12** and **21** becomes clear thus indicating that the phase transition has taken place.

The imaging composition between electrodes **12** and **20** remains in a light scattering condition. Hence, an image comprising a clear image area on a light scattering background is formed. A holding field is then applied across the entire liquid crystalline film by adjusting potentiometers **18i** and **18b** to establish across the entire film an electric field having a strength below the phase transition field strength range of the imaging composition.

When the imaging composition is placed in the imaging sandwich or member, and if the member is observed between crossed polarizers (as image enhancing devices), the imaging layer frequently appears colored of black, i.e., the imaging composition is initially in its predominantly Grandjean or disturbed texture. When an electrical field having field strength in the field-induced texture change range is placed across the liquid crystalline film, for example, by adjusting potentiometer **18i** illustrated in Figure 2.11b, the electrical field is thereby placed between the imagewise substantially transparent electrode **21** and the opposite electrode **12**, and the imaging composition comprising cholesteric liquid crystals predominantly assumes its focal-conical or undisturbed texture in the field-affected areas and the texture change is observable because the liquid crystalline film becomes white in these areas.

Similarly, the background image areas or indeed any area in such a liquid crystalline imaging member may be placed predominantly in the focal-conic or undisturbed texture by placing an electrical field having field strength within the field-induced texture change range across such an imaging composition comprising cholesteric liquid crystals. In this way, the texture of an entire liquid crystalline imaging member may be uniformly maintained in a substantially uniform electrical field action upon said composition.

Where field strengths typically higher than the field-induced texture change field strength range and above the cholesteric-nematic phase transition threshold field strength are used, the imaging composition comprising cholesteric liquid crystals exhibits the field-induced phase transition transforming the field-affected areas into the nematic liquid crystalline mesophase structure. This cholesteric-nematic phase transition is the primary imaging mechanism in the process whereby the imaging composition is changed from its typically translucent, milky white, opalescent appearance, into a substantially transparent film in the areas where the high electrical fields are present across the imaging composition. In this way, images, for example, images comprising transparent image areas on white or colored background areas, are produced by the advantageous imaging system of the process.

In the imaging system and image control system, it has been found to be particularly advantageous to maintain a substantially uniform holding field having field strengths below the cholesteric-nematic phase transition or the texture transformation threshold field strengths of the imaging composition comprising cholesteric liquid crystals, and to maintain such a holding field across the composition whenever the imaging member is being used, during both image "on" and image "off" periods. The holding field may be applied in the form of any of the various phenomena previously described with respect to effecting the cholesteric-nematic phase transition and the Grandjean to focal-conic texture transformation. Thus the holding field, or stimuli, may be a magnetic field, an electrical field, a thermal field, various types of pressure fields including static, dynamic and piezoelectric, photochemical action, surface effects and combinations thereof.

For purposes of illustrating the process, it will be described in detail with respect to the use of an electrical field as the holding field. By maintaining a substantially uniform field across the imaging composition, it has been found that the texture of the entire imaging composition layer can be maintained in a substantially uniform condition, as already described above. Furthermore, it has been found that after the imaging composition has been imaged by the cholesteric-nematic phase transition mechanism, and the phase transition imaging field is removed, that the application of the holding field of the process maintains the composition in a partially transformed or partially aligned state, which facilitates reimaging of the imaging composition.

It is believed that the previously phase-transition imaged areas of the composition are maintained in a partially transformed state, so that the subsequent application of a phase-transition imaging field causes the image to reappear or to be renewed in a shorter time than was formerly possible when the imaging composition was allowed to return to its equilibrium state in the absence of the holding field of the process. The holding field even makes possible the renewal of an image by briefly increasing the field strength across the entire surface area of the imaging composition to a field strength above the phase-transition threshold, as well as by the more conventional application of an imagewise phase-transition field.

It is therefore seen that the holding field controls the return of the imaged liquid crystalline composition to its equilibrium state. The transient return of the composition to its equilibrium state and the control of that transient response of the composition which is made possible by the process may itself be used as the imaging system.

The advantageous effects of the process where a substantially uniform electrical holding field below the phase-transition field strength threshold is maintained across the entire surface area of such a liquid crystalline imaging member is more clearly pointed out by the data shown in Figure 2.11c, which is representative for liquid crystalline imaging systems such as those used in this system. Figure 2.11c is a photomultiplier response versus time plot for a specific liquid crystalline imaging system, namely an imaging composition comprising cholesteric liquid crystals of about 60% cholesteryl chloride and about 40% cholesteryl nonanoate.

The data illustrated in Figure 2.11c show photomultiplier responses corresponding to the times of such responses, the times being measured from the time that a phase-transition imaged member has the transition-inducing field removed therefrom. The various response data lines represent the various responses of the particular liquid crystalline imaging system when used with holding fields having field strengths below the phase-transition threshold, and such fields are maintained across the entire surface area of the imaging member during various imaging sequences of the imaging cell.

In the system represented by the data in Figure 2.11c the substantially uniform holding field maintained across the entire surface area of the imaging member is shown to have been varied through the range between 0 and 47 volts, with the higher uniform field strengths attenuating the peak transmissivity until the transient response of the imaging cell, here represented by the photomultiplier response versus time data, asymptotically approaches the system's equilibrium condition, which is here represented by a photomultiplier response somewhere between 10 and 20 millivolts. The photomultiplier responses are proportional to the amount of light transmitted through the previously imaged (phase-transformed) areas of the imaging composition. The data in Figure 2.11c clearly illustrates the maintenance of the substantially uniform electrical holding field, of field strength below the phase-transition threshold field strength, across the entire surface area of such a liquid crystalline imaging member, controls the transient response of the imaging material in returning to its equilibrium state after the phase-transition inducing electrical fields have been removed or reduced in magnitude.

Another advantageous result from the use of such substantially uniform holding fields maintained across the entire surface area of such liquid crystalline imaging members is the resultant reduction in the amount of additional field strength necessary to phase-transition image desired image areas of the imaging composition. For example, if a substantially uniform field strength maintained across the entire surface area of an imaging member is about 50 volts, and the phase-transition threshold field strength for the imaging members is about 200 volts, it is necessary to place only an additional field of strength greater than about 150 volts across the desired image areas in order to produce the desired phase-transition image in the cholesteric liquid crystalline imaging composition.

In addition, the decreased imaging time can at least in part be explained because the presence of the substantially uniform holding field is believed to maintain the cholesteric liquid crystalline material predominantly in its focal-conic or undisturbed state thereby eliminating the increment of the response time typically necessary to transform the cholesteric liquid crystalline material from its original predominantly Grandjean or disturbed texture state into its predominantly undisturbed state before continuing through the cholesteric-nematic phase transition imaging which is the primary imaging mechanism in the system of the process.

The following examples define the process with respect to an imaging system and controlling an imaging system wherein an imaging composition comprising cholesteric liquid crystalline material is imaged by the application of electrical fields across the composition. The parts and percentages are by weight unless otherwise indicated.

Example 1: An imaging member is provided having a pair of parallel plate transparent electrodes each comprising a substantially transparent continuous coating of tin oxide on a glass substrate, and between the transparent coatings, a spacer gasket of Mylar film, a polyester resin film, enclosing an about 38 micron thick film of imaging composition here comprising a mixture of about 60% cholesteryl chloride and about 40% cholesteryl nonanoate. This imaging member is placed between polars, a polarizer-analyzer combination on each side of imaging member, and an incandescent light source is placed on the outside of the polarizer on the polarizer side of the imaging member and light source, polarizer, imaging member and analyzer, are aligned with a phototube detector which is outside the analyzer on the analyzer side of the imaging member.

The imaging member is provided with suitable electrical circuitry to provide an electrical field between the electrodes, and included in this circuitry is a switch for applying voltage to or removing it from the imaging member and simultaneously activating the sweep of an oscilloscope which meters the phototube response. The source-polarizer-imaging member-analyzer-phototube apparatus is a Leitz Ortholux polarizing microscope equipped with a hot stage and a 10-power objective which serves as the polarizer-analyzer combination; the imaging member described above placed on the hot stage; the phototube an RCA 7102 photomultiplier with an S-1 spectral response placed in optical alignment with the other elements of the system; and an incandescent light bulb.

The cholesteric-nematic phase transition threshold voltage for this composition is about 72 volts at about 25°C. This cell is typically imaged at about room temperature. A voltage of about 400 volts is applied to the imaging member or cell, and the imaging composition transforms to the nematic state in about 0.15 second. The transformed material appears dark when the imaged member is observed between polars in transmitted light. The field is removed by shorting the leads to the imaging member and the liquid crystalline composition relaxes back into the cholesteric state in a transient response which includes a time period of substantially increased light transmissivity. This maximum transmissivity occurs at about 4 seconds after shorting the imaging member, and the material continues to gradually assume the cholesteric state, and first, the predominantly focal-conic texture. An equilibrium is reached in about 16 to 20 seconds and upon further relaxation the sample will slowly return to its predominantly Grandjean texture. This latter and final relaxation may take quite a long time, ranging from minutes to hours.

A voltage of about 400 volts, (i.e., above the phase-transition threshold field strength) is reapplied, and the imaging cycle repeats itself. The material transforms from its predominantly Grandjean texture to its predominantly focal-conic texture before the cholesteric-nematic phase transition imaging mechanism occurs. Upon phase transition imaging, the imaging member is then returned to a holding voltage, below the cholesteric-nematic threshold, of about -30 volts. The maximum transmissivity in the transient response is now observed to occur about 13 seconds after removing the phase transition imaging field, and the magnitude of the maximum transmissivity is observed to be less than the magnitude of the maximum transmissivity observed when the phase-transition imaged member is entirely shorted as described above.

The data illustrated in Figure 2.11c is representative of transient responses of the imaging system of this example when the system has the phase-transition threshold imaging field removed therefrom while the imaging member is under the influence of holding voltage in the range between about 0 and -47 volts.

Example 2: The imaging member of Example 1 is provided with an imaging composition comprising about 60% anisylidene-p-n-butylaniline and about 40% cholesteryl chloride. This composition has a shorter natural relaxation time than the composition used as the imaging member in Example 1. This imaging member is imaged as described in Example 1 and is then shorted, without the inventive holding voltage, and the maximum transmissivity occurs about 6 milliseconds after shorting when operated at about 26°C.

The cholesteric-nematic phase transmission imaging is achieved by applying about 300 volts to the imaging member, which is subsequently reduced to a holding field voltage of about 110 volts, and the maximum transmissivity in relaxation is observed to occur at about 18 milliseconds after reduction of the imaging field to the holding field strength.

Additional studies with systems for transforming a cholesteric liquid crystalline material from its Grandjean or disturbed texture state to its focal-conic or undisturbed texture state by an applied electrical field, and an imaging system where a cholesteric liquid crystalline member is imaged in a desired image configuration by the electric field-induced texture transition system are described by *J.J. Wysocki, J.E. Adams and R.W. Madrid; U.S. Patents 3,642,348, February 15, 1972 and 3,704,056, November 28, 1972; both assigned to Xerox Corporation.*

AC Electrical Fields

W.E.L. Haas, J.E. Adams, Jr. and J.B. Flannery, Jr.; U.S. Patent 3,680,950; August 1, 1972; assigned to Xerox Corporation describe a system which transforms a liquid crystalline composition having cholesteric optical characteristics from the focal-conic or undisturbed state into the Grandjean or disturbed texture state by an applied AC electrical field and an imaging system where such a liquid crystalline member is imaged in a desired image configuration by the AC electrical field-induced texture transition system. Such transformed compositions exhibit colors and memory characteristics.

Display of Infrared Beams Using Encapsulated Crystals

P.J. Allen; U.S. Patent 3,604,930; September 14, 1971; assigned to the U.S. Secretary of the Navy describes a method and device for displaying a visual pattern of an incident beam of infrared radiation which includes a film of minute droplets of a liquid crystal composition encapsulated in gelatin. The nature of the composition is such that within a certain temperature range its optical properties change with a change of temperature so that it appears to change color. The film of liquid crystal is thermally biased to its critical temperature so that when incident infrared radiation is absorbed by the composition, the temperature thereof is raised which causes a color change in the zone of the film on which the radiation is incident.

Of the three general classes of liquid crystals, smectic, nematic, and cholesteric, it is the cholesteric phase which is of most interest. The molecules of a cholesteric substance are disposed in layers, each layer being formed of parallel rows of axially aligned molecules. From layer to layer there is an angular progression of the orientation of the molecular axis within each layer, resulting in a helical arrangement. It is this periodic, helical arrangement of molecules which is responsible for the color sensitive scattering observed with liquid crystals. The change in color of the scattered light with change in temperature of the liquid crystal is due to a slight change in the angular displacement between the molecules of adjacent layer, i.e., to a change in the pitch of the helical arrangement. This observable color change in liquid crystals makes a very sensitive visual indicator of temperature.

Liquid crystals often are sensitive to other influences including electric and magnetic fields, mechanical stress, ultraviolet radiation, and various vapors. It is the latter two sensitivities which are responsible for deterioration in the thermooptical sensitivity of exposed liquid crystals. This problem has restricted their measurement application mainly to short term use with freshly coated surfaces.

In initial efforts to develop an imaging detector for infrared laser beams, a fluid liquid crystal composition was applied to a thin Mylar sheet coated on the reverse side with flat black lacquer which provided the necessary dark background and also functioned to absorb the infrared radiation. Using thermal bias provided by visible and infrared incandescent lamps, some success was achieved in detecting a fraction of a kilowatt laser beam at 1.15 microns. However, the deterioration in performance of the liquid crystal on exposure to the atmosphere remained a problem.

It was found, however, that rapid deterioration of performance does not seem to be a problem with encapsulated liquid crystals, in which the liquid crystal material is sealed in minute gelatin capsules some 10 to 30 microns in diameter. These minute gelatin capsules are then suspended in a water-latex vehicle to form a slurry which, when spread over a surface and allowed to dry, leaves a randomly packed film or coating of the tiny capsules of liquid crystal which adhere because of the latex. The heat sensitive screen thus formed is ready for use as soon as drying is complete.

The temperature-color characteristic of the encapsulated liquid crystal used in the devices to be described is shown in Figure 2.12a. This particular material (NCR Encapsulated Liquid Crystal Type R-33) has a useful temperature range of 33° to 37°C, while other available types offer considerable choice of operating temperature and range. For example, materials can be obtained with a 1°C operating range.

Early success was obtained with a 4-inch diameter "tuna-can" infrared imaging detector shown in Figure 2.12b. This detector employs a 1-mil Mylar diaphragm sprayed on one side with ultraflat black Krylon lacquer as absorber, and coated on the outside surface with type R-33 encapsulated liquid crystal. The screen **10** thus prepared is stretched over the mouth of a can **12** and held taut using a plastic can cover **14** with the center portion removed at **16**.

Referring back to Figure 2.12a, the temperature-color characteristic of the encapsulated liquid crystal used is seen to have a maximum slope in the red-yellow-green region. It has been found that the best sensitivity can be achieved by thermally biasing the liquid crystal film to a temperature at which it first begins to change color, which will be referred to as the "critical temperature," so that a readily discernable color contrast can be obtained with relatively small changes of temperature, (i.e., small changes of laser beam intensity).

FIGURE 2.12: LASER INFRARED BEAM DISPLAY

a.

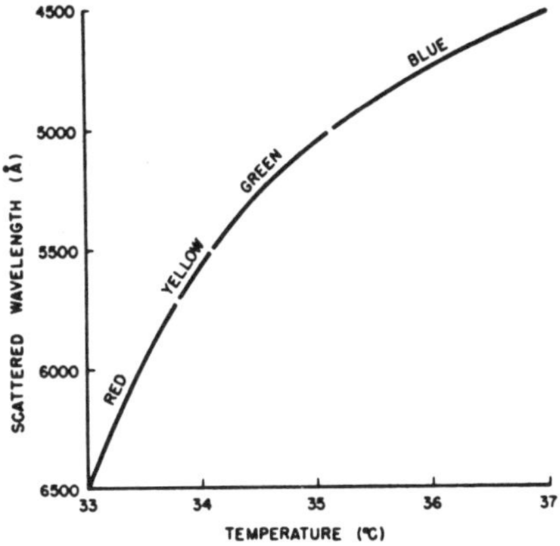

Color Change Versus Temperature Change

b. c. d.

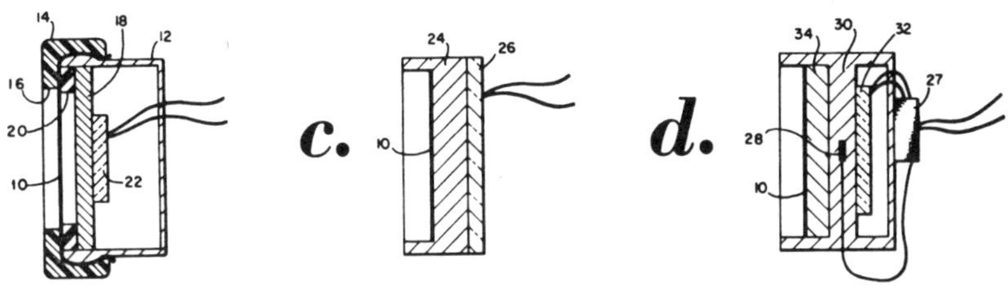

Elevational Views of Three Forms of the Device

e.

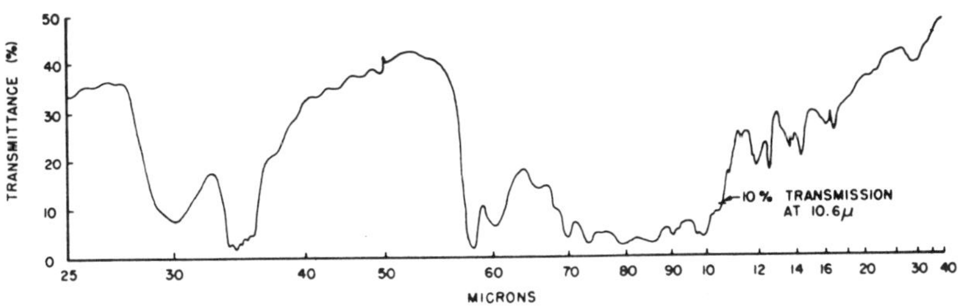

Transmittance for a Range of Wavelengths

(continued)

FIGURE 2.12: (continued)

f.

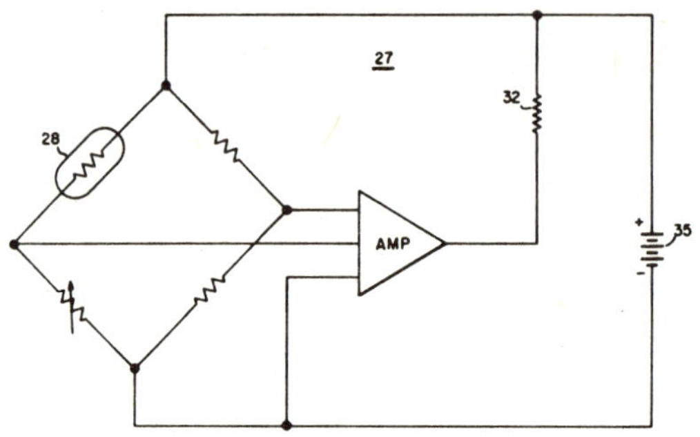

Heater Control

g.

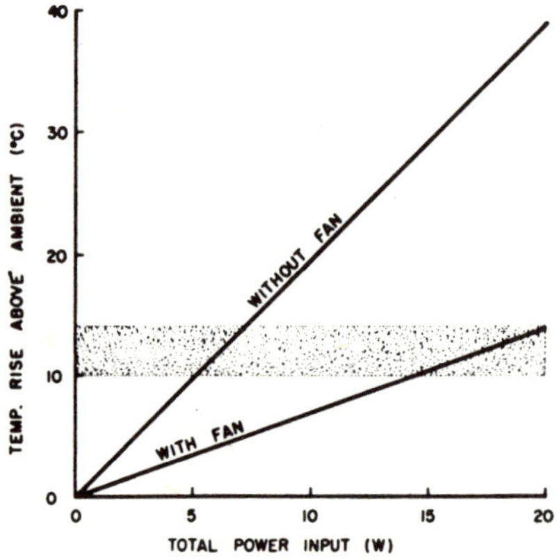

Air Flow Effect on Temperature Increase

Source: P.J. Allen; U.S. Patent 3,604,930; September 14, 1971

In this detector, the necessary thermal bias is provided by a heated baffle plate **18** air spaced an eighth of an inch behind screen **10** by means of spacer ring **20**. Less than 2 watts of electrical power to a 40-ohm resistor **22** epoxied to plate **18** are required to maintain the screen at proper operating temperature (about 33.5°C for the type encapsulated liquid crystal used). A regulated DC supply may be used to stabilize heater input power, but an automatic temperature control is preferred.

This detector will give a good visual indication with a 10-micron flux of 10 milliwatts per square centimeter incident on the screen, and a narrow 1-milliwatt beam at 1.15 microns has been detected with this device. The screen is sensitive enough to respond to a flashlight and to incandescent lamps placed close by as illumination for photographing the images. As used herein, sensitivity refers to the intensity of the incident flux needed to produce an image on the screen.

The detector is handicapped by a recovery time of several seconds and suffers from image blooming because of thermal spreading in the Mylar diaphragm. However, both of the factors are improved with a gentle steady flow of air over the face of the screen. To avoid screen damage, incident flux should not exceed about 250 milliwatts per square centimeter, average. This power density is easily exceeded, however, even with a low power laser if brought to a focus on the screen.

The detector just described is useful when milliwatts of power are involved, but there is greater need for a device with faster time constant which will image CO_2 laser beams of a watt and more. As used herein "time constant" is defined to mean the time that elapses between a change of the beam and a change of the image to correspond to the new beam form and intensity. Since adequate sensitivity is inherent in the liquid crystal mechanism, the problem is to provide good energy transfer from the laser beam to the liquid crystal while providing adequate heat sinking to dissipate the absorbed laser power.

A 2-inch diameter detector of another tube is shown in Figure 2.12c. Although limited to a continuous average power input of about 1 watt in still air, this unit will tolerate 5 watts per square centimeter for short periods. The detector employs a one-half inch thick black anodized aluminum body **24** which functions as a heat distributor and a heat sink. A heater resistor **26** is intimately attached to the rear of the unit and provides the necessary thermal bias with about 2 watts of electrical input. This particular unit is not equipped with automatic temperature regulation and consequently heater input power needs to be adjusted to compensate for laser beam heating and changes in ambient temperature. An automatic temperature regulator may, however, be added to obviate the need to monitor the heat sink temperature and adjust the heater input power.

In this type detector, screen **10** is formed by applying the encapsulated liquid crystal directly to the black anodized aluminum heat distributor **24** which also serves as the required black background. In this way, the time constant is minimized and resolution is maximized because the heat distributor, having high thermal conductivity, will conduct heat away from the zone of beam incidence. "Resolution," as used herein, refers to the precision of the correspondence between the size, shape and mode structure of the beam cross section and the image. The device relies on the infrared absorption of the encapsulated liquid crystal to convert laser beam energy to heat, directly in the encapsulated liquid crystal where it is most effective.

The degree of infrared absorption can be inferred from the dual beam spectrometer curve of Figure 2.12e which shows the transmission through a comparable film of encapsulated liquid crystal. Assuming modest reflection, absorption is seen to be high at 10.6 microns and probably is adequate for laser beam imaging over the entire 2.5 to 40 micron range shown in the curve, although sensitivity will be down where absorption is low. The sensitivity of the Mylar diaphragm detector does not drop off in the same way because infrared absorption also takes place in the black backing. Qualitative measurements at 10.6 microns indicate reflection from the face of the detectors is low at that wavelength.

A third example, shown in Figure 2.12d, employs a transistorized temperature control **27**, one form of which is shown in Figure 2.12f, to maintain essentially constant bias temperature. A small bead thermistor **28** embedded in the one-half inch aluminum plate heat distributor **30** senses temperature for automatic control. Heat is applied by two high wattage resistors **32** intimately attached to heat distributor **30**. A thermal insulating layer **34** is interposed between the screen **10** and the heat distributor to provide some design control over the time constant, sensitivity and image resolution of the device. Insulating layer **34** critically affects the image quality. The better it insulates, the more sensitive the detector but image resolution is then reduced because of thermal spreading, and the thermal time constant is increased because the encapsulated liquid crystal cannot cool as quickly through the insulation.

Generally, a compromise is required among these factors in any design. For many applications, however, satisfactory results are obtained by using black anodized aluminum for the heat distributor. The aluminum also acts as a thin insulating layer. To some degree, the encapsulant also insulates the liquid crystal from the heat distributor.

Certain improvements are obtained using a thermally anisotropic material such as pyrolytic graphite for the layer **34**. The material is made to have high thermal conductivity normal to the large area surface in contact with the heat distributor and the encapsulated liquid crystal film to provide a low thermal resistance therebetween, while thermal resistance parallel to the large area surfaces is high, thereby minimizing thermal spreading and preserving maximum resolution of the image.

Power is provided by a 24-volt DC supply **35**. Several similar detectors have been made using SCR control to permit operation directly from AC lines. In situations involving sensitive circuitry which might be affected by SCR generated noise, the transistor control may be preferable.

These detectors work well with an incident power density at 10.6 microns of between 0.25 and 5 watts per square centimeter. Much less than 250 millimicrons per square centimeter does not give satisfactory indication, and permanent scarring of the screen results at about 8 watts per square centimeter. Beams of higher power density can be examined after expanding through a lens. At 1 micron, the sensitivity of this detector is down by a large factor, althouth a 1-watt Nd:YAG laser beam has been detected. The principal advantage of this detector at 1 micron over the greater sensitivity of an infrared phosphor screen is the nonfading image.

A continuous beam power input of a watt or so can be accommodated with natural convention cooling, but higher beam input powers require augmented forced draft cooling so as to maintain proper thermal bias. Figure 2.12g shows the detector temperature rise (above ambient) as a function of input power to the electrical heater for natural, unassisted convention cooling, and with supplementary fan cooling. The shaded area represents the operating temperature range of encapsulated liquid crystal type R-33 (with an assumed ambient temperature of 23°C).

It is seen that dissipation can be increased threefold with very modest forced air cooling. A small axial fan might be attached to the rear of the detector. For higher beam powers and air cooling, there would also be advantage in employing an encapsulated liquid crystal that functions at a higher bias temperature. Power handling capability could be further increased using controlled temperature water cooling.

For many applications, a larger screen area is required because of uncertainty in beam location, or to accommodate higher beam powers and larger images. Fortunately, imaging detectors of this type can be made as large as required, and a number of 4-inch diameter detectors have been made.

Various methods were tried for applying the encapsulated coating, including spreading, brushing, and spraying, but usually with results which were less than satisfactory in one way or another. Attempts at spraying resulted in frequent clogging of the air brush used, but more important, the resulting screens did not provide good contrast and color brilliance. Brushing resulted in nonuniform and streaked screens. Best and most consistent results were obtained using a turntable technique in which the surface to be coated is rotated slowly during spreading and drying.

The unit to be coated was centered on a variable speed turntable, and speeds and amounts of encapsulated liquid crystal slurry were varied until good results were achieved. Thereafter, the proper amount of slurry was placed on the center of the detector face using a dropper, and with the table turning, the slurry could be spread in a matter of seconds using a toothpick gradually moved radially toward the circumference of the disc. The encapsulated liquid crystal slurry expands spirally to fill the area, and a fairly uniform coating can be achieved. Moderate nonuniformities in coating at this time often will not be apparent after the vehicle evaporates. Some smoothing occurs after application as the table continues to turn. It is essential that the turntable be level.

Two principal causes of screen blemish were encountered, "pimples" due to clumps of capsules and void spots which appear to be caused by the oil-like liquid crystal which had escaped its capsule, possibly due to rupture while shaking up the slurry which tends to separate in the bottle. Filtering the slurry through nylon stocking material eliminiates most of the pimple problems.

For the 2-inch screens, 10 drops of encapsulated liquid crystal slurry were applied to the center of the screen and were spread while it was turning at about 30 rpm. For the 4-inch screens, 40 drops of slurry were applied and spread while the unit was turning at about 8 rpm. Rotation should continue until the coating has dried.

Film type screens have been made using one-quarter, one-half and one-mil Mylar sheet. The one-half mil sheet is a good compromise between sensitivity and strength. The cleaned film was stretched over an embroidery hoop and sprayed on the inside with flat black lacquer. After drying, the hoop is fitted over the turntable which holds a dummy 4-inch plate to provide a flat surface against the underside of the Mylar film. The face of the screen is then coated with the encapsulated liquid crystal in the same manner as the aluminum plates. Some difficulty was experienced in wetting the Mylar film, but this was helped by washing the film with detergent and adding a small amount to the slurry. While still stretched on the hoop, the coated screen can be clamped between halves of a mounting ring, or it can be removed from the hoop and mounted on the detector as in Figure 2.12b. The excess Mylar film is trimmed with a razor blade after mounting.

Optical Storage Capability

W.E.L. Haas; U.S. Patent 3,806,230; April 23, 1974; assigned to Xerox Corporation describes liquid crystalline compositions having the optical properties of the cholesteric liquid crystalline mesophase and comprising at least one nematic liquid crystalline material and at least one optically active, nonmesomorphic material.

These liquid crystalline compositions are utilized in an imaging system by applying an electric field across a relatively thin film of the imaging composition. When the field is applied, light scattering is observed in the field-affected areas. The imaging compositions possess a memory capability whereby this light scattering effect persists after the electrical force is removed from the liquid crystalline film. According to a preferred form of the process, the electrical field may be applied across the liquid crystalline film in imagewise fashion whereby the image is formed immediately on the film. Alternatively, a uniform electrical field may be applied across the liquid crystalline film whereby the entire film becomes light scattering and subsequently the film may be "erased" in imagewise fashion to provide a visible image.

The imaging system advantageously provides a permanent image which can be utilized in a number of different ways, for example, providing the optical image input for a xerographic reproduction system or to expose a photographic medium thereby leading to the formation of a hard copy reproduction of the image.

Figure 2.13a illustrates a typical liquid crystalline electrooptic imaging cell. In this figure a typical liquid crystalline imaging member **10**, sometimes referred to as an electroded imaging sandwich, is shown in cross-section wherein a transparent pair of plates **11** having substantially transparent conductive coating **12** upon the contact surface comprise a pair of parallel substantially transparent electrodes. An imaging member wherein both electrodes are transparent is preferred where the imaging member is to be viewed using transmitted light; however, a liquid crystalline imaging member may also be viewed using reflected light thereby requiring only a single transparent electrode while the other may be opaque.

The transparent electrodes are separated by spacing member **13** which contains voids which form one or more shallow cups which contain the liquid crystalline film or layer which comprises the active element of the imaging member. A field is created between the electrodes by external circuit **15** which typically comprises a source of potential **16** which is connected across the two electrodes through leads **17**. The potential source may be either DC, AC, or combinations thereof.

In the system the mechanism by which images are formed in the liquid crystalline film or layer is a Grandjean to focal-conic texture transformation. A system for transforming a cholesteric, or optically negative liquid crystalline substance, from its Grandjean or disturbed texture state to its focal-conic or undisturbed texture state by an applied electric field is described in U.S. Patent 3,642,348.

Figure 2.13b illustrates a preferred form of the liquid crystal imaging member described in Figure 2.13a. In Figure 2.13b the desired image is defined by the shape of an electrode, and therefore by the shape of the corresponding electrical field. The imaging member here comprises transparent plates **11** separated by spacer gasket **13** having void area **20** filled with liquid crystalline material, and comprising substantially the entire area of spacer **13**. The desired image is defined by the shape of the substantially transparent conductive coating shown at **21**, which is affixed to the inner surface of one or both of the transparent support plates and is affixed only in the desired image configuration. The example illustrated in Figure 2.13b shows only one of the two electrodes in image configuration; however, it will be understood by those skilled in the art that both electrodes could be conveniently made in a matched pair to define the same desired image.

When the single image electrode configuration is used, the second electrode will comprise transparent plate **11** with substantially transparent conductive coating **12** upon the entire area of the inner surface of the transparent electrode. It is noted that a very thin, or substantially invisible conductor **22** is necessary to electrically connect the electrode in the desired image configuration to external circuit **15** which is similarly connected to the conductive coating of the opposite electrode. In operation this example will produce an electric field only in areas where there are parallel electrodes, i.e., between the electrode in the desired image configuration and the opposite electrode, whether or not the second electrode is also in the desired image configuration. Again here, one of the electrodes may be opaque if it is desired to observe the imaged member by reflected light rather than by transmitted light.

In Figure 2.13c an X-Y address system suitable for imaging a liquid crystalline imaging member is illustrated in exploded isometric view. The liquid crystalline imaging film is placed in void area **20** within the transparent and substantially insulating spacer-gasket **13**. The liquid crystalline film and spacer **13** are sandwiched between a pair of substantially transparent electrodes comprising transparent support plates **11** upon which strips of substantially transparent, conductive material **41** are coated. The substantially transparent electrodes are oriented so that conductive strips **41a** and **41b** on the respective electrodes cross each other in an X-Y matrix or grid. Each conductive strip in each set of parallel strips **41a** and **41b** is electrically connected to a circuit system **42** which is suitable for selective or sequential operation.

Through selection system **42** and external circuit **15** including source of potential **16**, an electrical field suitable for creating the Grandjean to focal-conic texture transformation can be created across selected points or a selected sequence of points in the illustrated imaging system. It will be understood that substantially transparent conductive strips **41** may vary in width from a very fine, wire-like structure to any desired strip width. In addition, one support plate **11** may be opaque where the imaging system is to be observed from one side using only reflected light.

Figure 2.13d describes another example of a liquid crystalline electrooptic imaging cell where one of the electrodes comprises a photoconductor and imaging is effected by applying a uniform potential across the entire area of the electrodes and subsequently exposing the photoconductor to an imagewise pattern of activating electromagnetic radiation corresponding to a desired image configuration.

FIGURE 2.13: IMAGING SYSTEM HAVING OPTICAL STORAGE CAPABILITIES

a.

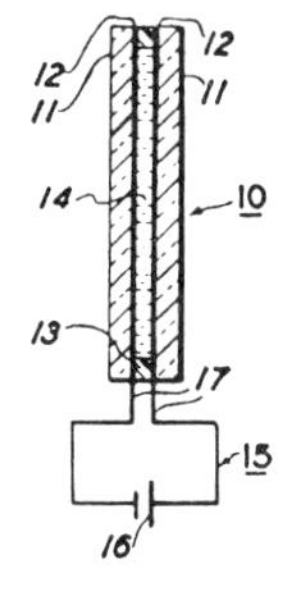

b.

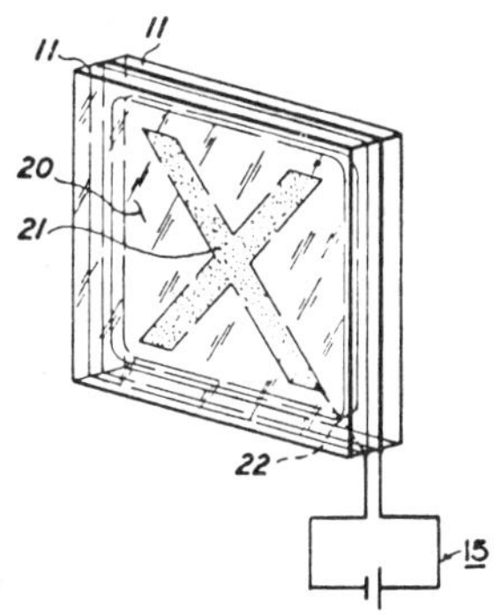

c.

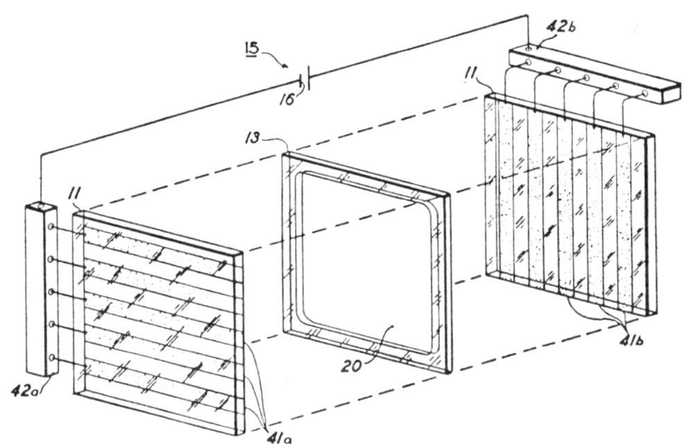

d.

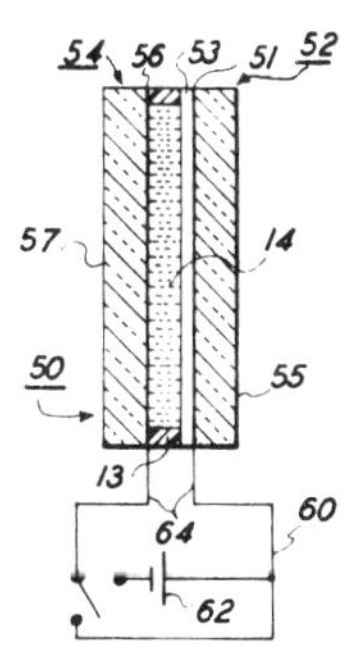

(continued)

FIGURE 2.13: (continued)

e.

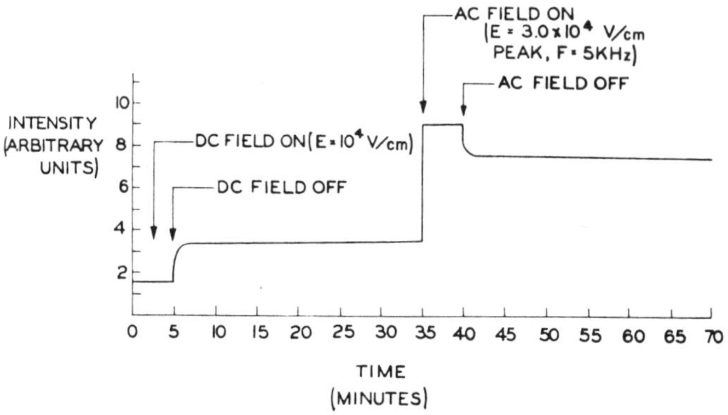

Source: W.E.L. Haas; U.S. Patent 3,806,230; April 23, 1974

In reference to Figure 2.13d, there is seen an electrooptic imaging cell, generally designated **50**, where a pair of plates, generally designated **52** and **54**, respectively, comprise a parallel pair of electrodes at least one of which is substantially transparent. For purposes of illustration both electrodes are shown as being transparent. Electrode **52** is made up of a photoconductive insulating material layer **53** overlying a conductive substrate which in this instance is shown as a substantially transparent conductive layer **51** deposited on a substantially transparent support member **55**. Electrode **54** is shown as a substantially transparent conductive layer **56** deposited on substantially transparent substrate **57**.

When it is desired to view the electrooptic cell using transmitted light, it is preferred that both electrodes be substantially transparent. Of course, in this instance there is required a photoconductive insulating material layer which is substantially transparent to the viewing electromagnetic radiation. Typical suitable substantially transparent photoconductive materials include, for example, relatively thin, e.g., about 5 microns, layers of selenium. However, the electrooptic imaging cell may also be viewed using reflected light thereby requiring only a single transparent electrode while the other may be opaque. In this instance one of the electrodes is preferably made up of an opaque photoconductive insulating layer deposited on an opaque substrate which may be any suitable conducting material such as a metallic layer.

The transparent electrodes are separated by spacing member **13** which contains voids which form one or more shallow cups which contain the liquid crystalline film or layer **14**. The electrodes are connected to opposite terminals of an external circuit **60** which typically comprises a source of potential **62** which is connected across the two electrodes through leads **64**. When a potential is applied to the conducting surface **54** of electrode **52** in the dark, no current will flow and no field will be established across the liquid crystal film since layer **53** is insulating under these conditions.

However, when the imaging cell is exposed to an image-wise pattern of activating radiation, the light-struck areas of photoconductive insulating layer **53** becomes conductive causing current to flow and establishing a field across the liquid crystal film in the light-struck areas. It is theorized that the current flow associated with the electric field causes the optical properties of the liquid crystalline material to change and the liquid crystal film which, in the Grandjean texture is substantially transparent prior to the application of the field, becomes "milky," i.e., focal-conic and scattering thereby effecting imaging. When the imaging cell is viewed by reflected light, the image will typically appear as milky-white image areas on a dark background. When the imaging cell is viewed using transmitted light, the image will typically appear as dark image areas with transparent nonimage or background areas. The following examples illustrate the process; all parts and percentages are by weight unless otherwise indicated.

Example 1: A liquid crystalline composition comprising about 10% of racemic menthol (an optically inactive species) and about 90% of p-ethoxybenzylidene-p'-n-butylaniline is prepared and observed under a Leitz Dialux polarizing microscope. The composition exhibits its typically nematic textures.

Another liquid crystalline composition is prepared by mixing together about 10% of l-menthol (an optically active species) and about 90% of p-ethoxybenzylidene-p'-n-butylaniline. The composition is observed under the polarizing microscope and is seen to be strongly optically active when sheared, a characteristic of cholesteric liquid crystalline materials in the Grandjean texture.

The optical response of both of the liquid crystalline compositions to electrical fields is then observed in the following manner. A layer of the liquid crystalline material is disposed between two transparent NESA glass electrodes separated by a 2-mil thick Mylar spacer. The intensity of the light transmitted through the cell is monitored as a function of time with a silicon photodiode, mounted on a polarizing microscope, and a Brush recorder. A tungsten light source is used. The cell initially appears transparent.

A DC electric field of about 10^4 v/cm is applied across the liquid crystalline film for about 5 minutes. For the composition containing the optically inactive racemic menthol light-scattering is observed when the field is applied across the liquid crystalline film and ceases when the field is removed. The cell is observed to return to its initial transparent state in about 200 milliseconds.

However, the behavior of the composition containing the optically active l-menthol contrasts sharply with that described above. For the composition containing l-menthol, the light-scattering is observed to persist for about 6 hours. The intensity of the light transmitted through the cell for 30 minutes after field removal is plotted as a function of time and shown in Figure 2.13e. Referring to this figure, it can be seen that after an initial relaxation, the intensity of the transmitted light increases very slowly in time. The pattern observable under the microscope is identifiable as a cholesteric focal-conic texture.

After about 35 minutes an AC electrical field of about 3.0×10^4 v/cm is applied across the liquid crystalline for about 5 minutes. It can be seen from Figure 2.13e that the application of the AC field clears the cell, i.e., returns the liquid crystalline imaging material to its transparent state. The time required to produce scattering or erasure is typically shorter than about 1 second. The 5 minute time period is selected for measurement purposes only.

Example 2: A liquid crystalline composition comprising about 5% of l-menthol and about 95% of p-ethoxybenzylidene-p'-n-butylaniline is prepared and observed in the same manner as described in Example 1. The composition is observed to be strongly optically active when sheared and the optical response of the composition to electrical fields is observed to be similar to that shown in Figure 2.13e.

Electric Field Sensitive Devices

J.L. Fergason; U.S. Patent 3,627,408; December 14, 1971; assigned to Westinghouse Electric Corporation describes an electric field sensitive device for providing a visual indication in response to an electric field incorporating liquid crystalline materials of the type exhibiting the cholesteric phase.

Such a liquid crystal device may be utilized for detection of a field with an indication by color of the value of a voltage or a current impressed thereon. In addition, the liquid crystal device may be utilized for display of a visual image. A variety of devices may be constructed to bring about and utilize this phenomenon. The optical property most readily utilized is that of selective scattering since it does not require polarization for observation. The change in transmission properties of light through the liquid crystal device may be utilized and controlled. This transmission shift is in the same direction and to the same extent as the scattering band. The following table gives a listing of specific mixtures of material by percent weight which exhibit the field effect. These materials are stable at room temperature. Various detection devices utilizing these mixtures are described in considerable detail in the patent.

Mixture	Color at Room Temperature
20% Cholesteryl chloride (CC) 80% Cholesteryl oleyl carbonate (OCC)	Blue
25% CC 75% OCC	Green
30% CC 70% OCC	Red
20% CC 20% Cholesteryl nonanoate (CN) 60% OCC	Blue
25% CC 75% Cholesteryl oleate (CO)	Green
25% Cholesteryl bromide (CB) 75% OCC	Green
30% CB 70% OCC	Red
25% CC 25% OCC 25% CO 25% CN	Green

Undisturbed Image on a Disturbed Background

B. Mechlowitz, J.E. Adams and W.E.L. Haas; U.S. Patent 3,666,948 and W.E.L. Haas, J.E. Adams, J.B. Flannery, Jr. and B. Mechlowitz; U.S. Patent 3,666,947; May 30, 1972; both assigned to Xerox Corporation describe an imaging system where an imaging member comprising a material having a cholesteric liquid crystalline phase in its Grandjean or disturbed texture state is thermally imaged by imagewise heating image portions of the imaging material to a temperature at least about the cholesteric liquid crystalline liquid isotropic transition temperature of the material and allowing the heated portions of the imaging material to return to a temperature within the cholesteric liquid crystalline mesophase temperature range of the material whereby the imaged areas assume the focal-conic or undisturbed texture state in the desired image configuration. Images provided by the system are erasible, and the imaging members are reusable.

In Figure 2.14a an imaging member **10** suitable for use in the process is illustrated where substrate **11** supports a layer of imaging composition comprising material which exhibits a cholesteric liquid crystalline mesophase. The supporting substrate may comprise any suitable material, and in various examples the substrate may take on any suitable form including the form of a web, foil, laminate or the like, strip, sheet, coil, cylinder, drum, endless belt, endless moebius strip, circular disc or other geometrical shapes. The layer of imaging composition **12** may comprise any suitable material which exhibits the cholesteric liquid crystalline mesophase.

The liquid crystal imaging layers or films suitable for use in the process are preferably of a thickness in the range of about 250 microns or less, although thicker films will perform satisfactorily in the system.

FIGURE 2.14: THERMAL IMAGING PROCESS

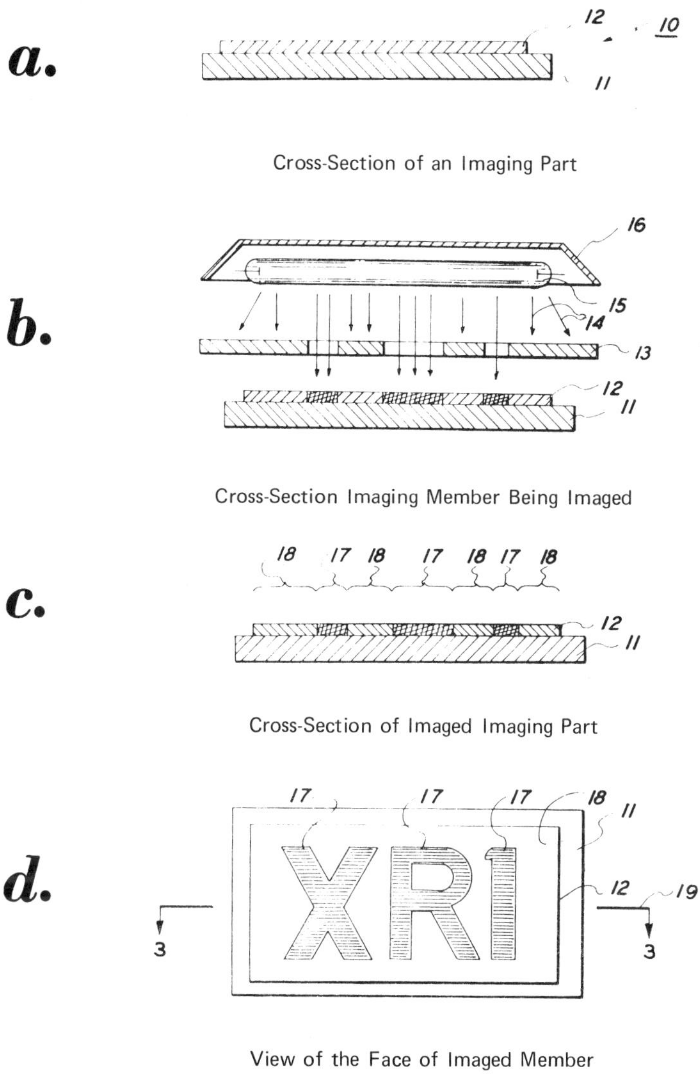

Source: B. Mechlowitz, J.E. Adams and W.E.L. Haas; U.S. Patent 3,666,948; May 30, 1972

Optimum results are typically achieved using layers in the thickness range between about 1 and 50 microns. When a layer of imaging composition **12** is prepared on a suitable substrate **11** by methods such as those described above, the material comprising the cholesteric liquid crystalline mesophase often assumes its Grandjean or disturbed texture state. However, in some instances the material may be made to assume its Grandjean texture state by mechanical disturbance such as by shearing or pressure, or by external forces such as electrical or magnetic fields, or by any other suitable means. The Grandjean texture is typically characterized by selective reflection of incident light around a wavelength λ_0 where $\lambda_0 = 2np$ where n equals the index of refraction of the composition layer and p equals the pitch of the liquid crystalline film, and is additionally characterized by optical activity for wavelengths of incident light away from λ_0. If λ_0 is in the visible spectrum, the composition layer comprising cholesteric liquid crystalline material appears to have the color corresponding to λ_0 for normally incident light and observation. If λ_0 is outside the visible spectrum, the composition layer typically appears colorless and nonscattering. The Grandjean texture of cholesteric liquid crystals is sometimes referred to as the disturbed texture.

An imaging member **10** which is prepared so that the composition layer comprising material in the cholesteric liquid crystalline mesophase is in its Grandjean or disturbed texture state is imaged by the imagewise application of thermal energy or energy which is capable of producing an imagewise thermal effect in the layer of imaging composition. One form of the system is illustrated, for example, in Figure 2.14b where the imaging composition layer **12** is shown being imagewise exposed through a stencil-like mask **13** to thermal radiation **14** which is emitted by a source **15** under shield **16**. Any source of thermal energy, for example, such as lasers, gas discharge lamps, and others, may be used as the source of energy in the imagewise exposure step.

In the system the thermal energy which is applied in imagewise configuration to the layer of imaging composition is applied in sufficient amounts so that the imaging composition comprising material having the cholesteric liquid crystalline mesophase is, in the imagewise exposed areas, heated to a temperature of at least about the liquid crystalline-liquid isotropic transition temperature of the imaging material having the cholesteric liquid crystalline mesophase characteristics. After the imagewise application of sufficient thermal energy to raise the temperature of the imaging composition in the imagewise exposed areas to a temperature at least about the isotropic transition temperature, the source of thermal energy is removed, and the imaging composition is allowed to cool into the cholesteric liquid crystalline mesophase temperature range below the isotropic transition temperature of the composition.

Upon cooling into the cholesteric liquid crystalline mesophase temperature range, the imagewise exposed areas of the imaging composition typically exhibit the focal-conic or undisturbed texture state, thereby providing an imaged member having image areas of the imaging composition comprising a material having the cholesteric liquid crystalline mesophase in the focal-conic texture state with background areas of the imaging composition comprising material having a cholesteric liquid crystalline mesophase in the Grandjean type texture state.

The focal-conic type texture is also typically characterized by selective reflection; but in addition, this texture state also exhibits diffuse scattering in the visible spectrum, whether λ_0 is in the visible spectrum or not. The appearance of the focal-conic texture state is typically milky-white when λ_0 is outside the visible spectrum. The focal-conic texture of cholesteric liquid crystalline materials is sometimes referred to as the undisturbed texture state.

The effect of thermally induced texture change in the imaged areas is to destroy polarization of transmitted light in the image areas whereas the background areas have little effect on polarization except in some small wavelength region. This small region, here designated $\Delta\lambda$, is centered about $\lambda_0 = 2np$ where n equals the index of refraction of the composition layer and p is the pitch of the cholesteric liquid crystalline material. Typically the ratio $\lambda_0/\Delta\lambda$ equals about 14. In this region, incident plane polarized light emerges from the composition layer comprising material having the cholesteric liquid crystalline mesophase, circularly polarized. Between crossed polarizers, therefore, the imaged areas typically appear white or light upon the black or colored background. Where an imaging member including a supporting substrate is viewed between crossed polarizers, the substrate material should be optically isotropic, i.e., it should not change the polarization of the light transmitted through the system.

Figure 2.14c illustrates in partially schematic, cross-sectional view an imaged member where the layer of imaging composition **12** exhibits imaged areas **17** in the focal-conic or undisturbed cholesteric liquid crystalline texture state, while the background areas **18** exhibit the Grandjean or disturbed cholesteric liquid crystalline texture state. The thermally induced texture transition imaging effect of the process is typically a bulk effect. The bulk effect is illustrated in Figure 2.14c where imaged areas **17** of the layer of the imaging composition are schematically illustrated as being transformed throughout the entire cross-section of the composition.

Figure 2.14d is a top view of the imaging member of Figure 2.14c (where the view of Figure 2.14c is a cross-section along line **19**) schematically showing imaged areas **17**, in the focal-conic texture state on background areas **18**, in the Grandjean texture state. Although the imaged areas **17** in Figure 2.14d are here illustrated as being the darker areas, the contrast and density of an imaged member produced by the system may vary from one example to another. Figure 2.14d is intended to be a representation of an imaged member wherein the imaged areas are clearly optically distinguishable from the background areas.

In addition to the imaging aspects of the system, it should also be noted that the imaging members of the process may have images like those produced by the system erased, for example, by the application of external forces such as pressure, shearing stresses, electrical fields, magnetic fields, of combinations thereof. After such an imaged member has the image erased, the member is typically suitable for reimaging by the system. In this way imaging members are reusable for large numbers of imaging and erasing cycles. The erasability of the system further enhances the utility of this one-step, immediately visible imaging system.

The following examples further define the process with respect to the thermally induced imagewise transformation of image portions of a layer of imaging composition comprising a material having cholesteric liquid crystalline characteristics, from the Grandjean or disturbed texture to an imaged, focal-conic or undisturbed texture. The parts and percentages are by weight unless otherwise indicated.

Example 1: An imaging member is provided by dissolving a solute mixture comprising about 57% cholesteryl formate and about 43% cholesteryl nonanoate in petroleum ether solvent, mixing the solution thoroughly, and depositing a layer of the solution on a thick, stable, black plastic substrate. The solvent is allowed to evaporate, leaving a layer of imaging composition comprising cholesteric liquid crystalline materials in a thickness of about 12 microns on the substrate. After solvent evaporation, the film of imaging composition is uniformly provided in the Grandjean texture state by shearing the film by dragging the edge of a glass slide across the surface of the film. The film exhibits a green color.

The imaging member is then imaged by the texture transition imaging system by placing the member, imaging composition side up spaced a few microns under an imagewise metal mask, and placing the masked member about 25 centimeters from a BH6-1 mercury arc lamp. The lamp is activated thereby exposing the member through the mask for an exposure time of about 10 seconds. The imagewise exposure heats the image areas to temperatures about or above the cholesteric liquid crystalline-liquid isotropic transition temperature of the imaging composition, and the imaged areas exhibit a texture transition.

A clearly optically distinguishable image, corresponding to the mask, of almost colorless imaged areas on a green colored background, appears almost immediately upon exposure. The composition is allowed to cool into its cholesteric liquid crystalline mesophase temperature range. The imaged areas exhibit the focal-conic or undisturbed texture state while the background areas of the cholesteric liquid crystalline imaging composition exhibit the Grandjean or disturbed texture state.

Example 2: An imaged member is provided by the method of Example 1. After imaging, this imaged member is covered with a transparent glass slide. The glass slide is slightly displaced with respect to the substrate of the imaged member. Displacing the cover slide erases the texture transformed image and uniformly provides the cholesteric liquid crystalline imaging composition in the Grandjean texture state.

Example 3: An imaging member is provided by mixing about 57% cholesteryl formate and about 43% cholesteryl nonanoate, placing the mixture in a crucible, and heating the mixture above the liquid isotropic transition temperature of the mixture and its components. The liquid mixture is thoroughly mixed in the crucible while the composition is in the liquid state. The liquid composition is applied to a transparent glass substrate, and squeegeed into a substantially uniform layer of imaging composition of a thickness of about 19 microns. The composition is allowed to cool into its cholesteric liquid crystalline mesophase temperature range, and the member is imaged as described in Example 1. The imaged member is examined under a microscope, and the image areas are seen to exhibit the focal-conic or undisturbed texture state and are colorless, while the background areas are seen to exhibit a green color and are in the Grandjean or disturbed texture state.

Molecular Orientation Technique for Xerography Process

A process described by *E. Forest and C.K. Keller; U.S. Patent 3,804,618; April 16, 1974; assigned to Xerox Corporation* relates in general to imaging and in particular to a xerographic system. In the art of xerography according to U.S. Patent 2,297,691, it is usual to employ the simultaneous application of electric field and a pattern of activating radiation on a photoconductive insulating member to form an electrostatic charge pattern otherwise known as an electrostatic latent image. This electrostatic latent image then is capable of being utilized such as, for example, developed by the deposition of electroscopic material thereon to form a visible image. Customarily, deposition of the electroscopic material is accomplished by cascading the electrostatic image with an electroscopic powder and a granular carrier as described in U.S. Patent 2,638,416.

Development by means of depositing electroscopic powders on the surface of photoconductive members has a disadvantage in that the softer photoconductors, particularly those based on selenium, are abraded by the action of the powders on the photoconductor surface. The abrasive action occurs at two stages of conventional xerography; first, during the initial cascading, and second, when the powder is removed by brushing the photoconductive member prior to the member's reuse.

In a display system, that is, where the developed image is viewed directly or where light information from the developed image is projected onto a screen or a fast developing film, the abrasion problem is more severe. In conventional xerography almost all of the electroscopic powder in the developed image is transferred to paper before the insulating surface is cleaned. In a display system, however, all of the electroscopic powder remains on the insulating surface until the cleaning step resulting in greatly increased abrasion.

It is an object of this process to provide a system for developing an electrostatic image which does not require physical contact between the developer and the photoconductive member. The process comprises placing a liquid crystalline substance in the field generated between an electrostatic latent image and a conductor, or in a field generated by a source of potential connected to shaped electrodes whereby the liquid crystalline substance is strained by the imagewise field in image configuration providing a visible image.

By strained is meant that the oriented molecules in the liquid crystal are reoriented. A visual difference between the field strained and the unstrained liquid crystals provides an image. The difference may be a difference in hue such as red, green, blue, etc. or in lightness, i.e., different shades of grays. The visible image may also result from a difference in reflectance, refraction, or transparency.

Alternatively, light information from the visible image may be projected onto a screen or onto a fast developing film. This process would be particularly useful in a rapid recording and display system such as that described in U.S. Patent 3,115,075. The insulating member is then removed from the system, solvent washed to remove any liquid crystal adhering to it, air dried, and stored for later use. Preferably, the liquid crystal is separated from the insulating member by a thin walled, opaque, insulating film, thereby preventing contact between the liquid crystal and the insulating member. This eliminates the requirement of solvent washing the insulating member followed by air drying.

It is also possible to effect reversal imaging, that is, to develop uncharged areas of the electrostatic latent image. Reversal imaging may be accomplished by charging the conductive member to the same potential as is present on the charged areas of the electrostatic latent image. In this case the resultant field is present only between the conductive member and the uncharged areas of the electrostatic image resulting in straining of the liquid crystalline material above uncharged areas of the electrostatic image providing a visible image which is the reverse of the electrostatic latent image.

The electrostatic latent image may be formed on either an insulating or photoconductive insulating surface. If a photoconductive insulating surface is to be used, care must be taken not to discharge the electrostatic latent image during display. That is, actinic radiation impinging on the surface of the photoconductive layer may discharge the electrostatic latent image before it has been developed and displayed. It may be necessary, therefore, to use light conditions during development and display which will not render the photoconductive layer conductive.

For example, red light would be suitable for selenium photoconductive layers since selenium is not affected by red light. Suitable lighting conditions for various photoconductive layers which will not render the photoconductor conductive may be determined by referring to graphs of the relative energy of response of photoconductors as a function of the wavelength of light used, avoiding wavelengths which result in photoconductor response. See, for example, *Xerography and Related Processes* by Dessauer and Clark, Focal Press Ltd., (1965). Preferably, however, the insulating film which is placed between the liquid crystal and the insulating member is opaque thereby eliminating the requirement of special lighting during development and display.

Figure 2.15 shows a sectional side view of a simple exemplary system for carrying out the process where the liquid crystalline material has been placed between a photoconductive insulating member and a grounded transparent electrode. Referring to Figure 2.15 a NESA glass plate **1** is prepared by placing on its surface an inert gasket **2**. The shallow cup formed by the plate and inert gasket is filled with a liquid crystalline material **3**. The depth of this cup, i.e., the depth of the liquid crystalline material used is a function of the composition of the liquid crystal, the temperature of the liquid crystal, the potential of the electrostatic latent image and the thickness and insulating properties of the thin film **4**.

FIGURE 2.15: XEROGRAPHIC TECHNIQUE

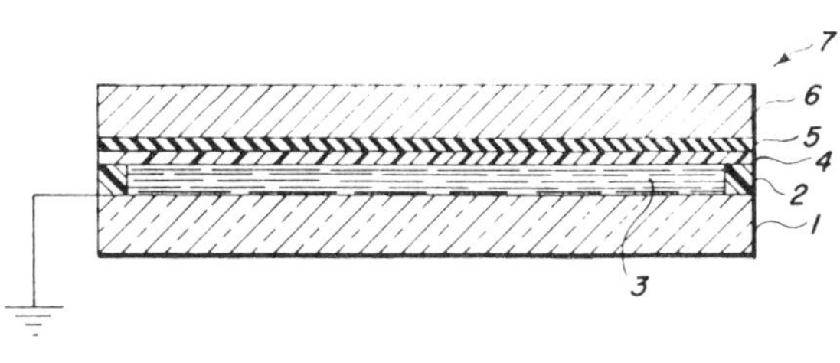

Source: E. Forest and C.K. Keller; U.S. Patent 3,804,618; April 16, 1974

Insulating film **4** is placed in contact with the liquid crystal and inert gasket. Xerographic plate **7** comprising a layer **5** of photoconductive material on a conductive substrate **6** is sensitized by corona charging. Plate **7** is then exposed to a pattern of light and shadow which causes the charge to drain away from those areas on which light impinges. The remaining surface charge is an electrostatic latent image. The xerographic plate is brought into contact with film **4**. The molecules in the liquid crystal are strained in field areas providing a visible image viewable through the NESA glass plate.

The following examples illustrate the process. The parts and percentages are by weight unless otherwise indicated. In the following examples the charge density of the latent image is such that a field of about 2,000 to 4,000 volts is developed across the liquid crystalline layer.

Example 1: A 2-inch square by one-eighth inch thick NESA glass plate is prepared by placing on its surface a Teflon gasket having over-all dimensions about 2 inches square and a cross-sectional dimension about 0.1 millimeter. The NESA glass plate is then grounded. The NESA glass plate and Teflon gasket are then heated to a temperature of about 130°C and maintained at about that temperature throughout the experiment by an electrical radiant energy heater.

The 0.1 millimeter deep cup formed by the Teflon gasket and NESA glass electrode is filled with a mixture of about 55% cholesteryl benzoate and about 45% cholesteryl acetate which is prepared as follows. The mixture of liquid crystals is placed in an oven and heated to a temperature of about 137°C. The mixture is then removed from the oven and allowed to cool to a temperature of about 130°C. The mixture is then poured onto the NESA glass plate. An approximately 0.05 millimeter thick film of Pigmented Type 30 Tedlar is placed in contact with the liquid crystalline material and the Teflon gasket

A 2-inch square by one-eighth inch thick Teflon member is corona charged through a grounded metal stencil in image configuration resulting in an electrostatic latent image on the surface of the insulating member. The Teflon member containing the electrostatic latent image is then placed in contact with the insulating film. The image is then viewed through the NESA plate. This experiment may be carried out under ordinary incandescent light conditions because the Teflon member is not photoconductive, hence will not discharge an electrostatic image in the presence of ordinary light conditions.

Example 2: A 2-inch square by one-eighth inch thick NESA glass plate is prepared by placing on its surface a Teflon gasket having over-all dimensions about 2 inches square and a cross-sectional dimension of about 0.1 millimeter. The NESA glass plate is then grounded. The NESA plate and Teflon gasket are then maintained at a temperature of about 24°C throughout the experiment.

The 0.1 millimeter deep cup formed by the Teflon gasket and NESA glass electrode is filled with a mixture of about 30% cholesteryl butyrate and about 70% cholesteryl myristate which is prepared as follows. The mixture of liquid crystals is placed in an oven and heated to a temperature of about 45°C. The mixture is then removed from the oven, poured onto the NESA glass plate, and allowed to cool to a temperature of about 24°C. An approximately 0.05 millimeter thick film of Pigmented Type 30 Tedlar is placed in contact with the liquid crystalline material and the Teflon gasket.

A 2-inch square by one-eighth inch thick selenium xerographic plate comprising an 80 micron selenium photoconductive layer on an aluminum substrate is sensitized by corona charging. The xerographic plate is then exposed to a pattern of light and shadow which causes the charge to drain away from those areas on which light impinges. The remaining surface charge is an electrostatic latent image. The xerographic member containing the electrostatic image is placed under red light conditions in contact with the film. The image is then viewed through the NESA glass plate under conventional incandescent light conditions.

Example 3: A liquid crystalline substance made of a mixture of about one part cholesteryl crotonate and about one part cholesteryl oleate is dissolved in chloroform to make a free flowing solution. This solution is poured onto the NESA glass plate Teflon gasket arrangement used in Example 1 except that in this experiment all components are at about room temperature. The chloroform is allowed to evaporate in air leaving a viscous liquid crystal material on the surface of the NESA glass plate. The insulating film of Example 1 is placed in contact with the liquid crystal and the Teflon gasket. The NESA glass plate is then grounded. The xerographic member of Example 2 containing an electrostatic latent image is then under red light conditions placed in contact with the insulating film. No heating device need be utilized in this example since the liquid crystal used in this example exhibits liquid crystal characteristics at or near room temperature. The image is then viewed through the NESA glass plate under conventional incandescent light conditions.

Erasable and Reusable Imaging System

W.E.L. Haas and J.E. Adams; U.S. Patent 3,671,231; June 20, 1972; assigned to Xerox Corporation describe an imaging system where a liquid crystalline imaging material on a photoconductive layer is provided with a suitable overcoating. The system is imaged by providing an electrical latent image on the above described imaging member, and it is erased by uniformly electrically charging the overcoated surface of the imaging member, thereby making the imaging member immediately reusable. Although any liquid crystalline composition having cholesteric liquid crystalline characteristics is suitable for use in the process, compositions comprising cholesteryl oleyl carbonate have been found particularly useful.

The protective overlayer is designed to protect the liquid crystalline imaging material which is typically tacky, soft, viscous, or liquid from foreign matter such as dust, insects, etc. and this overlayer is typically transparent to allow viewing of the imaged liquid crystalline materials. Electrically insulating materials will typically perform this function, and some partially conductive materials may perform this function satisfactorily. Materials suitable for use as the overlayer include Tedlar, a polyvinyl fluoride; polyethylene film; polyvinyl chloride film; Mylar; mixtures thereof; and others. The overlayer is typically a thin transparent film and preferably of thickness not greater than about 2 mils. Optimum results are achieved using overlayer thicknesses in the range between about one-quarter and one mil.

The following examples further define the process with respect to the liquid crystalline imaging system having a liquid crystalline imaging member which is quickly erasable and immediately reusable. The parts and percentages are by weight unless otherwise indicated.

Example 1: A liquid crystalline imaging member is provided by using a xerographic plate from a Model D Processor (Xerox Corporation) which comprises a photoconductive coating of vitreous selenium on a conductive sheet. The exposed surface of the photoconductor is framed in a spacer gasket comprising an about one-half inch wide strip of Mylar film. The area of the surface of the photoconductor framed by the spacing gasket is coated with an imaging composition comprising cholesteric liquid crystalline material, here a mixture of about 20% cholesteryl oleyl carbonate, about 28% cholesteryl chloride, and about 52% cholesteryl nonanoate. The liquid crystalline composition is then overcoated with an about one-half mil thick film of transparent Tedlar.

This imaging member is then substantially uniformly electrostatically charged to a surface potential of about 700 volts and imagewise exposed to an image pattern of activating electromagnetic radiation, here light, in the Model D Processor. Upon charging and exposing an image corresponding to the original image from which the imagewise pattern of activating electromagnetic radiation was derived appears on the above described imaging member. The liquid crystalline composition exhibits green image areas on a substantially colorless background, which exhibits the color of the xerographic plate itself in the background areas.

The imaged member is then again uniformly electrostatically charged to a surface potential of about 700 volts in the Model D Processor, which destroys the image produced by the previous imaging steps, and the imaging member is then ready to be reimaged by the above recited process.

Example 2: The imaging process of Example 1 is performed using an imaging member as described in Example 1 with an imaging composition comprising cholesteric liquid crystalline material comprising a mixture of about 16% cholesteryl oleyl carbonate, about 30% cholesteryl chloride, and about 54% cholesteryl nonanoate. The imaging composition is overcoated with an about one-quarter mil thick film of transparent Mylar, a polyester resin film available commercially. When imaged by the process of Example 1, this composition exhibits red image areas. This imaged member is also uniformly electrostatically charged whereby the image is destroyed or erased, thereby providing the imaging member in condition for reuse.

Example 3: The imaging and erasing process of Example 1 is performed using the imaging member and apparatus of Example 1 with an imaging composition comprising cholesteric liquid crystalline material here comprising a mixture of about 25% cholesteryl oleyl carbonate, about 26.3% cholesteryl chloride, and about 48.7% cholesteryl nonanoate. This imaging member is imaged by charging and exposing, and then erased by recharging, as described in Example 1.

Ultraviolet Light

According to a process described by *W.E.L. Haas; J.E. Adams; J.H. Becker and J.J. Wysocki; U.S. Patent 3,655,971; April 11, 1972; assigned to Xerox Corporation* a film of cholesteric liquid crystalline material is exposed to ultraviolet radiation thereby producing a visible, color image corresponding to the areas of different total exposure. The following examples illustrate the process; parts and percentages are by weight unless otherwise indicated.

Example 1: A liquid crystalline film comprising a mixture of about 50% cholesteryl iodide and about 50% cholesteryl nonanoate is prepared by heating the liquid crystalline materials in a crucible to a temperature above the isotropic transition temperature of the mixture, mixing the heated materials thoroughly, and depositing the heated mixture onto a glass slide. Samples of approximately uniform thickness of about 2 mils are prepared in this manner by cutting a hole of the desired area in a Mylar spacer about 2 mils in thickness, placing the spacer on the glass slide, filling the hole in the spacer with the heated liquid crystal mixture and dragging a glass slide across the spacer and mixture thereby filling the hole in the spacer and removing excess mixture while shearing and smoothing the sample. The shearing action of the leveling slide aligns the molecules in the sample. The pitch of the sample is determined using the observation apparatus described below.

The plate having the film of liquid crystalline material thereon is then exposed to ultraviolet radiation less than about 3,000 A in wavelengths, produced from a General Electric BH-6 high pressure, short arc mercury lamp. Exposure time is controlled by a shutter. The film material is exposed while in its cholesteric liquid crystalline mesophase. The film material is in its aligned state and shows various reflected colors. During ultraviolet exposure, the color of the film material shifts toward the red, as observed in a fixed observation system. Using this exposure system, about 150 millijoules per square centimeter of ultraviolet radiation with wavelengths below about 3,000 A shifts the color peak about 1,000 A.

Samples prepared as shown above are exposed in incrementally increasing amounts of ultraviolet radiation, and incremental changes in colors of the samples changing from blue to red are observed with increasing total exposure.

The ultraviolet exposed liquid crystalline film is observed with a fixed observation apparatus having a source of monochromatic incident light which transmits the incident light beam through a collimating lens system and then to the exposed liquid crystal film which is displayed on a dark background on a spectrometer stage. Since the color of light scattered from a liquid crystal film is a function of the angle of incidence and the angle of observation, the light source and exposed sample are oriented so that the angle of incidence is fixed at about 60° from normal to the plane of the sample. Similarly, the angle of observed, reflected light is fixed at about 30° from normal to the plane of the sample where the reflected light is detected by a photodiode detector in conjunction with an oscilloscope read-out system. A mechanical chopper located in the line of the incident beam provides a reference signal for use with a lock-in amplifier.

Shifts in bright colors of the liquid crystalline material are observed with the eye. Data illustrating the color shifts corresponding to amounts of ultraviolet radiation are shown on line **50** in Figure 2.16. This data shows that the effect of ultraviolet radiation is to shift the color of the liquid crystalline sample toward the red.

Example 2: Liquid crystalline films comprising mixtures of 55% cholesteryl iodide and 45% cholesteryl nonanoate; 45% cholesteryl iodide and 55% cholesteryl nonanoate and 40% cholesteryl iodide and 60% cholesteryl nonanoate are exposed to ultraviolet radiation from a General Electric BH-6 mercury arc lamp in the system described in Example 1. Results similar to those of Example 1 are achieved, and the data illustrating the color shifts corresponding to the amounts of ultraviolet radiation are shown on lines **55**, **45** and **40** , respectively, in Figure 2.16.

FIGURE 2.16: COLOR CHANGE AS FUNCTION OF ULTRAVIOLET LIGHT EXPOSURE

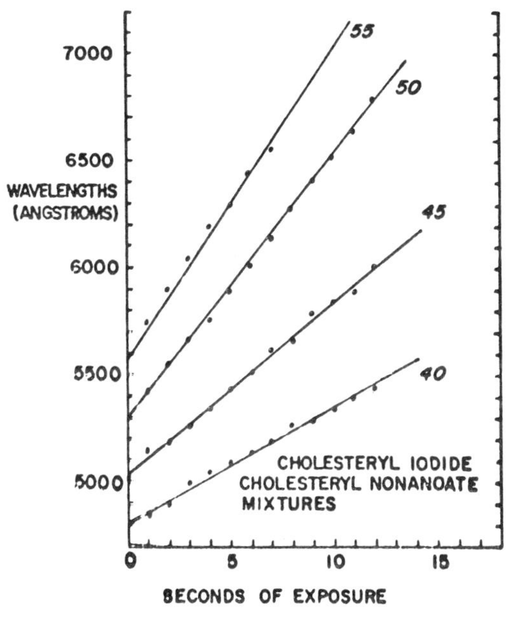

Source: W.E.L. Haas; J.E. Adams; J.H. Becker and J.J. Wysocki; U.S. Patent 3,655,971; April 11, 1972

Example 3: A liquid crystalline film comprising a mixture of 20% cholesteryl iodide and 80% cholesteryl nonanoate is prepared by dissolving the liquid crystals in petroleum ether in a concentration of about 1 gram per 10 cubic centimeters and coating the solvent mixture on a polished copper plate about 3 x 4 centimeters in size. The ether is then allowed to evaporate. The film has a thickness of about 5 microns and is an unaligned liquid crystalline film. The imaging plate is exposed to ultraviolet radiation through a steel mask using a xenon flashtube, Novatron-186, (Xenon Corporation) in the exposure process of Example 1. The liquid crystalline material in its originally unaligned state is almost colorless. No image pattern is immediately visible. The exposed film is slowly heated to about 40°C and the image becomes clearly visible.

OTHER APPLICATIONS AND DEVICES

Decorative Iridescent Compositions

W. Mahler and M. Panar; U.S. Patent 3,766,061; October 16, 1973; assigned to E.I. du Pont de Nemours and Company describe solid compositions having cholesteric properties, particularly iridescence which are made by intimately mixing esters of cholesterol or cholesterol-like sterols by melt or solution methods. The esters include a substantial proportion of hydrogen phthalate esters. The remaining esters preferably contain a carbocyclic ring. Sterols or sterol 3β-halides can also be present in the mixture. If cinnamate esters are employed, all or part of the hydrogen phthalate esters can be replaced with hydrogen isophthalate esters.

The compositions are particularly suitable for application as a decorative coating to articles such as paper cards, tile, leather, poromeric materials, wooden articles such as panels, furniture, plastics including thermoplastics and thermosetting resins, metals, glass and like solid objects. Particularly pleasing effects are obtained when the cholesteric glass compositions are applied over a smooth light absorptive ground, preferably black, which can be, for example, a suitable coating of paint, pigmented plastic or the like. If desired, a transparent coating may be applied to enhance soil resistance, water resistance or otherwise improve the properties of the coating.

For use as decorative coatings, the cholesteric solid compositions are preferably applied in a thickness of about 1 to 10 microns. A thickness of up to about 200 microns is practical. Below the lower limit, the color intensity produced is insufficient for practical applications.

The following examples illustrate the process; parts and percentages are by weight unless otherwise indicated. The normal color of the compositions was examined in most instances as a film spread on a black colored substrate. The following abbreviations are employed.

Normal Color	Abbreviation
Infrared	IR
Maroon	M
Red	R
Bronze	Br
Orange	O
Yellow	Y
Green	G
Violet	V

Examples 1 through 6: The compositions listed in the table below were prepared in the following manner. The ingredients were mixed together, then placed on a glass plate which was heated on a hot plate to a little above the melting point, about 175° to 200°C. The glass plate containing the melt was then removed from the heat, covered with another glass plate and allowed to cool. The colors formed by incident white light were viewed at a 90° angle to the plates.

Example	Composition	A	B	C	D	E	F
1	Cholesteryl hydrogen phthalate, percent	60	40	40	80		
	Cholesteryl hydrocinnamate, percent	20	20	0	10		
	Cholesteryl cinnamate, percent	20	40	60	10		
Resulting color		G	R	Y	B		
2	Cholesteryl hydrogen phthalate, percent	10	30	50	50	80	
	Cholesteryl benzoate, percent	65	35	25	50	20	
	Cholesterol, percent	25	35	25	0	0	

(continued)

	Composition	A	B	C	D	E	F
Resulting color		R	Y	YG	GB	B	......
3	Cholesteryl hydrogen phthalate, percent	30	40	45	60	80	......
	Cholesteryl benzoate, percent	35	40	10	20	10	......
	Cholesteryl cinnamate, percent	35	20	45	20	10	......
Resulting color		Y	YG	Y	BG	B	......
4	Cholesteryl hydrogen phthalate, percent	0	10	15	30	50	......
	Cholesteryl hydrogen isophthalate, percent	30	45	70	15	25	......
	Cholesteryl cinnamate, percent	70	45	15	55	25	......
Resulting color		YG	B	B	YG	BG	......
5	Cholesteryl hydrogen phthalate, percent	10	20	35	40	60	65
	Cholesteryl cinnamate, percent	45	60	35	60	40	25
	Cholesterol, percent	45	20	30	0	0	10
Resulting color		Y	R	R	Y	BG	IR
6	Cholesteryl hydrogen phthalate, percent	30	40	40	60	80	......
	Cholesteryl cyclohexanecarboxylate, percent	70	40	20	20	10	......
	Cholesteryl cinnamate, percent	0	20	40	20	10	......
Resulting color		R	G	G	G	B	......

Examples 7 through 21: In Examples 7 through 21 a mixture of 40 parts by weight of cholesteryl hydrogen phthalate, 40 parts of cholesteryl cinnamate and 20 parts of a benzoate ester of a similar sterol were dissolved in a solvent consisting of a mixture of 2 parts of toluene and 5 parts of methylene chloride to form a 25% solution. The solution was applied to black polyethylene terephthalate, evaporated and the film heated to 150°C, then cooled. The results are tabulated in the table below.

Example	Similar sterol	Normal color	$\lambda_{max.}$
7	7-dehydrocholesterol	B	4,300
8	Cholestanol	O-G	6,300
9	Stigmasterol	B-G	....
10	Ergosterol	B-G	....
11	6-ketocholestanol	O-G	5,900
12	Cholestan-5α,6α-epoxy-3β-ol		5,500
13	Lanosterol	V	....
14	Pregnenolone	B	4,200
15	β-Sitosterol	B	4,500
16	Isoandrosterone	B	4,300
17	5β-chloro-6β-nitrocholestan-3β-ol	O-G	6,300
18	6,6-difluorocholesten-3β-ol	B-G	5,100
19	5α-fluoro-6-ketocholestan-3β-ol	R	7,500
20	6-nitrocholesterol	R	....
21	5α-fluoro-6-nitriminocholestan-3β-ol	R	....

Examples 22 through 25: The compositions listed in the following table were prepared in the following manner. The ingredients were mixed in the indicated proportions and dissolved in toluene to give a 20% solution which was painted onto a black primer coated steel plate. The toluene was allowed to evaporate on standing at room temperature and the color, formed as the mixture dried, was observed.

	Composition	A	B	C	D	E
Example:						
22	Cholesteryl hydrogen phthalate, percent	40	60	70	70	90
	Cholesteryl hydrocinnamate, percent	40	40	15	0	5
	Cholesteryl cinnamate, percent	20	0	15	30	5
Resulting color		R	G	B	B	G
23	Cholesteryl hydrogen phthalate, percent	25	40	55	70	80
	Cholesteryl benzoate, percent	45	40	35	25	20
	Cholesterol, percent	30	20	10	5	0
Resulting color.		R	G	B	B	B
24	Cholesteryl hydrogen phthalate, percent	60	70	80	90	......
	Cholesteryl cinnamate, percent	20	30	0	5	......
	Cholesteryl benzoate, percent	20	0	20	5	......
Resulting color.		B	B	B	B	......
25	Cholesteryl hydrogen phthalate, percent	20	40	60	65	90
	Cholesteryl cinnamate, percent	60	50	40	25	10
	Cholesterol, percent	20	10	0	10	0
Resulting color.		R	R	B	G	BG

Tunable Liquid Crystal-Dye Laser

L.S. Goldberg and J.M. Schnur; U.S. Patent 3,771,065; November 6, 1973; assigned to the U.S. Secretary of the Navy describe a lasing medium having internal distributed feedback by virtue of the molecular structure of a cholesteric liquid crystal material. The medium comprises a mixture of a strongly fluorescent dye, 7-(diethylamino)-4-methylcoumarin, and a cholesteric liquid crystal solution of 40% cholesteryl oleyl carbonate, 30% cholesteryl chloride and 30% cholesteryl nonanoate.

Figure 2.17a shows in cross-sectional view a form of the process comprising a pair of supporting glass plates **10** and **12** separated by a thin film **14** of liquid, the plates being bounded by a liquid seal **16** such as a Teflon O-ring. A light blocking filter **18** is attached to the outside of one of the glass plates **12**. The spacing of the plates may be some 25 to 1,000 microns.

The liquid between the glass plates is the lasing medium and comprises a mixture of a strongly fluorescent dye, such as 7-(diethylamino)-4-methylcoumarin, and a cholesteric liquid crystal. The latter may, for example, consist of 40% cholesteryl oleyl carbonate, 30% cholesteryl chloride and 30% cholesteryl nonanoate.

The dye is pumped with light at its absorbing wavelength λ_0, which is about 3,400 A, and emits light in a band centered at a wavelength λ_1 of about 4,500 A. Pumping source may, for example, be a nitrogen laser. The light blocking filter **18** transmits light at a wavelength of λ_1 and blocks out light at a wavelength of λ_0 so that only lased light passes out of the device.

FIGURE 2.17: TUNABLE LIQUID CRYSTAL-DYE LASER

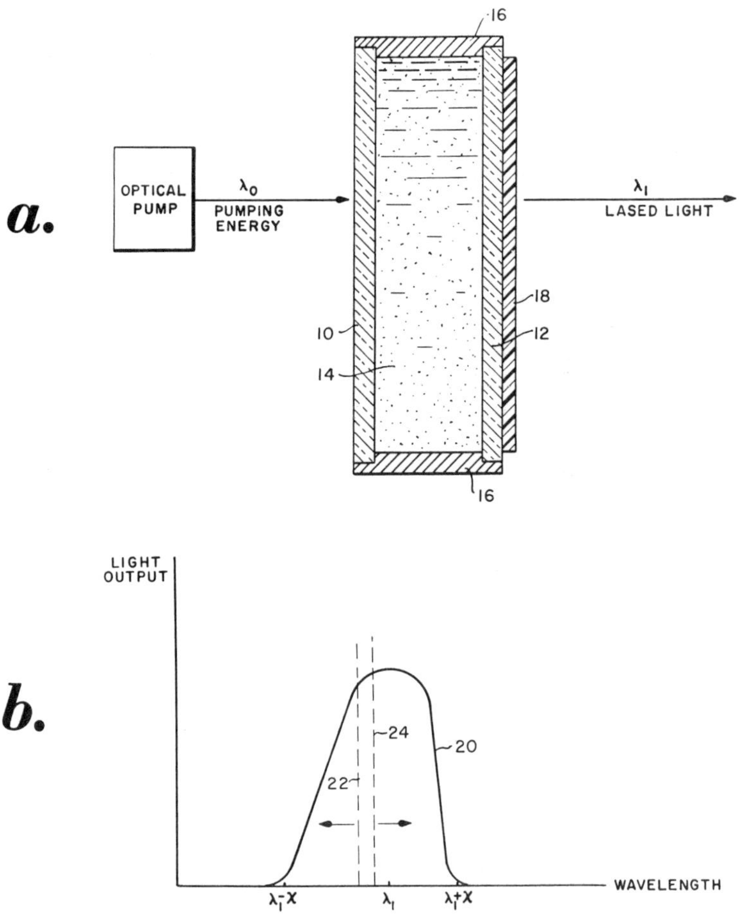

Source: L.S. Goldberg and J.M. Schnur; U.S. Patent 3,771,065; November 6, 1973

The system lases with circularly polarized output at wavelength λ_1 as a result of the internal reflection from the periodic molecular structure of the cholesteric liquid crystal material. Lasing occurs in the direction normal to the plane of the supporting glass plates. The wavelength of internal reflection is adjusted to overlap the fluorescent band of the dye by adjusting the pitch distance of the internal helical molecular structure. This pitch distance has been found to lie in the range of 3,000 to 15,000 A and has been found to be a function of temperature, electric field, pressure and composition.

Thus, in Figure 2.17b, which shows the fluorescent light output from the dye in terms of wavelength (curve **20**), the lased light output lies between the dotted lines **22** and **24** which might be, for example, a few A or less wide. This region can be tuned, that is, moved to the right or to the left (as indicated by the arrows) either by changing the temperature of the lasing liquid or by applying a DC electric field across the liquid and varying the voltage between 100 and 1,000 volts, for example.

To make the device efficient, the surfaces of the supporting glass plates must be highly polished and cleaned to give the dye-cholesteric solution a homogeneous, nonscattering texture. This involves preparation techniques such as cleaning with chromic acid and drying in water-free atmosphere.

The device can be used in spectroscopy as a tunable source of circularly polarized light. The compact thin-film form of the lasing medium makes the device useful as a multicolor display and projection system. Films can be arranged in single units or multiple units emitting different primary colors, and can be separately addressed or tuned at various spatial positions by external means. The device can be used as a compact multiwavelength channel communications source, as a remote temperature sensor by projection of color changes, and other applications requiring a tunable light source. Other fluorescent dyes may also be employed as shown in the following table to change the laser output center wavelength.

Organic Component	Laser Output Center Wavelength (A)
BBO or 2,5-dibiphenylyloxazole	4,100
Bis-MSB or p-bis(o-methylstyryl)benzene	4,200
POPOP or p-bis[2-(5-phenyloxazolyl)]benzene	4,300
Acridine	4,370
7-(Diethylamino)-4-methylcoumarin	4,500
3-Aminophthalimide	5,000
Pyronin B	5,760
3,3'-Diethyloxadicarbocyanine iodide	6,580

Analytical Methods

In a process described by *J.L. Fergason; U.S. Patent 3,409,404; November 5, 1968; assigned to Westinghouse Electric Corporation* the optical properties of a cholesteric liquid crystalline material are changed when the cholesteric material is contacted with another material. A variety of materials, particularly vapors, are identified by observing their effect on cholesteric liquid crystalline materials. The most convenient observable effect is a change in the color of the cholesteric material and, if necessary, comparing the change to the change effected by a known standard material. An analytical device may comprise one or more distinct elements of cholesteric liquid crystalline material. Suitable cholesteric liquid crystalline materials include a wide variety of compounds and mixtures thereof derived from cholesterol.

The process permits identification of a variety of materials by reason of their affect on cholesteric liquid crystalline materials. For example, reversible effects on the liquid crystal optical properties have been observed with common organic solvents, amines, simple alcohols and organic acids while irreversible effects have been observed with halogens, oxidizing agents, alcohols, amines, acids, bases and reducing agents.

As indicative of the results that can be achieved in accordance with the process, the following demonstration was made. All percents given are by weight. Ten liquid crystalline compositions were made, each consisting of a mixture of cholesteryl chloride and cholesteryl nonanoate. The amount of cholesteryl chloride in the compositions varied from 18 to 30%. Films, of substantially uniform thickness, of each composition were made by mixing the constituents in a solvent of 20% chloroform and 80% petroleum ether. The solutions were poured on polyethylene terephthalate film having a black coating on the opposite surface formed by spraying with a black acrylic lacquer, Krylon.

The solvent was then permitted to evaporate. Each film was then exposed at 27°C to a group of common organic solvents. The solvent concentration varied from about one part per thousand to about fifty parts per million. The concentration was sufficient to produce color changes readily apparent to the human eye. Where no color change was observed, the detecting element eventually became colorless due to a phase change. These solvents reversibly affected the film color, (i.e., the scattered waveband with the film exposed to white light).

Percent Cholesteryl Chloride in Cholesteryl Nonanoate

Solvents	18%	19%	20%	21%	22%	23%	24%	25%	27%	30%
Acetone	Red to blue	Red to green	Green to blue	Green to slightly red	Green to slightly red	Green to slightly red	Green to slightly red	Green to slightly red	Green to slightly red	Red (no change)
Butyl acetate	Red to blue	Red to blue	Green to blue	Green to blue	Green to blue	Green to blue	Green to blue	Green (no change)	Green to red	Red (no change)
Benzene	Red to blue	Red to blue	Green to blue	Green to blue	Green to blue	Green (no change)	Green (no change)	Green to red	Green to red	Red (no change)
Chloroform	Red to blue	Red to blue	Green to red	Green to red	Green to red	Green to red	Green to red	Green to red	Green to red	Red (no change)
Trichloro-ethylene	Red to blue	Red to blue	Green to blue	Green to blue	Green to red	Green to red	Green to red	Green to red	Green to red	Red (no change)
n-Heptane	Red to blue	Red to blue	Green to blue	Green to blue	Green to blue	Green to blue	Green to blue	Green to blue	Green to blue	Red to blue
Pyridine	Red to blue	Red to blue	Green to blue	Green to red	Green to red	Green to red	Green to red	Green to red	Green to red	Red (no change)

The data in the table illustrate that each solvent had a unique effect on the set of ten liquid crystals. Thus if an unknown solvent of the group described were exposed to the set of liquid crystals, it could be readily identified. For example, acetone is unique in producing a red to green shift in the 19% crystal; butyl acetate is unique in producing no change in the green color of the 25% crystal; benzene is unique in causing no change in the green color of the 23% or 24% crystals; chloroform is unique in producing a green to red shift in the 20% crystal; trichloroethylene is unique in producing different effects in the 21% crystal (green to blue) and the 22% crystal (green to red); n-heptane is unique in turning all of the liquid crystals blue and pyridine is unique in producing a green to red shift in the 21% crystal that is readily distinguished from the slightly red appearance of that crystal when exposed to acetone.

Consequently, an array of the seven compositions having from 19 to 25% cholesteryl chloride in cholesteryl nonanoate permits the specific identification of the abovementioned seven solvents.

The foregoing demonstrates the feasibility of forming an array of cholesteric liquid crystals that have a unique pattern in response to any of the materials that alter the optical properties of the liquid crystals, thus providing a "fingerprint" of each of those materials. Several additional analytical applications are described in the original patent.

Detection of X-Ray Radiation Using Iodine Compounds

J.L. Fergason and N.N. Goldberg; U.S. Patents 3,663,390, May 16, 1972; 3,657,538, April 18, 1972; and 3,776,927, December 4, 1973; all assigned to Westinghouse Electric Corporation describe a method of detecting x-ray radiation wherein there is prepared a composition that contains at least an effective amount of an iodine-containing cholesterogenic material. As illustrated below, such iodine-containing cholesterogenic material is substantially more sensitive to x-ray radiation, exhibiting for a given dosage of radiation a substantially greater shift in its color-play temperature range.

When iodine-containing cholesterogenic material is used in the above method, the iodine-containing compound used may be one that is itself already known, such as 3-iodocholestane, 5-iodocholestanol, 6-iodocholestanol, 5,6-di-iodocholestanol, 3,5,6-tri-iodocholestane, or 3-iodocholesterol, the last of these sometimes also being called cholesteryl iodide.

The iodine-containing compound may also be a compound not itself previously known, such as an aliphatic, alicyclic, aromatic or unsaturated mono- or di-basic acid ester of 5-iodocholestanol, 6-iodocholestanol, or 5,6-diiodocholestanol. When the acid is unsaturated, iodine might also be introduced into the side chain at the point of saturation, as in the making of 5,6-di-iodocholestanyl-9,10-di-iodostearate from cholesteryl oleate.

Other new compounds coming within the purview of the process are the carbonates that contain, in addition to suitable alkyl, aryl or alkaryl groups, a molecule portion that is cholesterogenic and contains iodine. Examples of material of this kind include 5-iodocholestanyl amyl carbonate, 6-iodocholestanyl phenyl carbonate, and 5,6-di-iodocholestanyl tolyl carbonate. If the alkyl, aryl or alkaryl group is unsaturated, other iodine-containing derivatives can be prepared. An example is the use of cholesteryl oleyl carbonate to prepare 5,6-di-iodocholestanyl-9,10-di-iodostearyl carbonate.

The manner of preparing the known compounds comprises essentially the treatment of cholestanol or cholesterol, suspended in a suitably unreactive organic solvent such as benzene, with an iodizing agent such as phosphorus pentiodide or triiodide, hydriodic acid, or iodine crystals. In most instances, the reaction proceeds smoothly at room temperature or at a moderately elevated temperature such as 80°C.

In most instances the iodine-containing product is substantially less soluble in the organic solvent than the cholesterogenic material reacted, so that it may be recovered from the organic solvent by filtration. In certain cases it is desirable or necessary to distill off a portion of the organic solvent to increase the quantity of iodine-containing cholesterogenic product material recovered. To be more specific, the 3-iodocholestane is made by using phosphorus pentiodide, and 5-iodocholestanol and 6-iodocholestanol may be made with the use of dry hydriodic acid. The 5,6-di-iodocholestanol and 3,5,6-tri-iodocholestane mentioned above are made by using iodine crystals.

The iodocholestanyl esters of aliphatic, alicyclic, aromatic or unsaturated mono- or di-basic acids are made by reacting the corresponding iodocholestanol or diiodocholestanol with the appropriate acid. A suitable organic solvent such as benzene and a suitable esterification catalyst such as concentrated sulfuric acid are used. Alternatively, it may in certain instances be possible to prepare the iodocholestane in situ by bringing cholesterol into contact at one time with both the iodizing agent and the acid with which the alcohol is to be esterified. It may also prove possible in certain instances to prepare the ester first from iodine-free cholesterogenic materials and then iodize it. Iodocholestanyl iodoesters are prepared from suitable starting materials such as cholesterol, iodine and oleic acid as shown in the following example.

Example 1: To 1 mol of oleic acid there is added 1 mol of iodine in benzene. When the solution becomes colorless, indicating complete absorption of the iodine by the oleic acid to form 9,10-di-iodostearic acid, there is added 1 mol of thionyl chloride and 10 cubic centimeters of dimethylformamide. The mixture is allowed to stand for 1 hour. Meanwhile, in a separate flask, 1 mol of cholesterol is reacted with 1 mol of hydriodic acid to produce 5-iodocholestanol. Then there are added to the reaction mixture in the first vessel or flask 1 mol of 5-iodocholestanol and 2 mols of pyridine. Pyridine hydrochloride is removed by filtration, and then methyl alcohol is added until a white crystalline precipitate forms, which is the product. The product is worked up in accordance with the conventional procedures and is characterized as 5-iodocholestanol 9,10-diiodostearate in a yield of over 70%, based upon the quantity of cholesterol charged.

Although it is possible in some instances to use either this iodine-containing cholesterogenic material or an iodocarbonate prepared in the same manner by itself or admixed with other iodine-containing cholesterogenic materials for the detection of x-ray radiation, it will be preferable in most instances to use such iodine-containing material as an additive to a composition composed principally of other cholesterogenic materials. Whether the iodine-containing cholesterogenic material is used by itself or is admixed with other cholesterogenic materials of better known characteristics, the procedure is the same.

The cholesterogenic materials are suspended in benzene, petroleum ether, or other suitable organic solvent to form a liquid that can be applied to a suitable substrate, with the organic solvent being permitted to evaporate to leave a thin film such as about 5 to 50 microns thick. In many instances, an iodine-containing cholesterogenic material sensitive to x-ray radiation does not itself have a color-play temperature range suitable for the intended use. Suitable color changes indicating that a film has been exposed to x-ray radiation are not obtained at temperatures far removed from those at which the cholesterogenic material customarily exhibits its color of play. Moreover, cholesterogenic materials differ, not only in respect to the temperature at which the color play takes place, but also in respect to the width of the temperature range over which color change takes place (this may be only 2° to 3°C or as much as 25° to 100°C) and in respect to the rapidity with which the material responds to an influence (some respond almost instantly and others take hours).

It is fortunate, therefore, that the various abovementioned compounds may be blended with other cholesterogenic materials of known performance as respects these properties. Substantial color changing effects can be obtained with, for example, about 1 to 10% of an iodine-containing cholesterogenic material blended with a suitable base composition of other cholesterogenic materials such as a base composition comprising a mixture of 40% by weight each of oleyl cholesteryl carbonate and cholesteryl nonanoate with 20% by weight of dicholesteryl sebacate, which exhibits a color play in the temperature range of about 30° to 40°C. Moreover, this base composition is one that is relatively sluggish in changing color, as is desirable for applications such as the making of dosimeters for the detection of radiation in the frequency range of 10^{17} to 10^{20} cycles per second.

Example 2: There was prepared on a substrate of glass a thin film of cholesteric-phase liquid crystal material that comprised, in percent by weight, 55% cholesteryl nonanoate, 35% oleyl cholesteryl carbonate, and 10% cholesteryl benzoate. Before having been irradiated with x-rays, this material exhibited a color-play temperature range of from 23°C (red) to 31°C (blue), with the color green being exhibited at 26°C.

When placed on a glass substrate provided with a suitable black background, as by spraying with black spray paint, and then irradiated, using a Norelco x-ray diffraction unit with a copper target and a beryllium window, at 35,000 electron volts and 20 milliamperes, there was obtained after an exposure of 10 minutes, a film that exhibited green at 25°C, red at 22°C, and blue at 30°C, i.e., a downward shift of 1°C in the color-play temperature range of the material. After 1 hour of exposure, the downward shift in color-play temperature range was 3°C, and after 5 hours of exposure, the shift was 10°C (red at 13°C, green at 16°C, and blue at 21°C).

Example 3: Example 2 was repeated except that there was used as the mixture of cholesterogenic material a mixture of 55% cholesteryl nonanoate, 35% oleyl cholesteryl carbonate, and 10% of cholesteryl 4-iodobenzoate. This was a temperature sensitive material, exhibiting a color play over a range of about 24° to 32°C with green being observed at 27°C.

After irradiation with x-rays of the same intensity as used in Example 2, the temperature for green was reduced to 26°C after 1 minute and to 19°C after 10 minutes. Comparison of these results with Example 2 reveals that with the iodine-containing cholesterogenic material of this example, an x-ray dosage detection means of substantially greater sensitivity is obtained. With the material of Example 2, 10 minutes of exposure lowered the temperature for green by 1°C, but with the iodine-containing cholesterogenic material of this example, the temperature for green was lowered by 8°C.

Example 4: Example 2 was repeated except that there was used as the cholesterogenic material a mixture consisting of 15% cholesteryl nonanoate, 60% oleyl cholesteryl carbonate, 10% cholesteryl 4-iodobenzoate and 15% cholesteryl chloride. This was a temperature-insensitive material, being green at room temperature and clearing at 45°C. After exposure to radiation as indicated in Example 2 at room temperature for 6 to 7 minutes, black was observed.

Conductive Ink Resistor for Heating

A process described by *T.L. Hodson, J.G. Whitaker and J.W. Jones; U.S. Patent 3,617,374; November 2, 1971; assigned to The National Cash Register Company* is directed to display devices having self-contained controlled resistor means for generation of heat upon electrical activation, which means can be formed in any desired configuration and thickness, and an encapsulated liquid crystal layer responsive to variations in heat to present a display, which can be polychromatic.

The resistor is comprised of conductive ink deposited on one major surface (usually the lower surface) of an opaque, substantially electrically nonconductive layer having a layer of encapsulated cholesteric liquid crystals in direct contact with at least a portion of the other major, (e.g., upper) surface of the opaque layer. Heating due to the resistor produces reversible color changes in the encapsulated liquid crystals resulting in the display.

In the process a low cost display is provided which is capable of almost infinite variation from one article to the next because of the adaptability of the encapsulated liquid crystals to deposition procedures enabling use of a high resolution and complex definition, e.g., silk screen and related printing procedures. In like manner, the conductive ink resistors can be printed in complex configurations by silk screening and related procedures thereby ensuring a highly variable display of portable nature, low unit price and high resolution.

The article offers great flexibility in that the color response in the liquid crystal capsules can be varied by varying the resistance of the conductive ink elements. The resistance of the conductive ink elements can be controlled in any one, more or all of the following manners: (1) by controlling the thickness of the conductive ink coating itself, (2) by controlling the total area of the coating, and (3) by varying the very composition of the conductive ink layer, e.g., by diluting the conductive ink with solvent (thinner) thereby reducing the intensity of the conductive pigments, and/or by introduction of adjuvant materials, such as plasticizers, in varying concentrations.

Also, the chromatic display effects can be varied by controlling the power input to the respective elements defining the conductive ink layer thereby varying the extent of thermal excitation. This latter factor can be utilized to produce a selective color response in an encapsulated liquid crystal capsular layer containing a profusion of capsules having liquid crystal compounds and mixtures possessing varying color response temperatures, thus exhibiting varying colors at the color advent temperatures and at temperatures above the advent temperatures. The composite articles can operate at low power input, can produce display characters and images of high visual resolution, and possess the ability of controlled color response achievable via a variety of readily controlled easily variable approaches. The following example illustrates the process; parts and percentages are by weight unless otherwise indicated.

Example: This example illustrates formation of an inverse-coated article having a polyester top layer, a capsular layer coated over the entire surface thereof, and a patterned printed coating of conductive ink. A liquid crystal mixture having the below indicated composition is encapsulated in a conventional manner using a standard two-way gelatin coacervation.

Liquid Crystal Component	Concentration (weight percent)
Cholesteryl pelargonate	63.8
Cholesteryl chloride	4.8
Oleyl cholesteryl carbonate	31.4

The encapsulation is conducted specifically as follows. Into an aqueous solution of 1 weight part of acid extracted pigskin gelatin (having a Bloom strength of 285 to 305 grams and an isoelectric point of pH 8 to 9) in 8.09 parts by weight of distilled water at 55°C, there are placed 14.0 parts by weight of the liquid crystal melt. The liquid crystal melt is milled with a shear agitator until the desired particle size is achieved, viz, from 15 to 30 microns.

While the milling progresses, an aqueous solution of 1 weight part gum arabic in 93.0 weight parts of distilled water is prepared in a separate container and maintained at a temperature of 55°C. When the desired particle size is achieved, the gelatin-liquid crystal emulsion is added slowly to the gum arabic solution. The pH is adjusted to 4.85 and the coacervate is permitted to cool to 27°C over two and one-quarter hours. The resultant capsules are cooled to below 15°C and hardened with 0.5 weight part of a 25 weight percent aqueous solution of glutaraldehyde for 12 to 15 hours. The resulting capsular slurry is then concentrated by filtration to a slurry having approximately 40 to 45 weight percent capsular solids.

Upon completion of encapsulation and subsequent concentration of the slurry, approximately 75 weight parts of the thus concentrated slurry is mixed with 25 weight parts of a 10 weight percent solution of commercially available polyvinyl alcohol (duPont 72-60) in water as a binder for the coating mixture forming the encapsulated cholesteric liquid crystal layer.

The above formulated liquid crystal capsules are then coated onto a "Cronar" sheet (duPont) using No. 12 nylon printing screen using two passes. When the capsular layer is dry, a coating of opaque, nonconductive "Zephyr R-M" black ink is applied to the capsular layer using a No. 12 nylon screen and the opaque background is allowed to dry.

Then a conductive ink pattern is printed on the dried opaque background layer using a No. 20 nylon printing screen. The conductive ink utilized is "Conductive Ink EL 796", having carbon black and titanium dioxide conductive particles in a conventional binder.

The printing operations are conducted at ambient room temperatures, which range from about 24° to 28°C, by screen printing in the following manner. The Cronar transparent support is laid on a flat surface. A nylon screen mounted on a soft white pine wood frame is positioned over the substrate. A supply of the encapsulated liquid crystal material is then poured onto the screen and a neoprene rubber squeegee is used to pull the supply of encapsulated liquid crystal coating formulation across the screen, at the same time pressing it through the open mesh of the screen.

The screen is then lifted from the substrate leaving the encapsulated liquid crystals affixed to the Cronar. The deposition of the black background ink and the electroconductive ink is done in similar fashion after the encapsulated liquid crystal layer dries. The electroconductive ink is deposited upon the dried opaque background layer in a pattern configuration of lines having a length of 4 inches, a width of one-sixteenth inch and a thickness of 0.0005 inch.

Upon drying of the conductive ink layer, the article is inverted thereby allowing a viewer to observe color changes occurring in the encapsulated liquid crystal layer when viewed by incident white light through the transparent Cronar top layer. The transparent Cronar layer originally serving as a transparent support for the deposition of the respective coatings then serves as a brightness enhancing and spectral purity (for color intensity) improving top layer in the manner noted previously.

The thus prepared display article having the integral conductive ink resistor means is then connected to a supply of electric power, and the voltage is adjusted until color is produced in the area of the encapsulated liquid crystal layer which is thermally activated by the printed conductive ink pattern. At a potential of 40 volts, the entire area overlying the printed conductive ink line is violet, indicating maximum color response at a low current input of less than 1 milliampere.

Another electroconductive display article containing encapsulated liquid crystals and electroconductive ink is produced in essentially the same manner except that the printed electroconductive ink lines have a length of 4 inches, a width of one-eighth inch and a thickness of 0.0005 inch. At a voltage of 25 to 30 volts, the area overlying the line appears violet at a noted current of between 1 and 2 milliamperes. Further articles were prepared by inverse coating wherein the display configuration was printed by use of encapsulated liquid crystal images and conductive ink images, respectively.

Detection System for Surface Flaws

J.E. Adams and W.E.L. Haas; U.S. Patent 3,744,920; July 10, 1973; assigned to Xerox Corporation system for detecting physical and/or electrical conductivity patterns in a surface of interest and more particularly a system which utilizes the optical characteristics of liquid crystal substances having optically negative properties.

A great variety of very sophisticated and complex articles of manufacture and apparatus are employed in modern technology. For example, there are known materials which can withstand tremendously high temperatures without being destroyed and thus are suitable for use in missile nose cones. The detection of flaws in these materials is extremely important since the presence of even a minute flaw can cause material failure during performance. For materials used in other applications, e.g., optics in precision instruments, a smooth surface free of physical deformation is essential. Thus, there exists a need for extremely sensitive detection systems which can be employed to identify the presence of such patterns in materials and which do not damage the material being analyzed in even the slightest manner.

Generally, the detection system makes use of a device comprising a thin layer of optically negative liquid crystal substances disposed on a supporting substrate which typically is a thin layer of a material having suitable physical properties to carry the liquid crystal layer and which may be flexible or rigid. Optionally, the top surface of the liquid crystal layer may be covered with a protective element which may be any suitable optically transparent material. The detection device is typically utilized by placing the support substrate against the surface to be analyzed. Subsequently, the physical deformation and electrical conductivity patterns present in the surface being analyzed can be quickly and efficiently identified.

Light Modulation System

J.E. Adams and W.E.L. Haas; U.S. Patent 3,726,584; April 10, 1973; assigned to Xerox Corporation describe a system for modulating light which comprises directing a beam of linearly polarized monochromatic light on an optically negative liquid crystal film and subsequently varying the pitch of the liquid crystal film in response to some stimulus to which the pitch of the film is sensitive.

The process takes advantage of the phenomenon associated with optically negative liquid crystalline substances to provide an advantageous system for modulating a monochromatic light beam. According to the process, a linearly polarized beam of monochromatic light is directed upon a thin film of an optically negative liquid crystalline substance and subsequently the pitch of the liquid crystal film is changed by applying thereto some stimulus to which the pitch of the film is sensitive.

As the pitch of the liquid crystal film is changed, the degree of rotation imparted to the incident light beam is varied. According to a preferred form of the process, a linear analyzer is positioned behind the liquid crystal film to detect the effect of the film upon the incident light beam. The method is preferably practiced by initially rotating the axis of polarization of the linear analyzer to a position where the incident light beam is completely extinguished and subsequently maintaining the linear analyzer in this position so that when some stimulus acts upon the liquid crystalline substance and changes the pitch of the substance thereby causing the degree of rotation imparted to the incident light to be changed, some light will be observed emerging from the linear analyzer. Thus, according to this process, the presence of a stimulus or a change in the amount of the stimulus present is characterized by a change from a condition where no light is observed emerging from the linear analyzer to one where some light is seen.

The light modulation system can be employed in any situation where it is desired to modulate an incident monochromatic light beam. Since the pitch of optically negative liquid crystalline films can be varied in response to a number of stimuli, the system can be employed in a number of specific uses such as, for example, to indicate temperature changes, and to provide an imaging method. An important practical advantage of the system is that the liquid crystalline materials employed are not expensive and the devices are relatively inexpensive since there is relatively little difficulty involved in the manufacture thereof. A further advantage of the system is that the devices can be very small, i.e., in the order of a few millimeters in thickness, and can be completely encapsulated by protective material. The following examples illustrate the process; all parts and percentages are by weight unless otherwise specified.

Example 1: An optically negative liquid crystalline composition is prepared by mixing together about 35% cholesteryl chloride and about 65% cholesteryl nonanoate. The composition has a λ_0 value of about 5,900 A. An approximately 5 to 10 micron thick film of the composition is placed on a glass slide and another glass slide is placed over the free surface of the liquid crystal film. The liquid crystal film is then placed in the path of a linearly polarized light beam having a wavelength of about 4,500 A provided by a suitable laser. A linear analyzer element is placed behind the liquid crystal film and rotated to a position where the incident light beam is completely extinguished. The liquid crystal film is heated slightly with a heat gun and light is observed passing through the linear analyzer element.

Example 2: A layer of the composition described in Example 1 is applied to the conducting surface of a NESA glass electrode. An approximately 1 mil thick Mylar film spacer is placed over the free surface of the liquid crystal film and a thin layer of photoconductive zinc sulfide is disposed on the free surface of the Mylar film. Subsequently a second NESA glass electrode is arranged over the zinc sulfide layer with the conducting surface of the NESA glass contacting the zinc sulfide layer. This combination is placed between a linear polarizer and a linear analyzer.

A uniform electric field is impressed across the zinc sulfide layer and liquid crystal film by applying voltage from a power source through the NESA glass electrodes. The device is flood illuminated with monochromatic light having a wavelength of about 6,500 A and the angle between the respective axes of polarization of the linear polarizer and linear analyzer is adjusted so as to extinguish the incident light. Ultraviolet activating radiation is then directed upon the device in imagewise fashion and a visible image is read out of the device.

NEMATIC COMPOUNDS

AROMATIC AZOXY COMPOUNDS

Substituted Azoxybenzene Compounds

R. Steinsträsser; U.S. Patent 3,773,747; November 20, 1973; assigned to Merck Patent Gesellschaft mit beschrankter Haftung, Germany has found that azo and azoxy compounds of the formula:

(1)

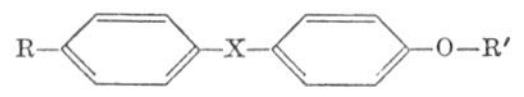

where X is azo or azoxy, R is straight-chain alkyl of 1 to 8 carbon atoms, and R' is straight-chain alkyl or straight-chain alkanoyl, each of 1 to 10 carbon atoms, in pure form and mixtures thereof with each other and with other nematic or nonnematic compounds, have nematic liquid-crystalline phases at low temperatures and exhibit wide nematic mesophases. In their nematic phase, their light scattering property varies proportionately and reversably to the voltage applied thereto, which renders them useful in the production of display screens in nuclear resonance and electron resonance spectroscopy and as solvents for stereospecific chemical reactions. Of the compounds of Formula (1), preferred are those where:

(1a) X is an azo group;
(1b) X is an azoxy group;
(1c) R is straight-chain alkyl of 2 to 4 carbon atoms;
(1d) R' is straight-chain alkyl of 1 to 6 carbon atoms;
(1e) R' is straight-chain alkanoyl of 4 to 8 carbon atoms;
(1f) Compounds defined by (1c) where R' is straight-chain alkyl of 1 to 6 carbon atoms or straight-chain alkanoyl of 4 to 8 carbon atoms;
(1g) Compounds as defined by (1f) were X is an azoxy group;
(1h) Compounds as defined by (1g) where R is n-butyl.

The process for the preparation of compounds of Formula (1) includes the cases in which: (a) A compound of Formula (2)

(2)

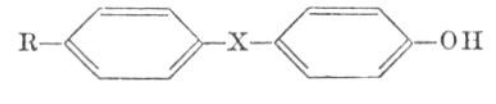

is reacted with a compound of the formula R'OH or a functional derivative thereof; or (b) A nitroso compound of Formula (3a) or (3b)

(3a) 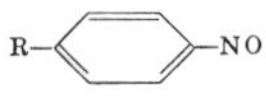 (3b) 

is reacted, respectively, with an amine of the Formula (4a) or (4b)

(4a)  (4b) 

or with a mixture of amino compounds of Formulas (4a) and (4b); or (c) A hydrazo compound of Formula (5)

(5) 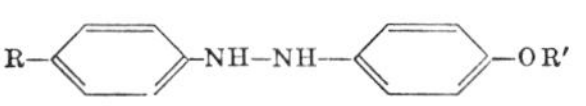

is oxidized with an oxidizing agent; or (d) A mixture of nitro compounds of Formulas (6a) and (6b)

(6a)  (6b) 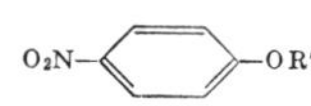

is reduced with a reducing agent; or (e) An amine of Formula (4a) or (4b) is diazotized and thereafter coupled, respectively, with a compound of Formula (7a) or (7b)

(7a) 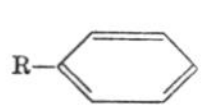 (7b) 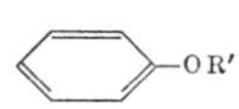

or (f) A nitroso compound of Formula (3a) or (3b) is reacted with a hydroxylamine of Formula (8a) or (8b)

(8a) 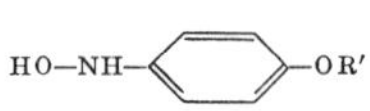 (8b) 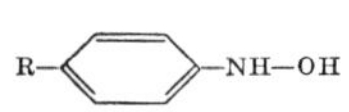

or, optionally, an azo compound of the Formula (1), i.e., where X is —N=N—, is converted, by treatment with an oxidizing agent, into the corresponding azoxy compound of Formula (1), i.e., where X is

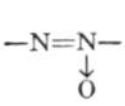

The compounds of Formula (1), especially the azoxy compounds, are obtained as substances having strictly defined melting points. Their melting points and their clear points are reproducible, although at least several of the azoxy compounds are not produced as a single compound but instead as a binary mixture of a p-R-p'-OR'-azoxybenzene and a p-OR'-p'-R-azoxybenzene.

Of the nematic mixtures of the process, preferred are those which contain at least two compounds of Formula (1). Mixtures which contain p-n-butyl-p'-methoxyazoxybenzene (or its isomer, p-methoxy-p'-n-butyl-azoxybenzene) preferably in an amount of 30 to 80 and especially 50 to 70% by weight, and at least one further compound of Formula (1), preferably in an amount of 20 to 70 and especially 30 to 50% by weight, exhibit particularly advantageous properties.

Example 1: (a) 51 grams p-n-butyl-p'-hydroxyazobenzene (MP 78°C; obtainable by the diazotization of p-n-butylaniline and subsequent coupling with phenol) is dissolved in 100 ml of 10% sodium hydroxide solution and agitated for 3 hours at room temperature together with 25 grams of dimethyl sulfate (or an equivalent amount of another methylating agent, such as methyl bromide, methyl iodide, methyl-p-toluenesulfonate, or trimethyl phosphate). During this time, p-n-butyl-p'-methoxyazobenzene precipitates quantitatively; MP 31°C, clear point 47°C (from ethanol).

(b) 32 grams of p-n-butyl-p'-methoxyazobenzene is dissolved in 80 ml of glacial acetic acid and mixed with 10 ml of 30% aqueous hydrogen peroxide solution. The mixture is heated for 3 hours to 80°C and then allowed to cool. The thus-separated oil is removed from the reaction mixture, dissolved in 50 ml of hot ethanol, and the solution is cooled to -15°C, thus obtaining a mixture of p-n-butyl-p'-methoxyazoxybenzene and p-methoxy-p'-n-butyl-azoxybenzene, MP 16.5°C, clear point 76°C (from ethanol).

Example 2: A suspension of 25 grams of p-n-butyl-p'-hydroxyazobenzene in 100 ml of absolute ether is mixed with a solution of 0.1 mol of diazomethane in 50 ml of absolute ether (produced from 28 grams of p-toluenesulfonylmethyl nitrosamide). Under the evolution of nitrogen, p-n-butyl-p'-methoxyazobenzene is quantitatively obtained, MP 31°C, clear point 47°C.

Example 3: 260 grams of p-n-butyl-p'-hydroxyazobenzene is dissolved in a solution of 82 grams of KOH in 300 ml of ethanol. Thereafter, 250 ml of ethanol is distilled off and the reaction mixture is mixed with 500 ml of acetone or ether, during which step the potassium salt of p-n-butyl-p'-hydroxyazobenzene is precipitated. The reaction mixture is filtered, dried and pulverized, 250 grams of the potassium salt are suspended, together with 84 grams of $NaHCO_3$, in 750 ml of acetone. At -15°C, a solution of 145 grams of enanthic acid chloride in 100 ml of acetone is added dropwise under agitation to this suspension.

The mixture is stirred for another 5 minutes and then mixed with 1 liter of ice water. Then, the mixture is agitated for 5 minutes, and the thus-precipitated p-n-butyl-p-heptanoyl-oxyazobenzene is filtered and recrystallized from ethanol until the melting point and the clear point are constant; MP 52°C, clear point 62°C. In place of the enanthic acid chloride, it is also possible to employ equivalent amounts of enanthic acid bromide or enanthic acid anhydride.

Example 4: 25.4 grams of p-n-butyl-p'-hydroxyazobenzene is dissolved in 100 ml of 4% sodium hydroxide solution, and 400 ml of ether is added. Under vigorous agitation, a solution of 15 grams enanthic acid chloride in 20 ml of ether is added dropwise thereto at -1°C. The reaction mixture is stirred for 1 hour under cooling, and at the same time 30 milliliters of 4% sodium hydroxide solution is added dropwise thereto so that a pH of 11 to 12 is maintained. The reaction mixture is filtered, the ether phase is separated, and the aqueous phase is washed several times with ether. The collected ether fractions are washed several times with water, dried, filtered, and concentrated. Under ice cooling, petroleum ether is added to the reaction mixture, thus precipitating p-n-butyl-p'-heptanoyloxyazobenzene, which is recrystallized several times from ethanol, MP 52°C, clear point 62°C.

Example 5: 5 grams of p-n-butyl-p'-hydroxyazobenzene is dissolved in 45 ml of dimethylformamide; 4.2 grams of calcined potassium carbonate is added thereto, and a solution of 3 grams of enanthic acid chloride in 20 ml of dimethylformamide is added to the reaction mixture under agitation at room temperature in a dropwise fashion. Thereafter, the reaction mixture is heated for 3 hours to 80°C, cooled, mixed with 150 ml of ether, and filtered. Under ice cooling, 90 ml of water is added thereto and the mixture thoroughly shaken several times. The phases are separated; the aqueous phase is washed with ether, the combined ether extracts are shaken several times with water, dried, and evaporated, thus obtaining p-n-butyl-p'-heptanoyloxyazobenzene, MP 52°C, clear point 62°C.

Example 6: In this example, mixtures of binary systems according to the process are set forth, along with their nematic ranges which are thermodynamically stable. The mixtures can readily be subcooled to temperatures ranging still below the disclosed melting point. The percentages listed are percent by weight. The mixtures set forth below are eutectic ones. Binary mixtures contain preferably 60 to 70% by weight of one component and 40 to 30% by weight of the second component.

(A) 62% p-n-butyl-p'-methoxyazoxybenzene (or p-methoxy-p'-n-butylazoxybenzene) and 38% p-n-butyl-p'-n-butyloxyazoxybenzene (or p-n-butyloxy-p'-n-butylazoxybenzene), MP -6°C and clear point 81°C.

(B) 65% p-n-butyl-p'-methoxyazoxybenzene (or p-methoxy-p'-n-butylazoxybenzene) and 35% p-n-butyl-p'-n-hexyloxyazoxybenzene (or p-n-hexyloxy-p'-n-butylazoxybenzene), MP –15°C and clear point 85°C.

(C) 67% p-n-butyl-p'-methoxyazoxybenzene (or p-methoxy-p'-n-butylazoxybenzene) and 33% p-n-propyl-p'-heptanoyloxyazoxybenzene (or p-heptanoyloxy-p'-n-propylazoxybenzene), MP -7°C and clear point 79°C.

(D) 68% p-n-butyl-p'-methoxyazoxybenzene (or p-methoxy-p'-n-butylazoxybenzene) and 32% p-ethyl-p'-valeryloxyazoxybenzene (or p-valeryloxy-p'-ethylazoxybenzene), MP -9°C and clear point 73°C.

Example 7: This example describes the use of these compounds. A thin film of one of the abovementioned compounds of Formula (1) or a mixture of two or more of these, e.g., where R is n-butyl, R' is methoxy and X is azoxy, or a composition of Example 6, is applied as a coating of a thickness of 5 to 20μ, between two light-permeable conductive glasses provided with contacts. Applying a direct voltage of, for example, between 5 and 20 volts to this unit, the previously transparent plate system becomes increasingly opaque with increasing voltage.

This process is reversible at will within the entire temperature range of the nematic phase of the compounds forming the coating. If one of the glass plates is provided with a conductive electrode raster in place of a conductive surface coating, this pattern becomes visible when the voltage is applied. Such an arrangement is illustrative of the principle of a flat image reproduction unit.

Azo Compounds Stable to Electrochemical Degradation

Y. Yamazaki; U.S. Patent 3,816,334; June 11, 1974; assigned to KK Suwa Seikosha, Japan has found that a nematic liquid crystal azo compound suitable for use in a display device has the general formula:

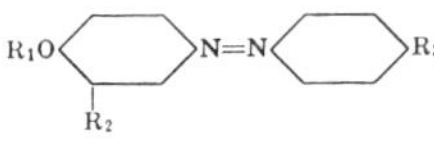

where R_1 is selected from the group consisting of an alkyl group having 1 to 8 carbons and an acyl group having 5 through 10 carbons, R_2 is selected from the group consisting of H and CH_3, and R_3 is selected from the group consisting of C_2H_5, n-C_3H_7, and C_4H_9, the compound being resistant to electrochemical decomposition.

The compounds having the above structure are strongly resistant to electrochemical degradation whether by hydrolysis, reduction or oxidation. This resistance is not affected by the presence of small quantities of water introduced during synthesis. The compounds can be mixed with each other or with other nematic liquid crystal compounds to form compositions having a mesomorphic range either near room temperature or including room temperature.

The compounds of this class may be divided into subclasses in accordance with the nature of the substituents R_1, R_2 and R_3. These will be described below and examples of each will be given. In addition, examples of mixtures with other nematic liquid crystal compounds will be given.

Subclass 1: R_1 is an acyl group of 6, 7 or 8 carbons, R_2 is H and R_3 is C_2H_5. Table 1 gives three examples of compounds of the subclass and the lower transition temperatures (A) and the upper transition temperatures (B). Examples 4, 5 and 6 are liquid crystal compositions comprising materials of subclass 1. The transition temperatures for these compositions are also shown, the tabulation making it possible to discern the drop in the lower transition temperature resulting from the addition of a second liquid crystal compound.

TABLE 1

Example	Composition	A(°C)	B(°C)	Ratio (mols)
1	p-(p'-ethylphenylazo) phenylhexanoate	38	55	—
2	p-(p'-ethylphenylazo) phenylheptanoate	44	52	—
3	p-(p'-ethylphenylazo) phenyloctanoate	46	57	—
4	mixture of p-(p'-ethylphenylazo phenylhexanoate and p-(p'-ethylphenylazo) phenylheptanoate	30	50	2:1
5	mixture of p-(p'-ethylphenylazo) phenylhexanoate and p-(p'-ethylphenylazo) phenyloctanoate	28	54	3:2
6	mixture of p-(p'-ethylphenylazo) phenylhexanoate and butyl p-(p'-ethoxyphenoxycarbonyl) phenylcarbonate	25	60	2:1

As can be seen from Table 1, Example 6, it is possible to achieve a lower transition temperature of 25°C, which is very close to normal room temperature. Generally, the nematic liquid crystals of the subclass are made by coupling phenol to the diazonium salt of p-ethylaniline, followed by esterification with the appropriate acid. In the manufacture of these compounds, recrystallization from alcohol is repeated until each transition temperature becomes constant. The liquid crystal materials of the subclass range from yellow to orange in color, this range being characteristic of azo compounds. When mixed with ester-bonded nematic liquid crystals such as butyl p-(p'-ethoxyphenoxycarbonyl) phenylcarbonate, the range of color is extended to light yellow.

Subclass 2: R_1 is an alkyl group having 1 to 4 carbons, R_2 is H and R_3 is C_4H_9: Compounds of this subclass are prepared by coupling phenol to the diazonium salt of p-butylaniline after which the product is converted to an ether by use of the proper alcohol. Crystallization is repeated until each phase transition temperature becomes constant. Table 2 presents Examples 7 through 12 showing suitable compounds of this subclass and mixtures employing compounds of this subclass.

TABLE 2

Example	Composition	A(°C)	B(°C)	Ratio (mols)
7	p-(p'-methoxyphenylazo) phenylbutane	32	48	—
8	p-(p'-ethoxyphenylazo) phenylbutane	44	75	—
9	p-(p'-butoxyphenylazo) phenylbutane			
10	Mixture of p-(p'-methoxyphenylazo) phenylbutane and p-(p'-ethoxyphenylazo) phenylbutane	23	70	2:1
11	Mixture of p-(p'-ethoxyphenylazo) phenylbutane and p-ethoxybenzylidene p'-butylaniline	18	72	1:2
12	Mixture of p-(p'-methoxyphenylazo) phenylbutane and p-(p'-ethylphenylazo) phenylhexanoate	10	50	1:2

As can be seen, mesomorphic temperature ranges including room temperature are achieved with mixtures containing members of this subclass.

Subclass 3: R_1 is an acyl group of 6, 7 or 8 carbons, R_2 is CH_3 and R_3 is C_4H_9. Compounds of this subclass are made by coupling o-cresol to the diazonium salt of p-butylaniline and esterifying with the appropriate acid. Recrystallization is repeated until the phase transition temperatures become constant. Table 3 shows Examples 15 and 16 which are compositions utilizing members of this subclass.

TABLE 3

Example	Composition	A(°C)	B(°C)	Ratio (mols)
13	p-(3-methyl-4-hexanoyloxyphenylazo) phenylbutane	20	12	—
14	p-(3-methyl-4-heptanoyloxyphenylazo) phenylbutane	18	10	—
14A	p-(3-methyl-4-octanoyloxyphenylazo) phenylbutane	15	10	—
15	Mixture of p-(3-methyl-4-hexanoyloxyphenylazo)phenylbutane and p-(p'-ethoxyphenoxycarbonyl) phenylcarbonate	5	30	1:2
16	Mixture of p-(3-methyl-4-hexanoyloxyphenylazo)phenylbutane and p-(p'-ethylphenylazo) phenylhexanoate	8	35	1:3

Subclass 4: R_1 is an acyl group having between 6 and 10 carbons, R_2 is H, and R_3 is C_4H_9. Compounds of this subclass are made by coupling phenol to a diazonium salt of p-n-butylaniline and esterifying with the appropriate acid. Recrystallization is repeated until each phase transition temperature becomes constant. Table 4 gives the properties of Examples 17 through 21; each of Examples 22 through 25 is a mixture containing at least one member of this subclass. The mesomorphic temperature range of Example 24 makes it particularly suitable for use at room temperature.

TABLE 4

Example	Composition	A(°C)	B(°C)	Ratio (mols)
17	p-(p'-hexanoyloxyphenylazo) phenylbutane	55	62	—
18	p-(p'-heptanoyloxyphenylazo) phenylbutane	49	56	—
19	p-(p'-octanoyloxyphenylazo) phenylbutane	46	61	—
20	p-(p'-nonanoyloxyphenylazo) phenylbutane	47	57	—
21	p-(p'-decanoyloxyphenylazo) phenylbutane	51	62	—
22	Mixture of p-(p'-octanoyloxyphenylazo) phenylbutane and p-(p'-nonanoyloxyphenylazo)phenylbutane	38	55	2:1
23	Mixture of p-(p'-heptanoyloxyphenylazo) phenylbutane and p-(p'-octanoyloxyphenylazo) phenylbutane	40	54	1:2
24	Mixture of p-(p'-octanoyloxyphenylazo) phenylbutane and methyl p-(p'-n-hexylphenoxycarbonyl) phenylether	20	50	3:1
25	Mixture of p-(p'-octanoyloxyphenylazo) phenylbutane and butyl p-(p'-ethoxyphenoxycarbonyl) phenylcarbonate	30	55	1:1

Subclass 5: R_1 is an acyl group of 6, 7 or 8 carbons, R_2 is H, and R_3 is n-C_3H_7. Compounds of this subclass are made by coupling phenol to a diazonium salt of p-propylaniline and coupling the product by esterification with the appropriate acid. Recrystallization is repeated until each phase transition temperature becomes constant. Table 5 gives the transition temperatures for members of this subclass and for mixtures containing members of this subclass. The mixtures are presented as Examples 30 through 33. Mixtures containing members of this subclass yield display devices showing colors ranging from light yellow to orange.

TABLE 5

Example	Composition	A(°C)	B(°C)	Ratio (mols)
26	p-(p'-n-propylphenylazo) phenylhexanoate	46	66	–
27	p-(p'-n-propylphenylazo) phenylheptanoate	40	63	–
28	p-(p'-n-propylphenylazo) phenyloctanoate	45	66	–
29	p-(p'-n-propylphenylazo) phenylpentanoate	49	62	–
30	Mixture of p-(p'-n-propylphenylazo) phenylheptanoate and p-(p'-n-propyl-phenylazo) phenyloctanoate	35	60	2:1
31	Mixture of p-(p'-n-propylphenylazo) phenylheptanoate and p-(p'-n-propyl-phenylazo)phenyloctanoate and p-(p'-n-propylphenylazo) phenylhexanoate	33	60	2:1:1
32	Mixture of p-(p'-n-propylphenylazo) phenylheptanoate and methyl p-(p'-n-hexanoylphenoxycarbonyl) phenylether	20	60	2:1
33	Mixture of p-(p'-n-propylphenylazo) phenylheptanoate and butyl p-(p'-ethoxyphenoxycarbonyl) phenylcarbonate	25	62	1:1

The transition points of the nonazo compounds used in some of the mixtures are given in Table 6.

TABLE 6

Compound	A(°C)	B(°C)
butyl p-(p'-ethoxyphenoxycarbonyl)phenylcarbonate	54	70
methyl p-(p'-n-hexylphenoxycarbonyl)phenylether	29	45
p-methoxybenzylidene p'-n-butylaniline	20	40
p-methoxybenzylidene p'-n-amylaniline	38	60
p-n-propoxybenzylidene p'-n-amylaniline	33	60

Azoxybenzene Compounds

M.T. McCaffrey and J.A. Castellano; U.S. Patent 3,816,336; June 11, 1974; assigned to RCA Corporation have found that certain azoxybenzene compounds are liquid crystals having a wide use temperature range and long life. The liquid crystal compounds comprise one or more azoxybenzene compounds of the formula:

(1)

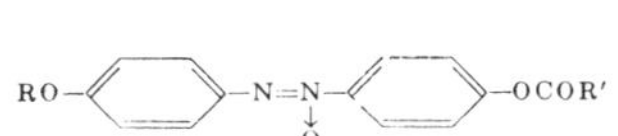

where R and R' are alkyl groups having from 1 to 7 carbons. R and R' can be the same or different and can be branched or straight chain alkyl groups. These compounds can be employed in electro-optic devices alone or mixtures of the azoxybenzene compounds can be employed. In addition, one or more of the compounds can be mixed with other azoxybenzene compounds.

The compounds can be prepared by reacting a p-alkoxy-substituted aromatic primary amine with nitrous acid in the cold and adding a phenol to form the corresponding p'-hydroxyaryl azo compound. The latter can be reacted with an anhydride in pyridine to esterify the hydroxy group and the resultant product reacted with formic acid and hydrogen peroxide to form the corresponding azoxybenzene compound. These reactions can be summarized by the equations presented below:

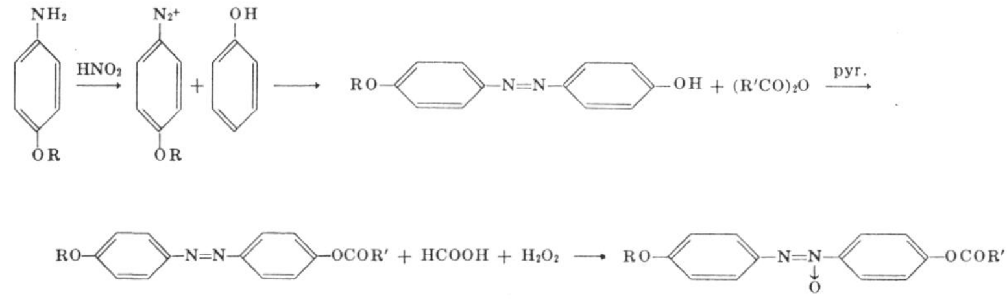

The following examples illustrate the process. In the examples, parts are by weight unless otherwise noted.

Example 1: To a solution of 15.1 parts of p-propoxyaniline in dilute hydrochloric acid (equal volumes of water and concentrated hydrochloric acid) cooled to 0°C was added a solution of 8 parts of sodium nitrite in 40 parts of water while maintaining the temperature below 10°C. The resultant mixture was added to a cold solution of 10.4 parts of phenol in 90 parts of 10% aqueous sodium hydroxide. Brown crystals of the p-propoxyphenylazo-p'-phenol product precipitated, were collected by filtration, washed with water and dried under vacuum.

2.6 parts of the azo product were mixed with 1.0 part of acetic anhydride in 50 parts by volume of pyridine and stirred for 2 hours at ambient temperatures. The resultant mixture was poured into 500 parts of ice water when orange crystals of p-propoxy-p'-acetoxyazobenzene precipitated. The precipitate was filtered, dried and recrystallized from isopropanol. A solution containing 1.5 parts of p-propoxy-p'-acetoxyazobenzene in 10 parts by volume of methylene chloride was added dropwise to a solution of 10 parts of formic acid (97%) and 8 parts by volume of hydrogen peroxide (30% stabilized) in 40 parts by volume of methylene chloride. The resultant mixture was refluxed for 2 hours, the solvent layer collected, washed with water, dried over sodium sulfate and the solvent stripped. The desired product, p-propoxy-p'-acetoxyazoxybenzene having the formula

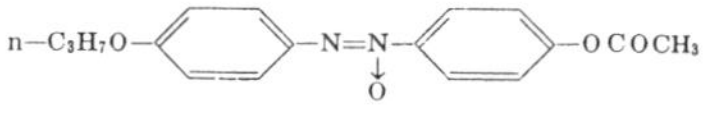

was recrystallized from isopropanol. This compound had a crystal-nematic transition temperature of 86°C and a nematic-isotropic liquid transition temperature of 123°C.

Examples 2 through 41: Other compounds were prepared in similar manner, except substituting the appropriate starting materials. Examples of the compounds are given in the following table, together with their crystal-nematic (C-N) transition temperatures and their nematic-isotropic liquid (N-L) transition temperatures. These compounds are mesomorphic or liquid crystalline in the temperature range between these two transition temperatures.

In the following table, R and R' have reference to the Formula (1), and are straight chain alkyl groups.

Example			C-N,°C.	N-L,°C.
2	R=CH_3,	R'= CH_3	125	133
3	do.	C_2H_5	82	127
4	do.	C_3H_7	78	130
5	do.	C_4H_9	87	116
6	R=CH_3,	R'= C_5H_{11}	73	120
7	do.	C_6H_{13}	77	110
8	do.	C_7H_{15}	72	114
9	R=C_2H_5,	R'= CH_3	98	151
10	do.	C_2H_5	99	152
11	do.	C_3H_7	84	151
12	do.	C_4H_9	69	136
13	do.	C_5H_{11}	81	138
14	do.	C_6H_{13}	66	129
15	do.	C_7H_{15}	77	130
16	R=C_3H_7,	R'= C_2H_5	89	118
17	do.	C_3H_7	77	127
18	do.	C_4H_9	70	117
19	do.	C_5H_{11}	69	119
20	do.	C_6H_{13}	67	111
21	do.	C_7H_{15}	68	115
22	R=C_4H_9,	R'= CH_3	81	130
23	do.	C_2H_5	82	134
24	do.	C_3H_7	79	135
25	do.	C_4H_9	66	118
26	do.	C_5H_{11}	64	125
27	do.	C_6H_{13}	64	120
28	do.	C_7H_{15}	63	122
29	R=C_5H_{11},	R'= CH_3	83	118
30	do.	C_2H_5	71	124
31	do.	C_3H_7	66	126
32	do.	C_4H_9	58	117
33	do.	C_5H_{11}	59	119
34	R=C_6H_{13},	R'= CH_3	82	123
35	do.	C_2H_5	68	128
36	do.	C_3H_7	66	128
37	do.	C_4H_9	68	119
38	do.	C_5H_{11}	66	122
39	R=C_7H_{15},	R'= CH_3	77	117
40	do.	C_2H_5	75	119
41	do.	C_3H_7	61	123

Examples 42 through 49: Certain higher molecular weight compounds exhibit a crystal-smectic (C-S) transition temperature and a smectic-nematic (S-N) transition temperature as well as the nematic-isotropic liquid (N-L) transition temperature. Such compounds are given in the following table where R and R' have the meanings given in Examples 2 through 41.

Example			C-S,°C.	S-N,°C.	N-L,°C.
42	R=C_5H_{11}	R'= C_6H_{13}	54	58	115
43	do.	C_7H_{15}	58	67	117
44	R=C_6H_{13}	C_6H_{13}	60	72	119
45	do.	C_7H_{15}	62	80	121
46	R=C_7H_{15}	R'= C_4H_9	69	71	117
47	do	C_5H_{11}	68	78	120
48	do	C_6H_{13}	61	85	117
49	do	C_7H_{15}	63	90	118

SCHIFF BASES

p-[N-(p-Methoxybenzylidene)Amino] Phenyl 2-Ethylhexanoate

S. Matsumoto and M. Kawamoto; U.S. Patent 3,738,942; June 12, 1973; assigned to Tokyo Shibaura Electric Co., Ltd., Japan describe a liquid crystal material which, when impressed with the prescribed electric field, presents a light scattering phenomenon over a wide range of temperatures around room temperature, and is particularly adapted for use with electro-optical apparatus such as a panel display, light switch and light valve device. The liquid crystal material comprises a mixed liquid crystal composition prepared by mixing at least one nematic liquid crystal material with p-[N-(p-methoxybenzylidene)amino]-phenyl-2-ethyl hexanoate. The liquid crystal material prepared by mixing the abovementioned constituents has a low transition temperature at which it changes from solid phase to nematic phase intermediate between solid and liquid phases and displays the desired electro-optical effect around room temperature. Therefore, an electro-optical device using the liquid crystal material need not be provided with heating or heat controlling means, offering many practical advantages due to its simple construction and freedom from handling difficulties.

Example: There was prepared a liquid crystal composition from raw materials consisting of butyl p-(p-ethoxy phenoxy carbonyl)phenyl carbonate (abbreviated as BEPCPC), p-[N-(p-methoxybenzylidene)amino]phenyl-2-ethyl hexanoate (abbreviated as MBAPH, p-[N-(p-methoxybenzylidene)amino]phenyl butyrate (abbreviated as MBAPB), p-[N-p-methoxybenzylidene)amino]phenyl acetate (abbreviated as MBAPA), and p-[N-(p-methoxybenzylidene)-amino]-n-butylbenzene (abbreviated as MBABB) in the proportions given in the following table.

	Proportion (parts by weight) of—					Temperature (° C.) range of nematic phase
	BEPCPC	MBAPH	MBAPB	MBAPA	MBABB	
Sample No.:						
1	80.6	19.4				23-60
2		19.1	80.9			21-67
3	36.1	27.1	36.8			3-42
4	25.2	10.8	48.8	15.2		−9-78
5		10.1			89.9	10-31
6		19.6			80.4	−1-13
Control No.:						
a	80.6	0				55-87
b		0	80.9			52-114
c	36.1	0	36.8			32-98
d	25.2	0	48.8	15.2		15-96
e		0			89.9	21-41
f		0			80.4	21-41

The weighed out raw materials were crushed and well mixed, followed by heating until the mixture was converted to an isotropic liquid. The liquid was further stirred under the isotropic liquid state to obtain a liquid crystal material. Determination was made of the temperatures at which the liquid crystal material thus prepared presented the nematic phase, the results being given in the above table.

By way of comparison, similar determination was made on the controls, the results being also shown in the table. The determination was carried out in the following manner. The samples of a liquid crystal material converted to the abovementioned isotropic liquid were fully cooled for solidification in a freezer. Thereafter there were observed the changing phases which the liquid crystal samples indicated during the heating process, using a microscopic melting-point measuring apparatus. Thus there were defined the transition temperatures at which the phase of the samples changed from solid to nematic and from nematic to isotropic liquid.

As apparent from the above table, the mixed nematic liquid crystal composition of this process containing p-[N-(p-methoxybenzylidene)amino]phenyl-2-ethyl hexanoate exhibits a nematic phase at far lower temperatures than a liquid crystal material prepared from the components alone. Namely, when impressed with an electric field, the nematic phase of the liquid crystal composition of this process readily presents a light scattering phenomenon over a wide range of temperatures around room temperature.

N-(o-Hydroxy-p-Methoxybenzylidene)-p-n-Butylaniline

M.M. Labes; U.S. Patent 3,827,780; August 6, 1974; assigned to Temple University describes stable anil-type nematic liquid crystals of the formula

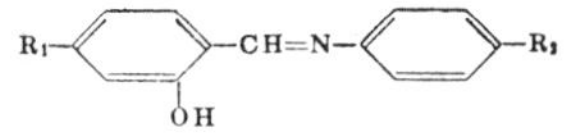

where R_1 and R_2 are each an alkyl, alkoxy, alkylester, aryl, aryloxy, arylester, aralkyl, carboxylic acid, halogen, nitro, cyano group or the like. The liquid crystals of this process, because of the excellent stability, are especially useful in light valves, optical display devices and similar applications which involve the modulation of light.

The anil-type nematic crystal compounds and compositions of the process, have excellent chemical stability as compared with the compounds previously employed, especially in comparison with anil-type compounds such as those described in French Patent 1,537,000. The reasons for the exceptional stability of the anil-type compounds of this process is believed to be due to the presence of the hydroxy group in the ortho position. The hydroxy group hydrogen bonds with the nitrogen atom of the anil linkage to form a compound which can be represented by the formula:

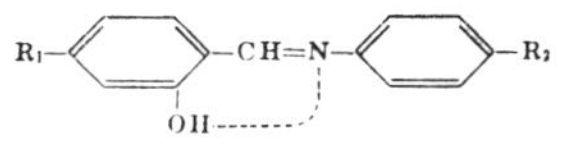

The hydrogen bonded ring thus formed is chemically stable. When the o-hydroxy group is not present in the ring, the compound is readily broken at the anil linkage and the nematic crystalline properties of the compound are destroyed. The compounds are prepared in a conventional manner. A para-substitued o-hydroxybenzaldehyde is prepared in accordance with the general procedure described in Zemplen et al, *Acta Chim, Hung* 22, 449 (1960). Equimolar quantities of the para-substituted o-hydroxybenzaldehyde and a parasubstituted aniline are dissolved in absolute ethanol, and then blended and heated at the boiling point of the mixture for 30 minutes. The mixture is cooled to 0°C at which point a crystalline solid forms. The solid is recovered and recrystallized from either ethanol or another suitable solvent such as hexane. The process is shown by the following equation:

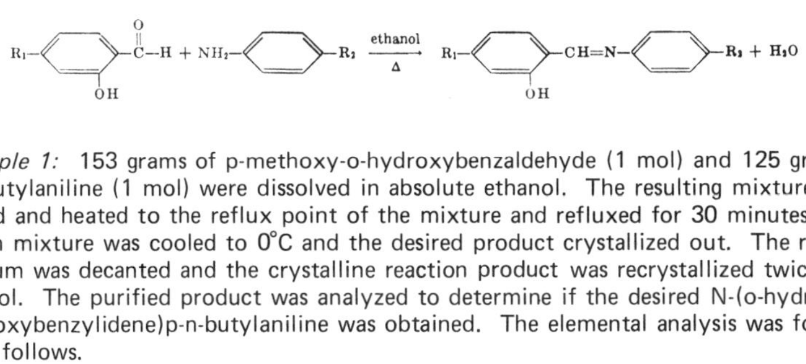

Example 1: 153 grams of p-methoxy-o-hydroxybenzaldehyde (1 mol) and 125 grams of p-n-butylaniline (1 mol) were dissolved in absolute ethanol. The resulting mixture was stirred and heated to the reflux point of the mixture and refluxed for 30 minutes. The reaction mixture was cooled to 0°C and the desired product crystallized out. The reaction medium was decanted and the crystalline reaction product was recrystallized twice in ethanol. The purified product was analyzed to determine if the desired N-(o-hydroxy-p-methoxybenzylidene)p-n-butylaniline was obtained. The elemental analysis was found to be as follows.

	Found	Theoretical
% C	76.36	76.29
% H	7.54	7.46
% N	5.21	4.94

The crystals thus obtained upon cooling from the isotropic phase remained nematic to <8°C in a capillary melting tube. In a thin film such as would be used in an electro-optical device crystallization occurred at about 26°C.

N-(o-hydroxy-p-methoxybenzylidene)-p-n-butylaniline was evaluated for the dynamic scattering properties by measuring the transmission as determined by a thermopile reading of a 24μ layer sandwiched between two tin-oxide coated quartz plates. Figure 3.1 shows the voltage transmission as well as the calculated contrast ratio which attains a value equal to 60. These results were obtained at 37°C. However, the material exhibits dynamic scattering effects down to 26°C at which point it solidified into crystals. This example was repeated numerous times and the results were reproducible.

FIGURE 3.1: PERFORMANCE DATA FOR STABLE ANIL-TYPE NEMATIC LIQUID CRYSTAL

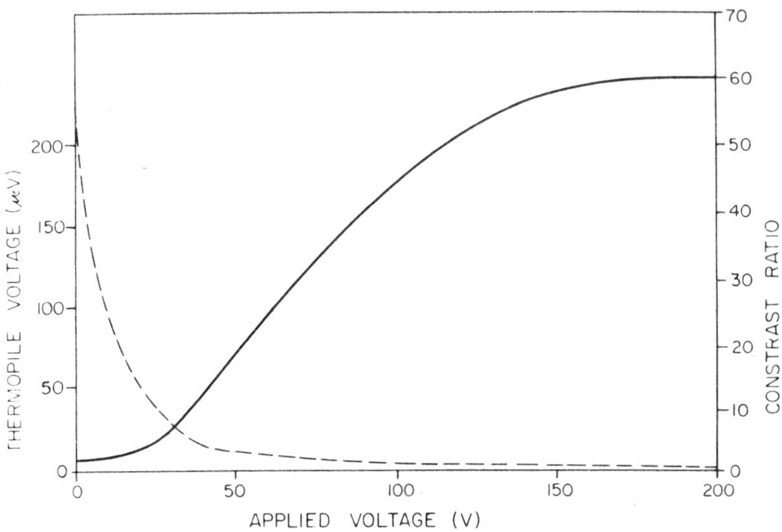

Source: M.M. Labes; U.S. Patent 3,827,780; August 6, 1974

The high purity and reproducibility was further established by blending materials obtained from several experiments and it was found the results were still reproducible. In comparison the N-(p-methoxybenzylidene)-p-n-butylaniline, which is the corresponding compound without the stabilizing ortho hydroxy group, was difficult to purify and the results were considerably less reproducible. This material also gradually decomposed during testing which further affected the reproducibility.

Example 2: The process described in Example 1 was repeated except that compounds of the following formula were prepared using other starting materials:

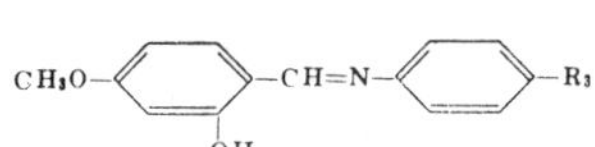

R_3	Nematic range, °C.
—CN............	141-147
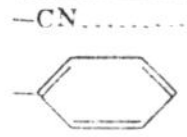	165-197

Both of the above compounds were evaluated and found to be suitable for use in electro-optical devices due to the relative stability and nematic crystal characteristics.

Alkoxy and Acyloxybenzylidene Anils

J.E. Goldmacher and G.H. Heilmeier; U.S. Patent 3,499,702; March 10, 1970 and J.A. Castellano; U.S. Patent 3,597,044; August 3, 1971; both assigned to RCA Corporation describe an electro-optical light valve which comprises a nematic liquid crystal composition comprised of a mixture of p-n-ethoxybenzylidene-p'-aminobenzonitrile with at least one member of the group consisting of p-n-butoxybenzylidene-p'-aminobenzonitrile and p-n-octoxybenzylidene-p'-aminobenzonitrile. The light valve includes means for applying an electric field to the liquid crystal composition.

In order to substantially lower the crystal-nematic transition temperature, mixtures of from 10 to 90 weight percent of p-n-butoxybenzylidene-p'-aminobenzonitrile with p-n-ethoxybenzylidene-p'-aminobenzonitrile or mixtures containing from 15 to 70 weight percent of each of the three compounds are preferred. The lowest crystal-nematic transition temperatures have been found with a mixture comprised of equal weights of p-n-ethoxybenzylidene-p'-aminobenzonitrile or a mixture comprised of equal weights of p-n-ethoxybenzylidene-p'-aminobenzonitrile, p-n-butoxybenzylidene-p'-aminobenzonitrile, and p-n-octoxybenzylidene-p'-aminobenzonitrile. The following table shows the comparison between crystal-nematic transition temperatures and the nematic-isotropic transition temperatures of some of the particular prior art compounds and that of the p-n-alkoxybenzylidene-p'-aminobenzonitrile compounds and the mixtures thereof.

Composition	Crystal-Nematic Transition Temperature (°C)	Nematic-Isotropic Transition Temperature (°C)
p-Azoxyanisole	116	135
4,4'-Diethoxyazoxybenzene	137	167
4,4'-di-n-Pentoxyazoxybenzene	79	121
p-n-Ethoxybenzylidene-p'-aminobenzonitrile	105	130
p-n-Butoxybenzylidene-p'-aminobenzonitrile	63	106
p-n-Octoxybenzylidene-p'-aminobenzonitrile	79	94.5
p-n-Ethoxybenzylidene-p'-aminobenzonitrile and p-n-butoxybenzylidene-p'-aminobenzonitrile (equal weights)	41	109
p-n-Ethoxybenzylidene-p'-aminobenzonitrile p-n-butoxybenzylidene-p'-aminobenzonitrile p-n-octoxybenzylidene-p'-aminobenzonitrile (equal weights)	50	91

Threshold voltages for a one-half mil thick cell containing the mixtures are approximately 10 volts AC or DC. Switching time for switching of a one-half mil cell at 100 volts DC is in the order of 10 milliseconds for the rise time and about 30 to 100 milliseconds for the decay time. By rise time, it is meant the time necessary to align the domains. By decay time, it is meant the time it takes after the removal of the applied field for the liquid crystals to return to the random state. The mixtures are comprised of anils having the structural formula:

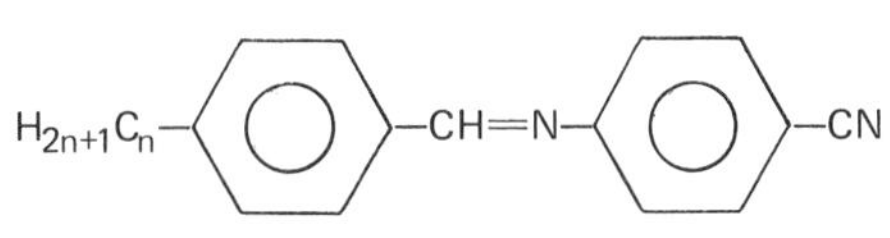

Generally, these materials can be prepared by the condensation of p-aminobenzonitrile with the appropriate p-alkoxybenzaldehyde in a refluxing benzene solution with benzenesulfonic acid as a catalyst and azeotropic removal of water. Purification of the compound is generally accomplished by repeated recrystallization from hexane solution to a constant nematic-isotropic transition temperature.

For example, the compound p-n-ethoxybenzylidene-p'-aminobenzonitrile can be prepared in the following manner. A solution of 0.10 mol of p-aminobenzonitrile, 0.10 mol of p-ethoxybenzaldehyde, 0.1 gram of benzene sulfonic acid in 500 ml of benzene is refluxed for about 3 hours. A Dean-Stark trap is used to collect the water which is azeotropically removed. After refluxing, the solvent is removed by evaporation under reduced pressure in a rotating film evaporator. The residue is collected and recrystallized several times from hexane. The p-n-butoxybenzylidene-p'-aminobenzonitrile compound is prepared in a similar manner except that p-butoxybenzaldehyde is used rather than the ethoxybenzaldehyde stated above. In a like manner, p-n-octoxybenzylidene-p'-aminobenzonitrile is prepared by substituting p-octoxybenzaldehyde for ethoxybenzaldehyde in the above example.

In order to prepare the mixtures, the pure crystalline compounds are weighed into a vial in the desired weight proportion and the mixture heated to the isotropic melting point so as to form an isotropic solution. The isotropic solution is allowed to cool to 0°C and becomes solidified during this cooling. The resulting solid mass can be broken off into small pieces and can be placed in a capillary tube or on a glass slide for melting point determination. The mixtures may be inserted in the liquid crystal device by having the planar supporting elements either separated by means of a shim or by sealing three sides together with an epoxy cement leaving a space between the elements. The liquid crystal composition is then heated to a temperature greater than its crystal-nematic transition temperature and injected into the space between the planar elements, for example, by means of a syringe.

N-(p-Alkanoyloxybenzylidene)-p'-n-Butylaniline

E.L. Steiger and H.J. Dietrich; U.S. Patent 3,784,513; January 8, 1974; assigned to Owens-Illinois, Inc. describe the preparation of liquid-crystal compositions of the structure:

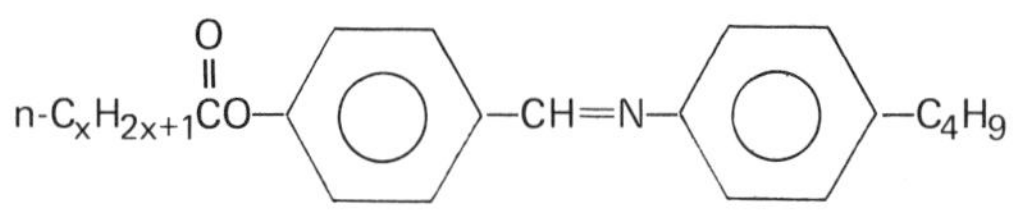

where x equals 1 to 10, typically 1 to 5. The compounds are prepared by the reaction of para-n-butylaniline and para-n-alkanoyloxybenzaldehyde. The following equation is representative of the reaction of para-n-alkanoyloxybenzaldehyde and para-n-butylaniline:

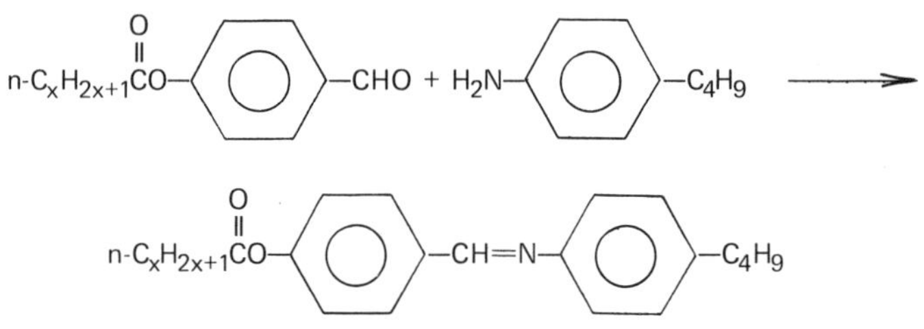

The para-substituted aromatic amines are commercially available and may be typically used without further purification. Schiff bases are prepared by refluxing equimolar quantities of the p-substituted benzaldehyde and aniline in anhydrous ethanol for 4 to 6 hours. The solvent and water are removed and the residue recrystallized several times from ethanol until the transition temperatures remain constant. The crude yields ranged from 70 to 90%. The infrared spectra show a strong band at 1,629 cm^{-1} corresponding to the carbon-nitrogen double bond in Schiff base compounds. Other absorptions are compatible with the expected structures. Transition temperatures are determined on a Leitz Ortholux polarizing microscope using a Mettler FP-2 heating stage.

N-(p-n-propionyl oxybenzylidene)-p'-n-butylaniline (where x is 2 in the basic structure) was prepared. The structure was analyzed at 77.40% by weight carbon, 7.51% by weight hydrogen, and 4.23% by weight nitrogen compared with calculated theoretical analysis values of 77.64% by weight carbon, 7.49% by weight hydrogen, and 4.53% by weight nitrogen. The compound changed from the crystalline to the nematic phase at 65.0°C and from nematic to isotropic at 71.2°C.

N-(p-n-butyryloxybenzylidene)-p'-n-butylaniline (where x is 3 in the basic structure) was prepared. The structure was analyzed at 77.65% by weight carbon, 7.81% by weight hydrogen, and 4.46% by weight nitrogen compared with calculated theoretical analysis values of 77.99% by weight carbon, 7.79% by weight hydrogen, and 4.33% by weight nitrogen. The compound changed from the crystalline to the nematic phase at 67.7°C and from nematic to isotropic at 72.1°C. The liquid-crystal compounds may be utilized in display devices, especially of the electronic type.

N-(p-n-Alkoxycarbonyloxybenzylidene)-p'-n-Butoxyaniline

E.L. Steiger and H.J. Dietrich; U.S. Patent 3,769,314; October 30, 1973; assigned to Owens-Illinois, Inc. describe the preparation of liquid-crystal compounds of the structure:

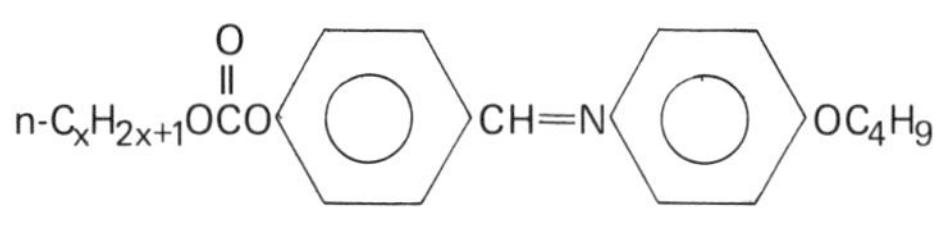

where x is an integer of 1 to 10. The compounds are prepared by the reaction of para-n-butoxyaniline and para-n-alkoxycarbonyloxybenzaldehyde. The following equation is representative of the reaction of para-n-butoxyaniline and para-n-alkoxycarbonyloxybenzaldehyde:

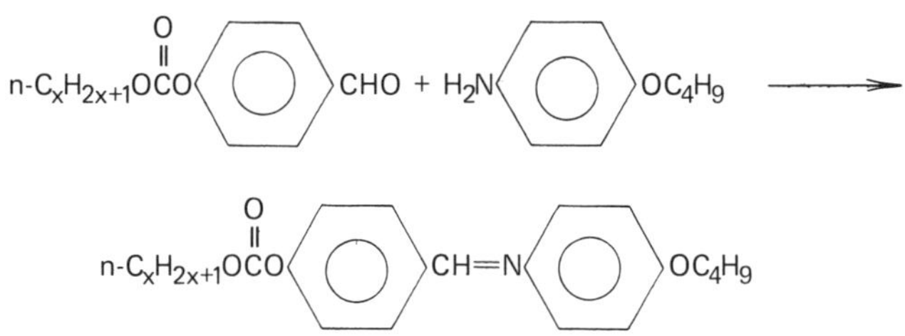

N-(p-ethoxycarbonyloxybenzylidene)-p'-n-butoxyaniline (where x is 2 in the basic structure) was prepared in accordance with this process. The structure was analyzed at 70.57% by weight carbon, 6.91% by weight hydrogen, and 4.18% by weight nitrogen compared with calculated theoretical analysis values of 70.36% by weight carbon, 6.79% by weight hydrogen, and 4.10% by weight nitrogen. The compound changed from the crystalline to the nematic phase at about 98.5°C and from nematic to isotropic at about 118.7°C.

N-(p-2-Methoxyethoxybenzylidene)-p-n-Butylaniline

H.J. Dietrich and E.L. Steiger; U.S. Patent 3,751,467; August 7, 1973; assigned to Owens-Illinois, Inc. describe the preparation of liquid-crystal compounds of the structure:

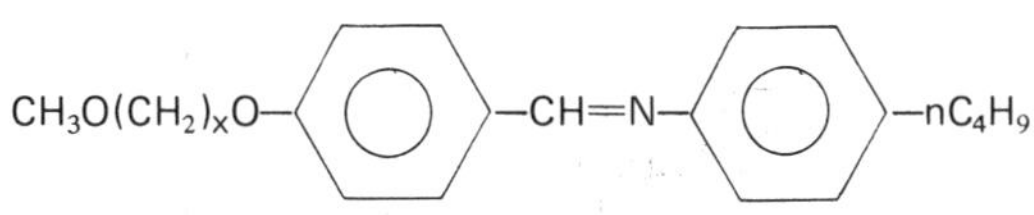

where x equals 2 to 10, typically 2 to 5. The compounds are prepared by the reaction of para-n-butylaniline and para-methoxyalkyleneoxybenzaldehyde.

N-(para-2-methoxyethoxybenzylidene)-para-n-butylaniline (where x is 2 in the basic structure) was prepared in accordance with this process. The structure was analyzed at 76.84% by weight carbon, 8.14% by weight hydrogen, and 3.77% by weight nitrogen compared with calculated theoretical analysis values of 77.14% by weight carbon, 8.09% by weight hydrogen and 4.49% by weight nitrogen. The compound changed from the crystalline to the nematic phase at 34.5°C and from nematic to isotropic at 56.6°C.

Mixtures and Derivatives of p-Methoxybenzylidene-p'-n-Butylaniline

W.E.L. Haas, J.B. Flannery, Jr., B. Mechlowitz and J.E. Adams; U.S. Patent 3,816,113; June 11, 1974; assigned to Xerox Corporation describe compositions having extended nematic mesomorphic temperature ranges which are stable at relatively low temperatures and their use in electro-optic applications.

The sensitivity of the optical properties of nematic liquid crystals to electric fields makes these materials extremely useful for electro-optical applications. For example, dynamic scattering is common to many materials which exhibit nematic liquid crystalline mesophases. The so-called dynamic scattering state occurs when an ionic current flow is set up through the nematic liquid crystalline material and the nematogenic material assumes a nonselective light-scattering appearance. This phenomenon may be utilized in electro-optical devices such as, for example, display systems.

Previously a serious drawback with respect to potential uses of nematogenic materials in electro-optic devices has been the relatively high temperatures at which the majority of the known nematic materials become mesomorphic. Typically, electro-optic devices are operated at or near room temperature. Thus devices utilizing such nematic liquid crystalline materials would require additional apparatus to maintain the temperature of the liquid crystalline material within its mesomorphic range thereby undesirably complicating the overall device configuration.

Recently there was described in the art a nematogenic material which is liquid crystalline at room temperature. H. Kelker and B. Scheurle, in *Angew. Chem.,* Internat. Ed. 8, 884 (1969), reported the preparation of p-methoxybenzylidene-p'-n-butylaniline (ABUTA) which has the chemical formula:

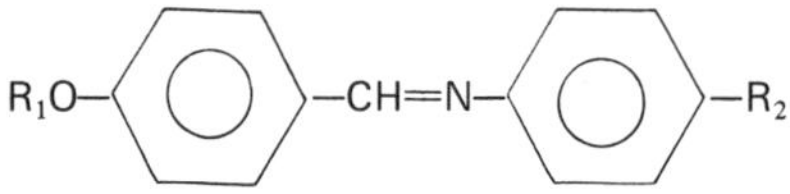

where R_1 = CH_3 and R_2 = C_4H_9 and has a nematic mesomorphic range of from 20° to 41°C. While this compound has been found to be very useful in various electro-optic techniques nevertheless there exists a continuing need for liquid crystalline materials which have a nematic mesophase at or near room temperature.

This process provides liquid crystalline compositions which are formed by mixing together liquid crystalline compounds having the general formula shown above and which have a nematic mesophase which is stable at relatively low temperatures. Liquid crystalline compounds which belong to the class of compounds having the general formula shown above and where R_1 and R_2 are particular combinations of alkyl groups are listed in the following table.

It can be seen that the compounds described in the table each possess a nematic mesophase which is stable at relatively low temperatures, i.e., near room temperature. These compounds have been found to be highly advantageous for use in various electro-optic applications such, as, for example, display devices.

It has been found that by utilizing some of the compounds listed in the table and other members of the same generic class it is possible to form compositions which have significantly larger nematic mesomorphic temperature ranges than those of the individual compounds. The liquid crystalline compositions of the process are: mixtures of p-methoxybenzylidene-p'-n-butylaniline (referred to hereinafter as ABUTA for simplicity) and p-n-butoxybenzylidene-p'-n-butylaniline (hereafter BOBUTA); mixtures of p-ethoxybenzylidene-p'-n-butylaniline ($R_1 = C_2H_5$ and $R_2 = C_4H_9$, hereafter EOBUTA) and ABUTA; mixtures of EOBUTA and BOBUTA; mixtures of p-n-butoxybenzylidene-p'-ethylaniline (hereafter BOETHA) and ABUTA; mixtures of BOETHA and EOBUTA; mixtures of p-n-pentyloxybenzylidene-p'-n-butylaniline (hereafter PENTOBUTA) and ABUTA; and mixtures of p-n-hexyloxybenzylidene-p'-n-butylaniline (hereafter HEXOBUTA) and ABUTA. As will be seen below an extremely important characteristic of these compositions is that they typically have nematic mesophases which are stable at room temperature or slightly below and in some cases significantly below typically extending to below 0°C.

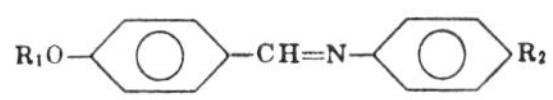

R_1	R_2	Smectic (° C.)	Nematic (° C.)	Name
C_3H_7...	C_4H_9	....	39–59	*p*-*n*-Propoxybenzylidene-*p'*-*n*-butylaniline.
C_4H_9...	C_4H_9	7–46	46–75	*p*-*n*-Butoxybenzylidene-*p'*-*n*-butylaniline.
C_5H_{11}...	C_4H_9	24–54	54–71	*p*-*n*-Pentyloxybenzylidene-*p'*-*n*-butylaniline.
C_6H_{13}...	C_4H_9	<25–70	70–78	*p*-*n*-Hexyloxybenzylidene-*p'*-*n*-butylaniline.
C_7H_{15}...	C_4H_9	29–76	76–77.5	*p*-*n*-Heptyloxybenzylidene-*p'*-*n*-butylaniline.
C_4H_9...	CH_3	*(49)–66	66–68	*p*-*n*-Butoxybenzylidene-*p*-toluidine.
C_4H_9...	C_2H_5	42–52	52–67	*p*-*n*-Butoxybenzylidene-*p'*-ethylaniline.
C_5H_{11}...	CH_3	35–45	45–64	*p*-*n*-Pentyloxybenzylidene-*p*-toluidine.

*Denotes monotropic transition.

The compositions are highly advantageous when employed in various electro-optic applications since they allow overall device considerations to be desirably simplified. The compositions may be conveniently utilized in any number of electro-optic applications such as for example, display cells since upon application of an electric field of sufficient magnitude, e.g., above about 3,000 volts per centimeter, the optical properties of the materials change and the liquid crystal material which is substantially transparent prior to the application of the field becomes frosted, i.e., exhibits dynamic scattering. For a detailed description of this phenomenon see G.H. Heilmeier et al, *Proc. IEEE* 56, 7 (1968).

Generally speaking the individual compounds used to form the compositions are prepared by condensation in refluxing ethanol of the respective p-n-alkoxybenzaldehyde and the appropriate p-n-alkylaniline. Refluxing is carried out for several hours and the solvent is then removed by distillation. The crystalline products are then purified by repeated recrystallization from an appropriate solvent such as methanol or petroleum ether.

The compositions are prepared by weighing appropriate fractions of the desired constituents, combining them in a vessel, e.g., a glass beaker, and heating above the isotropic transition temperatures of the respective constituents with mixing to ensure a homogeneous composition.

The temperatures at which transitions from the nematic mesomorphic state to the isotropic state occur, as indicated in Figures 3.2a through 3.2g for the respective compositions, are determined by differential thermal analysis employing a DuPont 900 Differential Thermal Analyzer equipped with chromel-alumel thermocouples. The thermograms are obtained at a heating rate of 10°C/min and the temperature at the beginning of the state transition endotherm is taken as the characteristic temperature. The temperature is determined by linear extrapolation of the leading edge of the endotherm to its intersection with the baseline.

Verification of the state assignments is obtained by polarized optical miscoscopy. Samples of these compositions are placed in the form of thin films, between glass cover slides and viewed through a Leitz Ortholux polarizing microscope having a calibrated hot stage. Visually observed texture changes have been found to coincide within 2°C with the phase transitions observed by thermal analysis.

Referring to Figure 3.2a there is seen a graphical plot illustrating the eutectic behavior of compositions prepared from varying mol fractions of BOBUTA and ABUTA. The nematic mesomorphic temperature range for BOBUTA is from 46° to 75°C and that for ABUTA from 20° to 41°C. It can be seen that the compositions possess extended nematic mesomorphic temperature ranges which generally increase progressively as the weight percent of the BOBUTA increases and which reach a maximum at from 55 to 65 weight percent BOBUTA. Figure 3.2b illustrates the eutectic behavior of compositions prepared from varying weight percentages of EOBUTA and ABUTA. The nematic mesomorphic temperature range for EOBUTA is from 38° to 82°C. The compositions possess extended nematic mesomorphic temperature ranges which reach a maximum at from 50 to 70 weight percent of ABUTA.

FIGURE 3.2: NEMATIC LIQUID CRYSTALLINE COMPOSITIONS WITH EXTENDED MESOMORPHIC RANGE

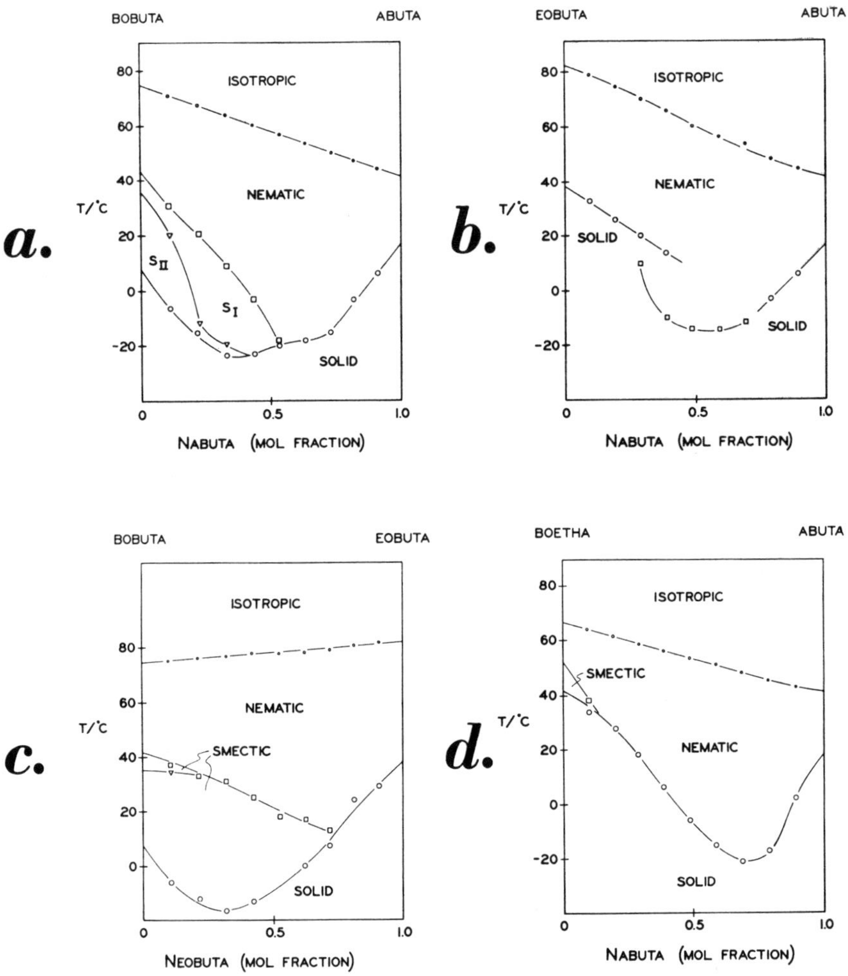

(continued)

FIGURE 3.2: (continued)

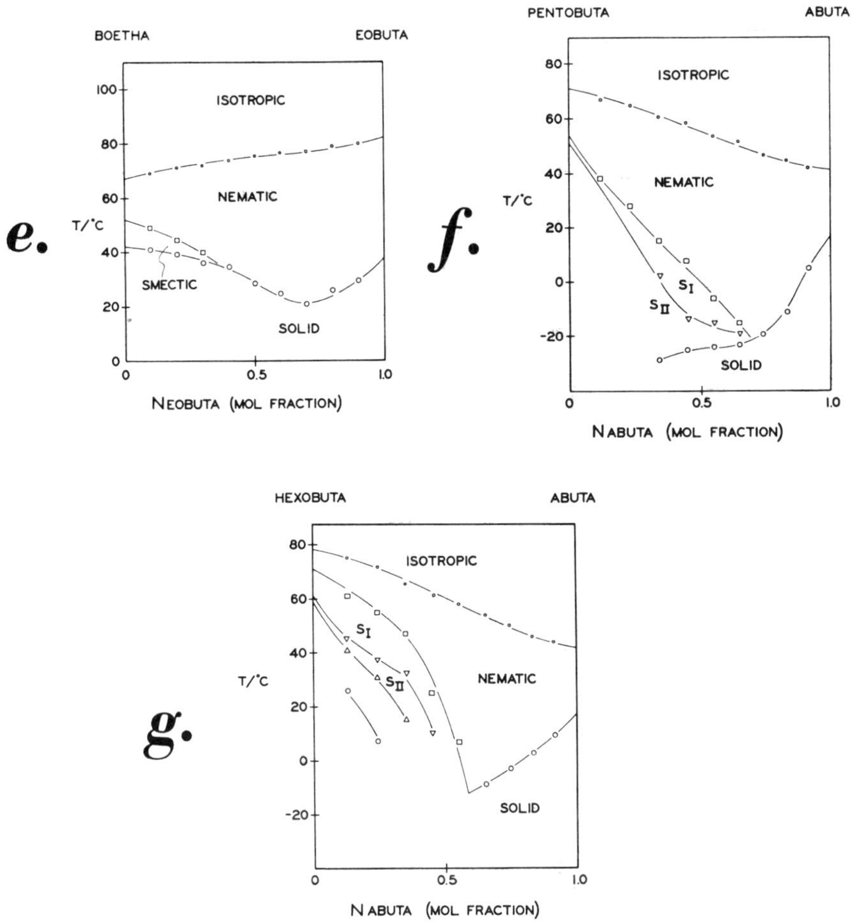

Source: W.E.L. Haas, J.B. Flannery, Jr., B. Mechlowitz and J.F. Adams; U.S. Patent 3,816,113; June 11, 1974

Figure 3.2c illustrates the eutectic behavior of compositions prepared from varying weight percentages of BOBUTA and EOBUTA. The compositions are seen to have extended nematic mesomorphic temperature ranges which reach a maximum at from 55 to 75 weight percent of EOBUTA.

The eutectic behavior of compositions prepared from varying weight percentages of BOETHA and ABUTA is shown in Figure 3.2d. BOETHA has a nematic mesomorphic range of from 48° to 62.5°C. The compositions have extended nematic mesomorphic temperature ranges which generally increase rapidly as the proportion of the ABUTA constituent increases and which reach a maximum at from 60 to 80 weight percent of ABUTA.

Figure 3.2e illustrates the eutectic behavior of compositions prepared from varying weight percentages of BOETHA and EOBUTA. The compositions have extended nematic mesomorphic temperature ranges which reach a maximum at from 70 to 90 weight percent EOBUTA.

Figure 3.2f graphically illustrates the eutectic behavior of compositions prepared from varying weight percentages of PENTOBUTA and ABUTA. PENTOBUTA has a nematic mesomorphic temperature range of from 54° to 71°C. The compositions have extended mesomorphic temperature ranges which generally increase rapidly as the proportion of the ABUTA constituent increases and which reach a maximum at from 50 to 80 weight percent of ABUTA.

The eutectic behavior of compositions prepared with varying mol fractions of HEXOBUTA and ABUTA is illustrated in Figure 3.2g. The compositions are seen to have extended nematic mesomorphic temperature ranges which generally become larger as the proportion of the ABUTA constituent increases and which reach a maximum at from 50 to 70 weight percent of ABUTA. The compositions may be utilized in a variety of electrooptical devices and provide simplified apparatus considerations.

p-Methoxyformyloxybenzylidene-p-n-Butylaniline

M.J. Rafuse; U.S. Patent 3,675,987; July 11, 1972; assigned to Sperry Rand Corporation describes room temperature nematic liquid crystal compositions suitable for use as scatterers of light in optical devices that include as a basic component p-methoxyformyloxybenzylidene-p-n-butylaniline. Such compositions may also include p-ethoxybenzylidene-p-n-butylaniline or p-butoxybenzylidene-p-n-butylaniline, along with certain additives including p-toluylidene-p-n-butylaniline. The compositions are employed in the form of thin layers within thin cells having transparent electrodes with means for applying electric fields across the layer. The display is formed by electrically controlling light-scattering properties of the liquid crystal layer.

The multiple component electro-optically active nematic liquid crystal compositions include as a common principal component an active electro-optic medium consisting of p-methoxyformyloxybenzylidene-p-n-butylaniline, which it is believed may be illustrated graphically by the formula:

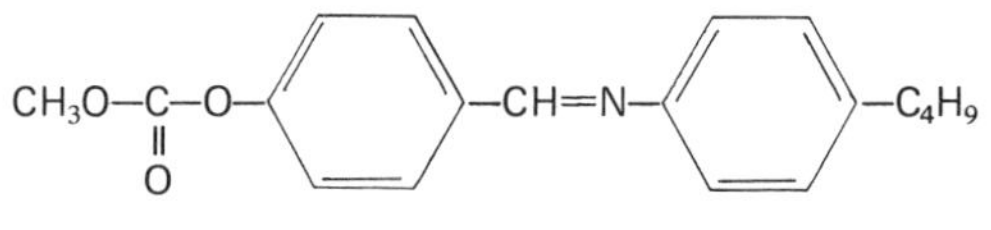

This principal component forms useful room temperature-operable nematic liquid crystal compositions when mixed with certain p-alkoxybenzylidene-p-n-butylanilines. Ionizable additives may also be added to such mixtures for purposes which will become apparent.

The above principal or primary component is prepared by refluxing the appropriate aldehyde and amine in dry alcohol. Commercially available benzaldehydes and p-n-butylaniline are used after purification. Specifically, p-methoxyformyloxybenzylidene-p-n-butylaniline is prepared by first preparing methyl p-formylphenyl carbonate. To a solution of commercially available p-hydroxybenzaldehyde (2.4 grams; 0.02 mol) in 15 ml of dry ether containing 4 ml of pyridine is added methyl chloroformate (2.5 ml), while cooling and stirring the solution. The reaction mixture is stirred substantially at room temperature for 2 hours and is then filtered. The solvent is removed by evaporation under reduced pressure. The residue is then induced to crystallize by chilling it.

The crystals are next redissolved in hexane and again recrystallized. A typical yield using the above proportions is 2.50 grams (a 70% yield) of white methyl p-formylphenyl carbonate crystals having a melting point lying between 34° and 36°C. The infrared spectrum of this material in CCl_4 included wave numbers 2,820, 2,730 and 1,710 cm^{-1} stemming from the aldehyde portion of the generated molecule and 1,760 cm^{-1} from the carbonate portion. The spectrum confirms the chemical constitution of the resultant material.

Methyl p-formylphenyl carbonate crystals (1.8 grams; 0.01 mol) are then refluxed with p-n-butylaniline (1.5 grams; 0.01 mol) in 10 ml of dry ethanol for 2 hours with stirring. The cool reaction mixture is filtered and the precipitate is recrystallized from hexane. With the above proportions, there is a yield of 2.48 grams (an 80% yield) of pale yellow crystals of the desired p-methoxyformyloxybenzylidene-p-n-butylaniline. The observed infrared spectrum in CCl_4 confirms the chemical constitution of the product, indicating the loss of nitrogen-hydrogen bonds in the material and the absence of the aldehyde portions.

Tests of the material yield an operational temperature range between 55° and 61°C displaying nematic characteristics. When cooled, the material forms a monotropic nematic liquid crystal state which converts to a crystalline solid state at 32°C. In the optical scattering condition, the material has a white appearance under white light. It is believed that the benzylidene portion of the molecule is active in conferring liquid crystal characteristics upon the material, the general geometry of the molecule being such as otherwise to permit the liquid crystal phase to exist. It is believed that the presence of the methylformyloxy portion of the molecule contributes to the high opacity of this primary material when in its optically scattering state and to its high resistivity and consequent longer lifetime of the state.

The relatively high electro-optically active temperature range of the principal material p-methoxyformyloxybenzylidene-p-n-butylaniline is reduced, according to the process, by the addition of a known alkoxybenzylidene-p-n-butylaniline, such as identified by the graphic formula:

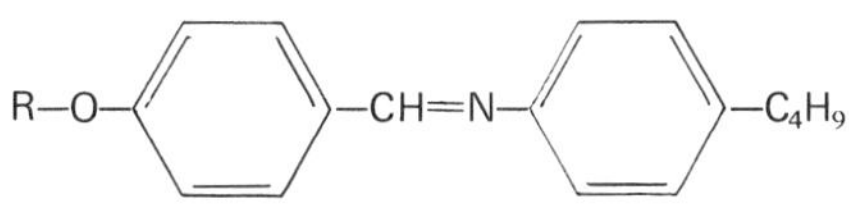

The added or secondary material may be a p-ethoxy material where R is C_2H_5, or a p-n-butoxy material, where R is C_4H_9.

The p-ethoxybenzylidene-p-n-butylaniline material is previously known to those skilled in the art as a relatively weak dynamic scattering nematic liquid crystal operable between 35° and 75°C. In the example, equal weights of the primary p-methoxyformyloxybenzylidene-p-n-butylaniline material and of the temperature lowering secondary p-ethoxybenzylidene-p-n-butylaniline crystals are weighed out and mixed thoroughly while warming until mutually dissolved and all of the mixture is in its liquid isotropic state. The material is cooled into the nematic liquid crystal state with continued agitation to ensure continued adequate mixing.

The dual component composition is utilized by placing it in a cell between closely spaced glass surfaces with a transparent electrode on at least one glass surface, as in the prior art. Application of an electric field across the thin film is accomplished for controlling the transparent and opaque states of the film, as in the prior art. The equal component mixture is found to demonstrate the desired liquid crystal display effects over the wide temperature range of −12° to +60°C. This range is particularly advantageous, extending considerably above and below ambient room temperature, and making the composition useful in many environments for which prior art liquid crystal materials are entirely unsuited. This beneficial result is achieved while still maintaining a contrast ratio (between the transparent and turbulent states) of substantially 5 to 1.

Other than equal proportions by weight of the primary and alkoxybenzylidene-p-n-butylaniline materials may be employed. For example, Figure 3.3a illustrates the behavior of the dual component mixture employing the primary material p-methoxyformyloxybenzylidene-p-n-butylaniline and the secondary p-ethoxybenzylidene-p-n-butylaniline in various ratios.

FIGURE 3.3: ROOM TEMPERATURE NEMATIC LIQUID CRYSTAL COMPOSITIONS

a.

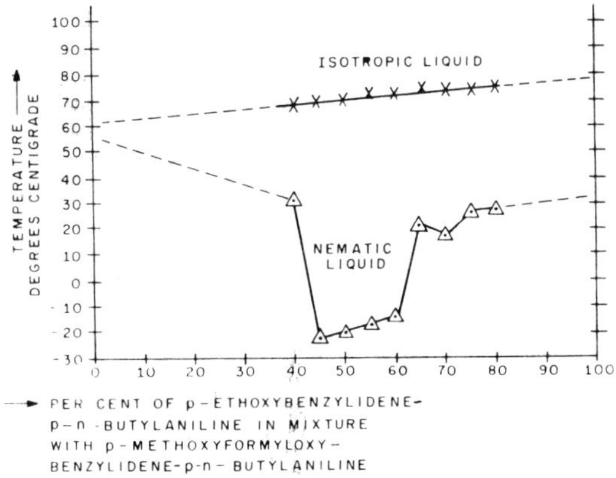

b.

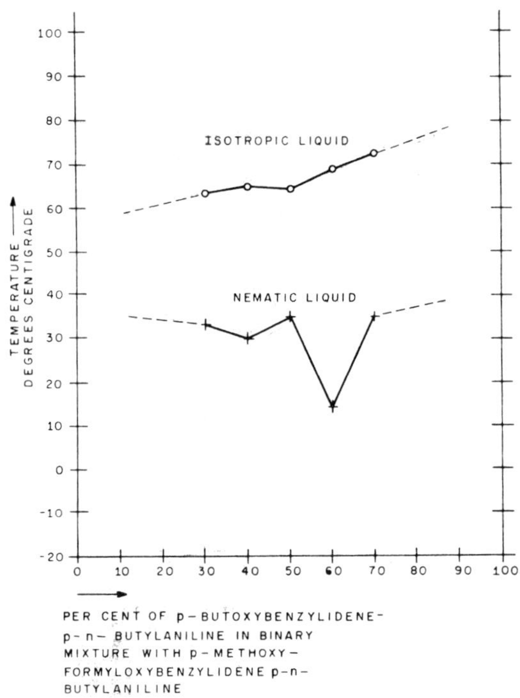

(continued)

FIGURE 3.3: (continued)

c.

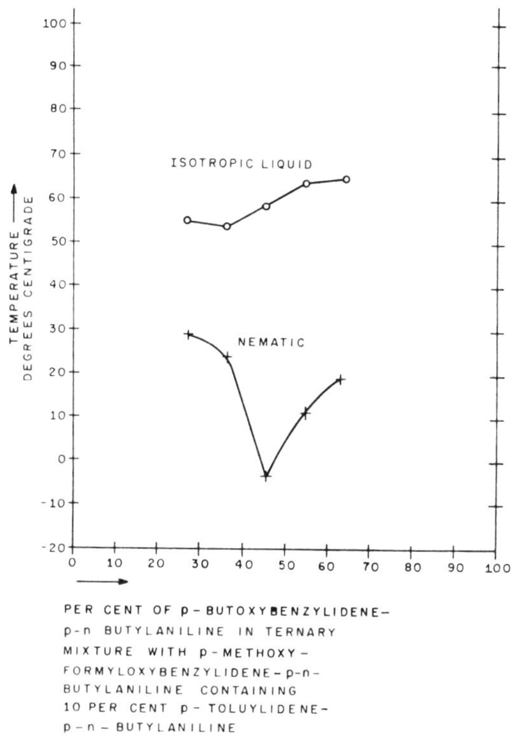

Source: M.J. Rafuse; U.S. Patent 3,675,987; July 11, 1972

It is observed that percentages of the secondary to primary materials in the measured range of substantially 42.5 to 63.3% permit operation at 10°C, of substantially 43.5 to 62.5% permit operation at 0°C, and of substantially 45 to 60.5% permit operation at -10°C. It is seen that the range of possible choice of percentages remains wide, though, decreasing somewhat, for example, from 20° to -10°C. In other words, changes in relative proportions of the dual mixture have little influence on the desired operation of the composition over the temperature range of +20° to -10°C. A mixture using about 45% of the secondary material operates at a temperature even as low as substantially -20°C. It is therefore a material of especially important advantages for use in special relatively low temperature applications. The appearance of the material in the cell switches from transparent to milky with none of the yellow cast observed in prior art materials.

p-Butoxybenzylidene-p-n-butylaniline may be mixed with the primary material p-methoxyformyloxybenzylidene-p-n-butylaniline to form a second electro-optically active material of desirable low temperature nature. The mixture is formed in the general manner employed in forming the previously discussed mixture employing the secondary material p-ethoxybenzylidene-p-n-butylaniline. From Figure 3.3b, it is observed that percentages of the secondary material p-butoxybenzylidene-p-n-butylaniline to the primary material p-methoxyformyloxybenzylidene-p-n-butylaniline in the measured range of substantially 30 to 70% by weight permit operation at substantially 35°C, and of substantially 55 to 65% by weight permit operation at substantially 25°C. A mixture using about 60% of the secondary material is found to operate at 15°C. This second binary composition is therefore an advantageous electro-optically active material for use in relatively low temperature applications.

It has furthermore been found that the operating temperature range of the binary material containing p-butoxybenzylidene-p-n-butylaniline may be desirably reduced by the addition of a relatively small quantity of a temperature depressant such as p-toluylidene-p-n-butylaniline so that a ternary composition is generated. The latter temperature depressant may be descriped graphically by the formula:

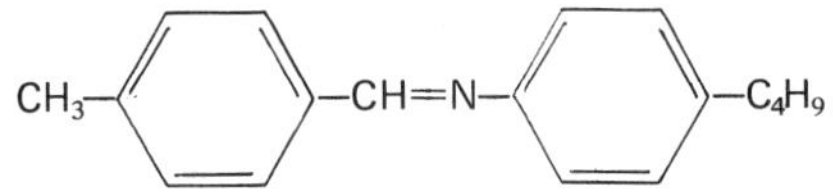

This material may be used in other binary or ternary compositions and has the general geometric symmetry of the kind which often provides a molecule having liquid crystal characteristics. However, it has no dipole characteristics at right angles to the long axis of the molecule. Therefore, the intermolecular interactions are too weak to confer liquid crystal properties. Because of its long, thin shape, the p-toluylidene-p-n-butylaniline molecule will fit compatibly between other molecules actually having good liquid crystal properties, weakening such intermolecular attraction and consequently lowering the operating temperature range of the multicomponent composition.

p-Toluylidene-p-n-butylaniline may be generated by the following method. Commercially available p-tolualdehyde (24.03 grams; 0.2 mol) and commercial p-n-butylaniline (29.85 grams; 0.2 mol) are refluxed for 2 hours in 25 ml of absolute ethanol. The reaction mixture, diluted with benzene, is washed with water, then with 10% sodium hydroxide, then again with water to neutrality before the dried solution is evaporated to remove the solvent. The residue is distilled twice under reduced pressure to yield 34.59 grams of a constant boiling fraction (BP 139° to 141°C at 0.06 mm of mercury) of p-toluylidene-p-n-butylaniline (a 69% yield). The refractive index of the pale yellow liquid is 1.6085 at 24°C.

One preferred example of a ternary liquid crystal composition employs 45% by weight of the primary p-methoxyformyloxybenzylidene-p-n-butylaniline material, 45% by weight of the secondary p-butoxybenzylidene-p-n-butylaniline material and 10% of the p-toluylidene-p-n-butylaniline material. This ternary composition displays excellent nematic liquid crystal properties over a range of temperatures from less than 0° to +47°C. Equal parts of these primary and secondary materials, without the additive demonstrate an operating temperature range of 35° to 75°C.

The ternary material is clear when unactivated in an optical cell and in the turbulent state has a white milky appearance in white light, again lacking the yellowish appearance general in prior art nematic compositions. The general range of operating temperatures may be altered arbitrarily by the use of a greater or lesser portion of the p-toluylidene-p-n-butylaniline material. Also, other than equal proportions of the primary p-methoxyformyloxybenzylidene-p-n-butylaniline and of the secondary p-butoxybenzylidene-p-n-butylaniline material may be employed.

For example, Figure 3.3c illustrates the behavior of such a ternary material when the proportions of the primary and secondary materials are altered while using p-toluylidene-p-n-butylaniline as an additive. As seen in Figures 3.3b and 3.3c, the additive significantly lowers the operating temperature range of the original binary composition. It is observed that percentages of the secondary butoxy material to the primary material in the measured range of substantially 37.5 to 62.5% by weight permit operation at substantially 20°C, and of substantially 41 to 59% by weight permit operation at substantially 10°C. Operation at substantially 5°C is possible over the percentage range from substantially 42.5 to 50 by weight. It is also observed that this ternary material operates at about –4°C for a 45% p-butoxybenzylidene-p-n-butylaniline mixture, and that it is therefore adapted to operate at much lower temperatures than the binary p-butoxy material lacking the p-toluylidene-p-n-butylaniline additive.

In related work *R.A. Soree and M.J. Rafuse; U.S. Patent 3,756,694; September 4, 1973; assigned to Sperry Rand Corporation* describe thin flat panel display devices which utilize optical polarization elements to control the amount and location of light transmitted through liquid crystal media having electric field controllable birefringence properties. Nematic liquid crystal compositions are provided, particularly adapted for use in such display systems, that demonstrate electric field controllable birefringence properties, for example, from above room temperature to about -30°C. These compositions include binary mixtures of p-toluylidene-p-n-butylaniline with p-n-butoxybenzylidene-p-n-butylaniline or pentyl p-anisylidene-p-aminophenyl carbonate with p-ethoxybenzylidene-p-n-butylaniline. Ternary compositions may be formed by adding ionizable materials.

N-(p-n-Alkoxycarbonyloxybenzylidene)-p'-Aminophenyl Acetate

H.J. Dietrich and E.L. Steiger; U.S. Patent 3,769,313; October 30, 1973; assigned to Owens-Illinois, Inc. describe the preparation of liquid crystal compounds of the structure:

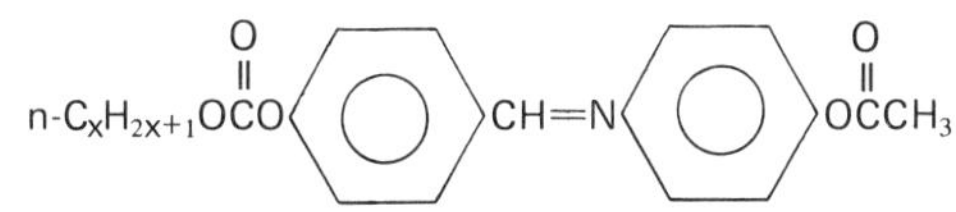

where x is an integer of 1 to 10. The compounds are prepared by the reaction of para-acetoxyaniline and para-n-alkoxycarbonyloxybenzaldehyde. The following equation is representative of the reaction of para-acetoxyaniline and para-n-alkoxycarbonyloxybenzaldehyde:

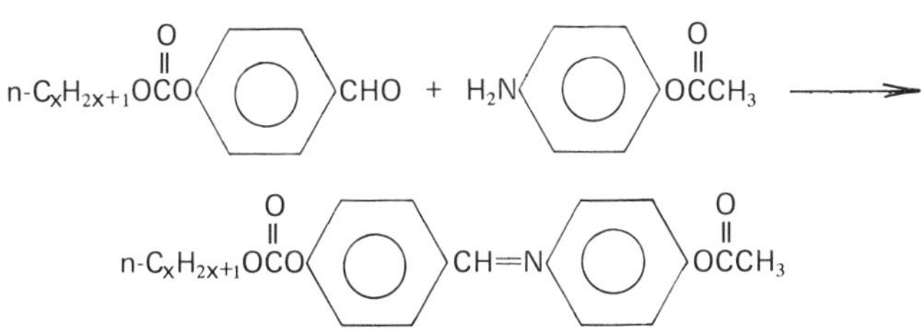

The para-acetoxyaniline, also known as para-aminophenyl acetate, is commercially available. Schiff bases are prepared by refluxing equimolar quantities of the p-substituted benzaldehyde and aniline in hydrous ethanol for 4 to 6 hours. The solvent and water are removed and the residue recrystallized several times from ethanol until the transition temperatures remain constant. The crude yields ranged from 70 to 90%. The infrared spectra show a strong band at 1,629 cm^{-1} corresponding to the carbon-nitrogen double bond in Schiff base compounds. Other absorptions are compatible with the expected structures. Transition temperatures are determined on a Leitz Ortholux polarizing microscope using a Mettler FP-2 heating stage.

N-(p-ethoxycarbonyloxybenzylidene)-p-aminophenylacetate (where x is 2 in the basic structure) was prepared. The structure was analyzed at 65.76% by weight carbon, 5.25% by weight hydrogen, and 4.32% by weight nitrogen compared with calculated theoretical analysis values of 66.05% by weight carbon, 5.23% by weight hydrogen, and 4.28% by weight nitrogen. The compound changed from the crystalline to the nematic phase at about 115.1°C and from nematic to isotropic at about 123.9°C. The liquid crystal compounds may be utilized in display devices, especially of the electronic type.

p-(Anisalamino)Phenyl Acetates

J.E. Goldmacher and J.A. Castellano; U.S. Patent 3,540,796; November 17, 1970; assigned to RCA Corporation describe a family of nematic liquid crystal compositions which is represented by the formula:

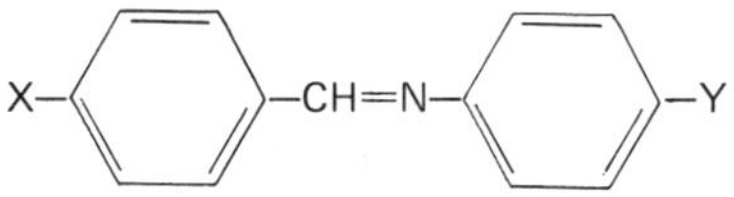

where X and Y are radicals chosen from the group consisting of saturated alkoxy radicals having from 1 to 9 carbons and saturated acyloxy radicals having from 2 to 5 carbons such that, when X is a saturated alkoxy radical, Y is a saturated acyloxy radical and vice versa. The saturated alkoxy radical has at least 3 carbons when the saturated acyloxy radical has only 2 carbons. The composition may include up to 50 weight percent of p-(anisalamino)-phenylacetate, based on the total weight of the composition.

One of the features characterizing the electrooptical devices employing this family of compositions is the relatively low minimum operating temperature due to the low crystal mesomorphic transition temperatures of members included in the family of compositions. Mixtures have been found which have a crystal mesomorphic transition temperature below room temperature. Another feature is the wide temperautre range in which the devices can be used. One example, included in the family of compositions, is the compound having the formula:

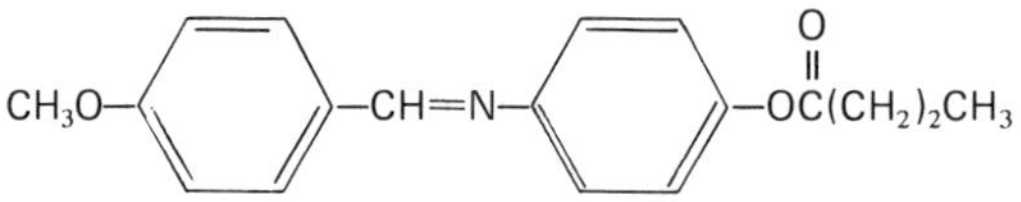

which has a crystal mesomorphic transition temperature of about 50°C and a mesomorphic isotropic liquid transition temperature of 113°C. Another example included in the family of compositions is an equal weight mixture of: p-(anisalamino)phenylacetate,

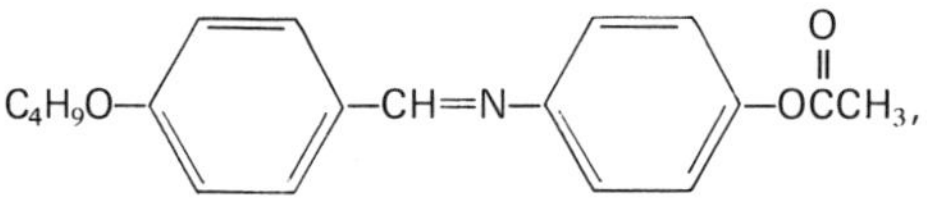

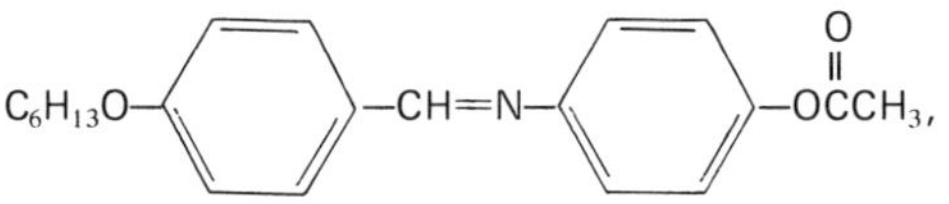

and

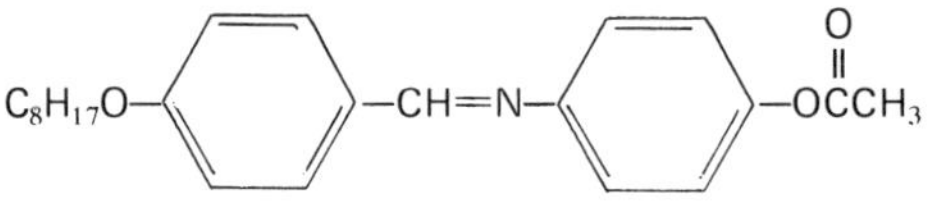

This mixture has a mesomorphic state from 39° to 104°C. Examples of known nematic crystals are p-(azoxyanisole) and p-(anisalamino)phenylacetate. These compounds have mesophase ranges from 117° to 136°C and from 81° to 110°C, respectively.

In general, the compounds can be prepared, for example, by the condensation of a p-acyloxy-phenylamine with the appropriate p-alkoxybenzaldehydes. The reaction is carried out in a refluxing benzene solution with benzenesulfonic acid or acetic acid as a catalyst and facility for azeotropic removal of water. The compounds can be purified by recrystallization from hexane solution. The recrystallization is repeated until a constant mesophase isotropic transition temperature is reached, followed by one additional recrystallization during which the hexane solution is filtered through a micropore filter.

Example 1: A solution of 1.51 grams of p-aminophenylacetate, 1.78 grams of p-butoxybenzaldehyde and 0.1 gram of benzenesulfonic acid in 50 ml of benzene is refluxed for about 3 hours. A Dean-Stark trap is used to collect the water which is azeotropically removed. After refluxing, the solvent is removed by evaporation under reduced pressure in a rotating film evaporator. The residue is collected and recrystallized three times from hexane. The yield is about 2 grams (64%) of colorless plates of:

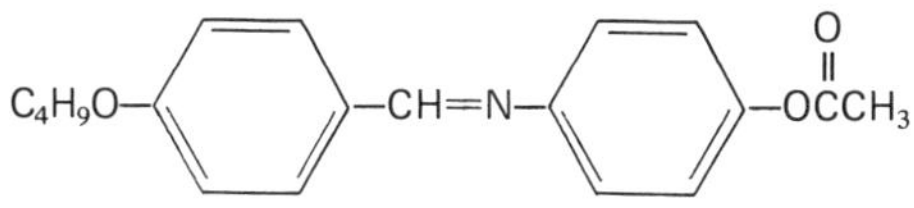

This product is then recrystallized once more, filtering the hot solution through a 0.2 micron filter prior to crystallization. Analysis of the product shows 73.02% carbon, 6.77% hydrogen and 4.78% nitrogen. This can be compared to the theoretical calculated amounts based on the formula $C_{19}H_{21}NO_3$ which are 73.29% carbon, 6.80% hydrogen, and 4.50% nitrogen.

Example 2: A solution of 1.23 grams of p-anisidine, 1.78 grams of p-formylphenylpropionate and 0.10 gram of benzenesulfonic acid in 50 ml of benzene is refluxed for about 4 hours. The same procedure as in Example 1 is used to isolate and purify the product. The theoretical calculated analysis for the product

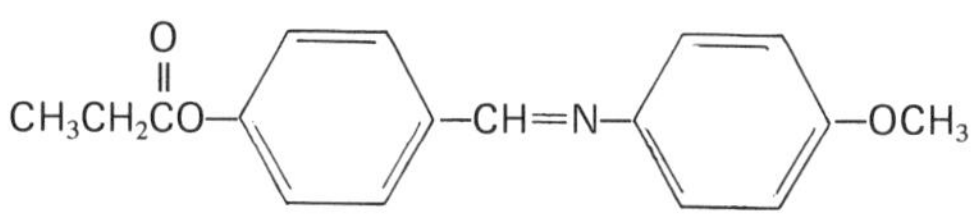

is 72.06% carbon, 6.05% hydrogen and 4.95% nitrogen. The amount found by experimentation was 72.71% carbon, 5.85% hydrogen, and 5.17% nitrogen.

Example 3: A solution of 5.4 grams of p-aminophenol, 8.9 grams of p-butoxybenzaldehyde and 0.1 gram of benzenesulfonic acid in 200 ml of benzene is refluxed for 4 hours. The same procedure as in Example 1 is used to isolate and purify the resulting yellow plates. Recrystallization of these plates from benzene results in white crystals of p-butoxybenzylidene-p-aminophenol. A mixture of 2.8 grams of this material in 10 ml of a cold (-5° to 10°C) 5 N sodium hydroxide solution and 100 ml of benzene is then treated with 1.4 grams of propionic anhydride. This mixture is stirred for about 1 hour. The mixture is filtered with suction and its resulting two-phase filtrate is transferred to a separatory funnel.

The lower aqueous phase is removed and the benzene layer is washed twice with water. The benzene solution is then dried over anhydrous sodium sulfate. The benzene is then evaporated and the residue is recrystallized from hexane to yield colorless crystals of p-butoxybenzylidene-p-aminophenylpropionate:

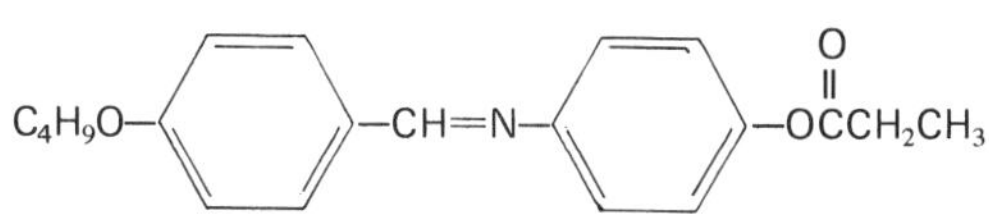

Example 4: A solution of 54.5 grams of p-aminophenol, 68.0 grams of anisaldehyde and 0.1 gram of benzenesulfonic acid in 200 ml of benzene is refluxed for 4 hours at which time about 9.0 ml of water is collected in a Dean-Stark trap. The product is collected as in Example 1 and recrystallized from a 50:50 ethanol:benzene solution to give colorless crystals. A mixture of 2.1 grams of these crystals, 50 ml pyridine and 1.9 grams of valeric

anhydride is stirred at room temperature overnight. The solvent is then evaporated and the residue triturated with hexane to yield colorless crystals of p-methoxybenzylidene-p-aminophenylpentanoate:

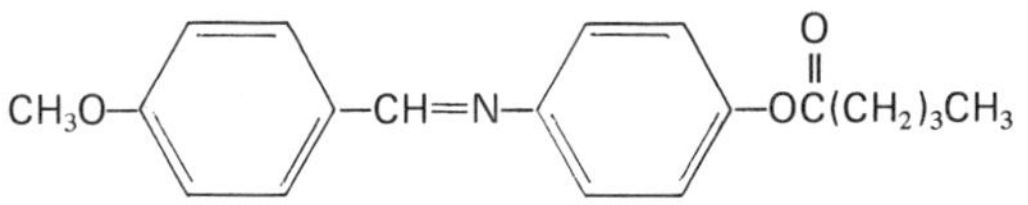

Olefinic-Substituted Benzylidene Structures

E.L. Steiger and H.J. Dietrich; U.S. Patent 3,742,053; June 26, 1973; assigned to Owens-Illinois, Inc. describe the preparation of mesomorphic compounds of the structure:

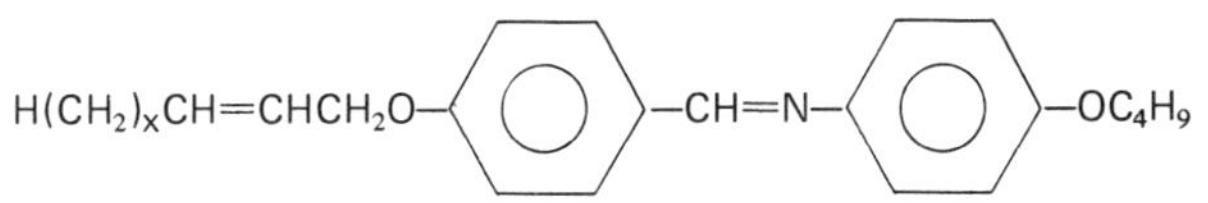

whre x is an integer of 0 to 10. The compounds are prepared by the reaction of para-n-butoxyaniline and para-n-alkenyloxybenzaldehyde. The following equations are representative of the reaction of para-n-butoxyaniline and para-n-alkenyloxybenzaldehyde:

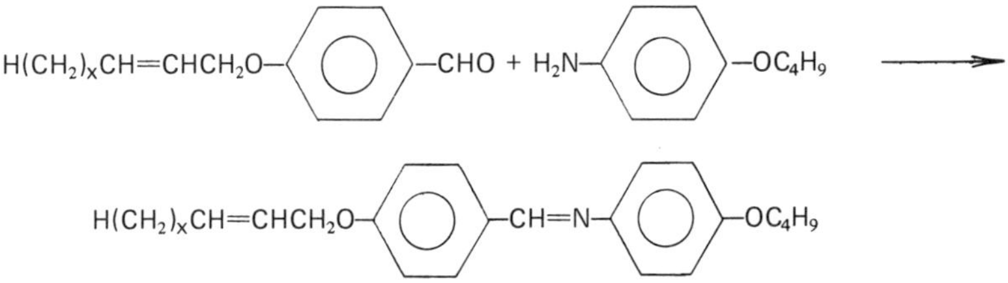

The appropriate alkenyl halides are commercially available. Also the para-n-butoxyaniline is commercially available. Both may be typically used without further purification.

Schiff bases are prepared by refluxing equimolar quantities of the p-substituted benzaldehyde and aniline in anhydrous ethanol for 4 to 6 hours. The solvent and water are removed and the residue recrystallized several times from ethanol until the transition temperatures remain constant. The crude yields ranged from 70 to 90%. The infrared spectra show a strong band at 1,629 cm^{-1} corresponding to the carbon-nitrogen double bond in Schiff base compounds. Other absorptions are compatible with the expected structures.

Transition temperatures are determined on a Leitz Ortholux polarizing microscope using a Mettler FP-2 heating stage. N-(p-allyloxybenzylidene)-p'-n-butoxyaniline (where x is 0 in the basic structure) was prepared. The structure was analyzed at 77.37% by weight carbon, 7.50% by weight hydrogen, and 4.64% by weight nitrogen compared with calculated theoretical analysis values of 77.64% by weight carbon, 7.49% by weight hydrogen, and 4.53% by weight nitrogen. The compound changed from the crystalline to the nematic phase at about 114.0°C and from nematic to isotropic at about 119.6°C. The liquid crystal compounds may be utilized in display devices, especially of the electronic type.

In related work *H.J. Dietrich and E.L. Steiger; U.S. Patent 3,743,681; July 3, 1973; assigned to Owens-Illinois, Inc.* describe the preparation of liquid crystal compounds of the structure:

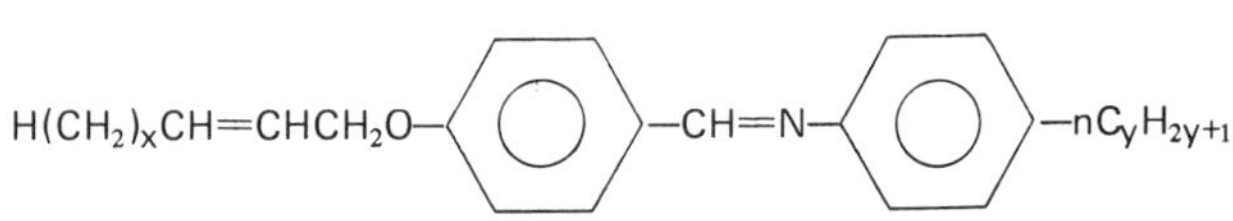

where x is an integer of 0 to 10 and y is the same or different integer of 1 to 10. The compounds are prepared by the reaction of para-n-alkylaniline and para-n-alkenyloxybenzaldehyde.

N-(p-n-Alkanoyloxybenzylidene)-p'-Aminophenyl Acetate

E.L. Steiger and H.J. Dietrich; U.S. Patent 3,799,971; March 26, 1974; assigned to Owens-Illinois, Inc. describe the preparation of mesomorphic compounds of the structure:

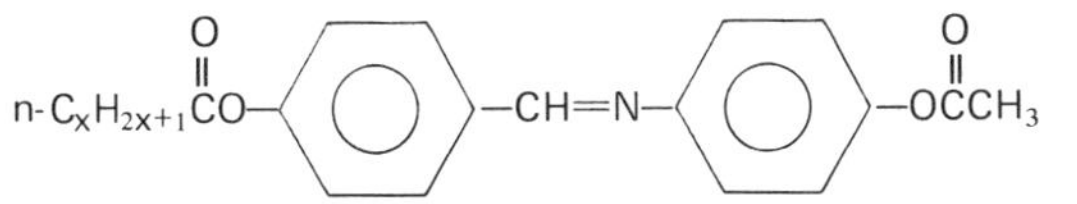

where x is an integer of 1 to 10. The compounds are prepared by the reaction of para-acetoxyaniline and para-n-alkanoyloxybenzaldehyde. The following equation is representative of the reaction of para-acetoxyaniline and para-n-alkanoyloxybenzaldehyde:

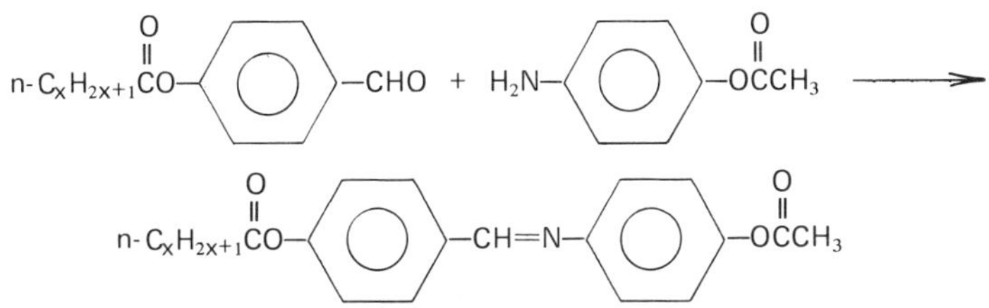

The para-acetoxyaniline, also known as para-aminophenylacetate, is commercially available. It may be typically used without further purification. Schiff bases are prepared by refluxing equimolar quantities of the p-substituted benzaldehyde and aniline in anhydrous ethanol for 4 to 6 hours. The solvent and water are removed and the residue recrystallized several times from ethanol until the transition temperatures remain constant. The crude yields ranged from 70 to 90%. The infrared spectra show a strong band at 1,629 cm^{-1} corresponding to the carbon-nitrogen double bond in Schiff base compounds. Other absorptions are compatible with the expected structures. Transition temperatures are determined on a Leitz Ortholux polarizing microscope using a Mettler FP-2 heating gauge.

N-(p-n-butyryloxybenzylidene)-p-aminophenylacetate (where x is 3 in the basic structure) was prepared in accordance with this process. The structure was analyzed at 70.3% by weight carbon, 5.81% by weight hydrogen, and 4.66% by weight nitrogen compared with calculated theoretical analysis values of 70.78% by weight carbon, 6.24% by weight hydrogen, and 4.13% by weight nitrogen. The compound changed from the crystalline to the smectic phase at about 92.3°C and from smectic to isotropic at about 135.1°C. The liquid crystal compounds may be utilized in display devices, especially of the electronic type.

AROMATIC ESTERS

Aryl Ester Mixtures

Y. Katagira and Y. Miyata; U.S. Patent 3,770,653; November 6, 1973 and Y. Katagira, Y. Miyata and O. Nagasaki; U.S. Patent 3,770,654; November 6, 1973; both assigned to KK Daini Seikosha and KK Hattori Tokeiten, Japan describe nematic liquid crystal compositions having a liquid crystal range including ambient temperature. The compositions include at least one compound having the general formula

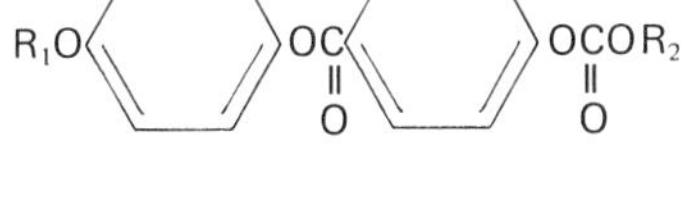

where R_1 and R_2 are alkyl groups such as $H(CH_2)_n$, and at least one compound having the general chemical formula

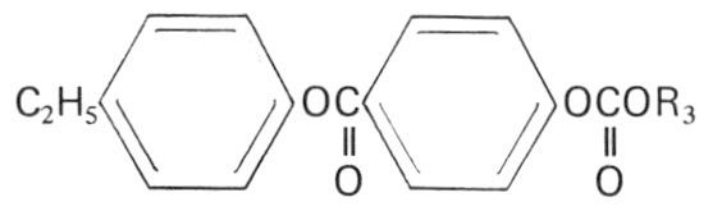

where R_3 is alkyl group such as $H(CH_2)_n$. The two types of compounds are strongly soluble in each other, making it possible to prepare compositions having the desired mesomorphic temperature range. Moreover the compounds are almost colorless and are transparent, making them highly suitable for use in digital devices. The way in which compounds are prepared is exemplified by synthesis of

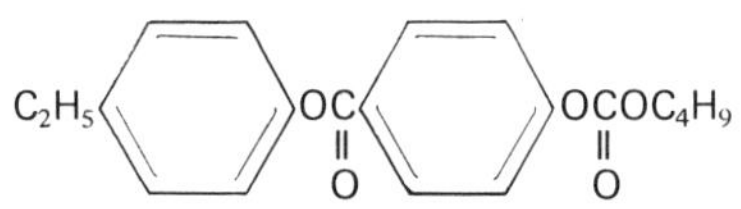

which is prepared as follows. p-Hydroxybenzoic acid (1 mol) was dissolved in 1 liter of 2 M aqueous sodium hydroxide. To this mixture was added normal butyl chloroformate in slight excess under vigorous agitation at room temperature. The mixture was then acidified with acetic acid. Benzoic acid p-n-butyl carbonate was separated off, washed and dried. Thionyl chloride (1.2 mols) was added to the solid. After reaction the excess thionyl chloride was eliminated. After cooling, p-ethylphenol (1 mol) and pyridine (1 mol) were added. The mixture was poured into 1 liter of water and the solid ester was recrystallized from ethyl alcohol. The following examples illustrate the process.

Example 1:

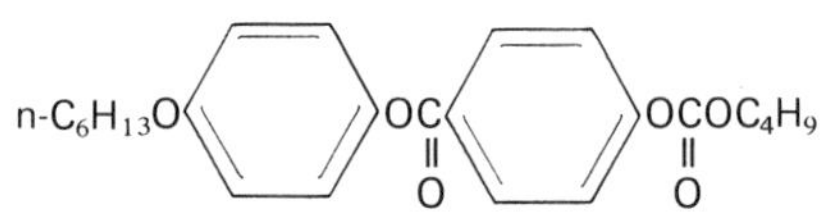

having a measured nematic mesomorphic temperature range of 38° to 75°C, is combined with

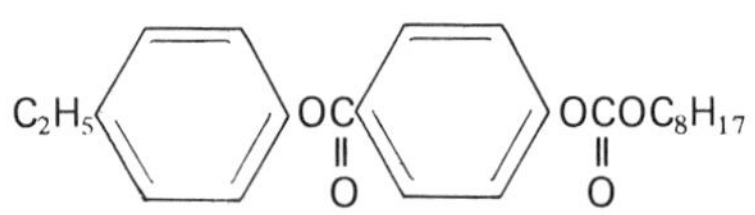

having a measured nematic mesomorphic temperature range of 34° to 40°C, at a mol ratio of 1:1. The product was a transparent nematic liquid crystal stable at 9° to 52°C. The product was displaced between Nesa glass slides and subjected to a direct current electric field of 10^4 v/cm. Strong light scattering was observed.

Example 2:

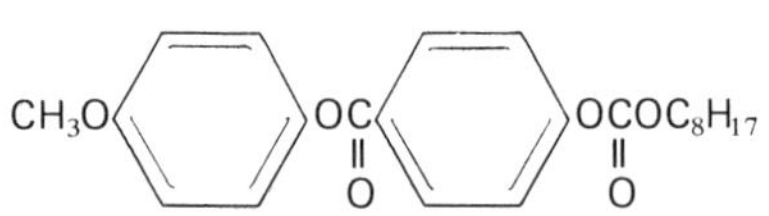

having a mesomorphic range of 37° to 65°C, is combined with

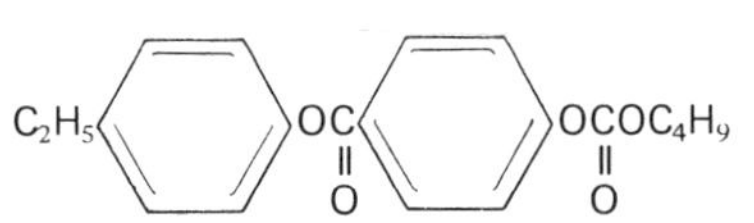

having a mesomorphic temperature range of 39° to 42°C at a mol ratio of 2:1. The nematic liquid crystal product was stable at 12° to 56°C. When tested as described in Example 1 strong light scattering was seen.

Example 3:

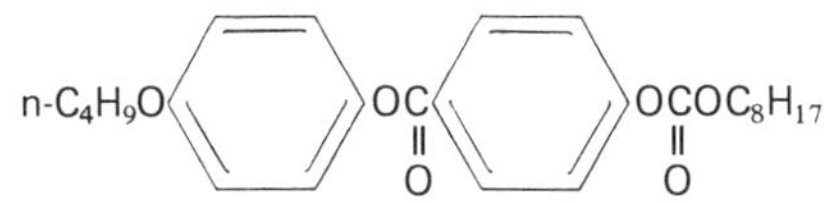

having a mesomorphic temperature range of 35° to 59°C, is combined with

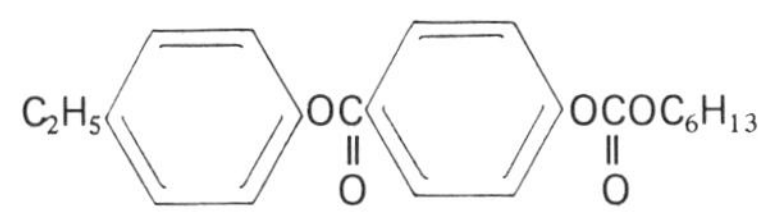

having a mesomorphic temperature range of 48° to 50°C at a mol ratio of 1:1. The nematic liquid crystal product was stable at 13° to 54°C. Strong light scattering again was observed when the product was tested as in Example 1.

In related work *Y. Katagira and Y. Miyata; U.S. Patent 3,790,498; February 5, 1974; assigned to KK Daini Seikosha and KK Seikosha, Japan* describe memory liquid crystal compositions containing from 80 to 95% by weight of one or more nematic crystal liquid compounds of the formula

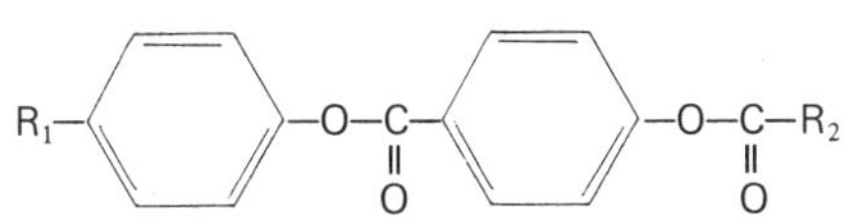

where R_1 and R_2 are alkyl and alkoxy and may be the same or different, and 5 to 20% by weight of a carboxylic acid ester of cholesterol, sitosterol or stigmasterol.

Positive Dielectric Anisotropy

D.C. Green and W.R. Young; U.S. Patent 3,836,478; September 17, 1974; assigned to International Business Machines Corporation describe nematic liquid crystal materials which are operatively nematic in a relatively wide range of temperatures which encompasses room, or close to room temperatures.

In considering the uses of liquid crystals, in one of their predominant types of display applications, i.e., the dynamic scattering cell, nematic liquids which have negative dielectric anisotropies are required. To this end, hydrolytically unstable Schiff bases, various azoxybenzene derivatives, and trans-chlorostilbene derivatives, all with room temperature nematic phases, have been employed.

In another and also relatively general type of display device employing nematic materials, operation is dependent upon the electric field alignment of the nematic liquids, i.e., field effect application. In this type of application, operation is not dependent upon current flow (and associated electrochemical reactions), as is the case in dynamic scattering type devices. Consequently, there is presented the possible advantage of extended operating lifetimes for the field effect devices.

These field effect devices are operable with nematic liquids which have a positive dielectric anisotropy. In this connection, three classes of room temperature operative classes of nematic liquids with positive dielectric anisotropy are known, viz, 4,4'-dialkylazoxybenzenes which are of yellow color and have a nematic range at the widest of 22° to 65°C for single material and a highest NI (nematic-isotropic) clearing point of 72°C, carbonate esters which are colorless and whose best nematic formulation (tertiary mixture) has a nematic range of 24° to 76°C, and diphenylacetylenes which are colorless, have a highest NI clearing point of 76°C and whose best nematic formulation (binary mixture) has a nematic range of 13° to 64°C.

While these known nematic liquids which have a positive dielectric anisotropy are operative in liquid crystal field effect electrooptical devices, they suffer from the deficiencies in that their nematic operating range is not sufficiently wide, in that some are not colorless, and in that some are not stable in the presence of visible and ultraviolet light, air, water and electric fields. This process provides nematic liquid compounds and mixtures of these compounds where the compounds have the following general structural formula

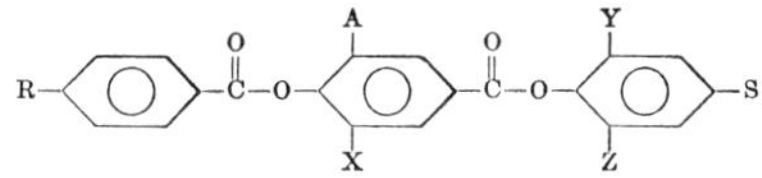

where R is a radical selected from the group consisting of alkyl radicals containing from 1 to 12 carbons, alkoxy radicals containing from 1 to 8 carbons, alkanoyloxy radicals containing from 1 to 8 carbons, and alkoxy carbonyloxy radicals containing from 1 to 6 carbons, S is a radical selected from the group consisting of alkyl radicals containing from 1 to 12 carbons and alkoxy radicals containing from 1 to 8 carbons, and A, X, Y and Z are radicals selected from the group consisting of hydrogen and methyl radicals with at least one of the radicals A, X, Y and Z in a compound being the methyl radical.

Also, there is provided a method for preparing the nematic liquid compounds having the above-set forth general structural formula. This method comprises the steps diagrammatically set forth in the following synthesis

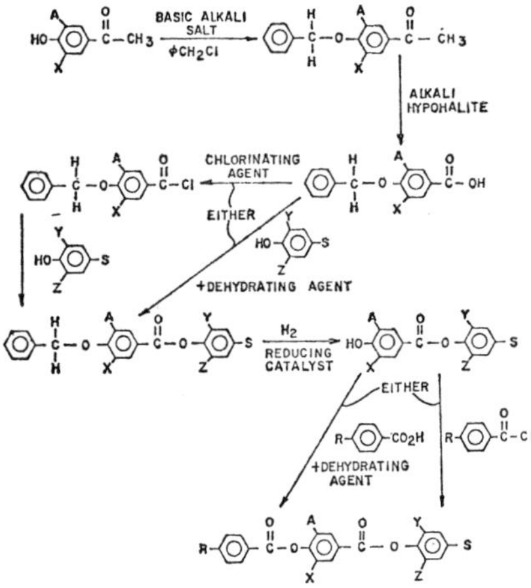

where R, S, A, X, Y and Z have their previous significance.

In one group of esters, the radical R is selected to be an alkoxy radical such as ethoxy, 2-methylpentoxy and hexoxy, an alkanoyloxy radical such as hexanoyloxy and an alkoxycarbonyloxy radical such as butoxycarbonyloxy and an alkyl radical such as butyl and pentyl, and the radical S is selected to be an alkyl radical such as butyl and pentyl, and an alkoxy radical such as ethoxy. The radicals R and S are so selected in analogy with the low melting points and high nematic CMD (clearing point/melting point difference) for

Schiff bases, acetylenes and chlorostilbenes. In the following table (Table 1), there are set forth the physical properties of mesomorphic (intermediate between crystalline and isotropic) derivatives of nematic esters according to the process where R is H_5C_2O and S is C_4H_9. The esters are numbered 1 through 8.

TABLE 1

Ester No.	A	X	Y	Z	Transition	Temp. (° C.)	ΔH (kcal./mole)	ΔS (e.u.)
1	H	H	CH_3	H	C→N	113	8.36	21.7
					N→I	138	0.25	0.55
2	CH_3	H	H	H	C→N	91	7.32	20.1
					N→I	179	0.39	0.87
3	H	H	CH_3	CH_3	C→N	143	8.82	21.2
					N→I	145	0.29	0.69
4	CH_3	H	CH_3	H	C→I	160	10.97	25.3
					N→I*	143	0.36	0.88
5	CH_3	CH_3	H	H	C→N	122	9.83	24.9
					N→I	129	0.51	1.26
6	CH_3	H	CH_3	CH_3	C→I	195	-------	-------
					N→I*	101	-------	-------
7	CH_3	CH_3	CH_3	H	C→I	138	10.41	25.9
					N→I*	103	0.50	1.32
8	CH_3	CH_3	CH_3	CH_3	C→I	137	-------	-------
					N→I*	79	-------	-------

In the above table, the * denotes a monotropic transition. In Table 2 set forth below, there are set forth physical properties of a mesomorphic derivative of the above esters where R is H_9C_4 and S is OC_2H_5. The ester is numbered 9.

TABLE 2

Ester No.	A	X	Y	Z	Transition	Temp. (° C.)	ΔH (kcal./mole)	ΔS (e.u.)
9	CH_3	H	H	H	C→N	96	6.28	17.0
					N→I	176	0.40	0.90

In the above tables, C→N denotes crystalline to nematic and N→I signifies nematic to isotropic. The esters 1 through 9 in Tables 1 and 2 exhibit intense vibrational absorption in the infrared region at 1,735 cm^{-1} (carbonyl stretching) and at 1,505 cm^{-1} and 1,606 cm^{-1} (phenyl ring vibrations). In addition, at least five intense vibrational bands occur in the C—O stretching region (1,000 to 1,300 cm^{-1}) for each ester. In the NMR spectra taken in deuteriochloroform, sharp singlets for the absorption of the methyl group on the central aromatic ring (A and/or X) appear at 2.3 ppm downfield from TMS, while the absorption for methyl groups on the terminal ring appears as a sharp singlet at about 2.2 ppm.

Illustrations of where there are no methyl groups in positions A, X, Y, Z, which thereby increase the melting point relative to the monomethyl derivatives 1, 2 and 9 in Tables 1 and 2, are set forth in Table 3.

TABLE 3

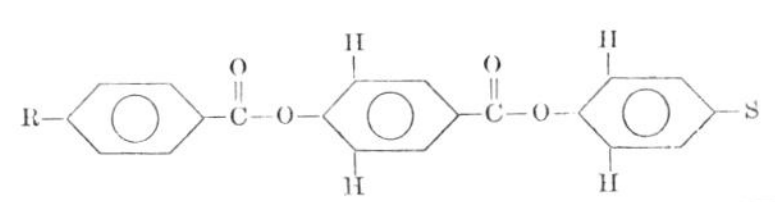

R	S	Transition	Temp. (° C.)	ΔH	ΔS
C_4H_9—	C_2H_5O—	C→N	138	7.63	18.6
		N→I	225	0.30	0.61
C_2H_5O—	C_4H_9—	C→N	124	9.15	23.0
		N→I	231	0.39	0.77

The esters are colorless materials and, in comparison with the nematic Schiff bases, azobenzenes, azoxybenzenes, nitrones and stilbenes, are highly transparent in the ultraviolet region. For example, in ethanol solution, ester No. 6 exhibits a λ_{max} at 262 nm (ϵ extinction coefficient = 34,000). Ester No. 6 does not absorb above 340 nm. The column headed Temperature in Tables 1, 2 and 3 indicates the melting points and clearing points for the respective esters.

With regard to the nematic-isotropic (N→I) transition temperatures, it is to be noted in Table 1 that the nematic-isotropic transition temperatures progressively decrease as the number of aromatic methyl group appendages increase. A possible explanation for such effects is that they result from increased molecular broadening with increased methyl substitution such that intermolecular interactions that are responsible for mesomorphism are decreased as the intermolecular separation increases.

With regard to the positional effects of methyl substitution, it is believed that the nematic-isotropic transition temperature, T_{NI}, is lowered by about the same amount when a single methyl group is placed in either of positions A or Y. The placing of two methyl groups on the same ring results in approximately twice the lowering of T_{NI} as compared to the lowering obtained by the addition of a single methyl group. However, when two methyl groups are affixed to different phenyl rings respectively, the lowering of the value of T_{NI} is somewhat less. When the molecule contains three or four methyl groups, the transition temperature drops more slowly with each additional methyl group which is added. This effect probably results from the fact that, with the molecule already broadened by two or more methyl groups, the addition of still more methyl groups has a progressively decreasing effect.

The two rightmost columns in Tables 1, 2 and 3 contain calorimetric data and pertain to entropies and enthalpies of the nematic-isotropic transition. In the examining of these columns, two facts are readily apparent. (1) Substitution on the central ring by one, and even more so by two methyl groups results in a significant increase in the entropy of the nematic-isotropic phase transition. Central ring substitution affects the heat of transition in the same direction, but a smaller extent. (2) Substitution on the end ring has no comparable effect.

These data are pertinent to the reduction in the transition temperature upon methyl substitution on the central ring. The transition temperature is determined by the heat and the entropy of the transition, the latter being two more fundamental quantities that are representative of the change in the interaction energy and in the order when the transition takes place. This transition is represented by the following equation: $T_{NI} = \Delta H_{NI}/\Delta S_{NI}$.

The decrease in the value of T_{NI} upon methyl substitution may be ascribed to a reduction in the energy of intermolecular interactions in the nematic phase, i.e., a lower value for ΔH_{NI}. This result is attributed to either increased intermolecular separation due to lateral substitution or to decreased molecular polarizability resulting from a steric loss of conjugation. While in some instances, the experimentally measured ΔH_{NI} is smaller for methyl-substituted derivatives than for the parent compound, in accord with previous postulates, this trend cannot be universally true. Methyl substitution in the A and X positions of the esters increases ΔH_{NI}. Consequently, the decrease in transition temperature is due to an increase in ΔS_{NI}.

To prepare the esters 1 through 9, set forth in Tables 1 and 2, there are utilized the 4-substituted benzoic acids, 4-ethoxybenzoic acid, which is known, and which is made by the Aldrich Chemical Company, and 4-n-butylbenzoic acid. The latter acid is prepared by the hydrolysis of 4-n-butylbenzonitrile which is prepared from 4-n-butylaniline available from the Aldrich Chemical Company by the method as described in the paper of H.T. Clarke and R.R. Read, *Org. Syn.*, Coll. Vol. I, 514 (1941). These acids are refluxed in thionyl chloride to provide 4-ethoxybenzoyl and 4-n-butylbenzoyl chlorides.

Also utilized in the synthesis of the esters 1 through 9 in Tables 1 and 2, are the 4-hydroxybenzoic acids, viz., 4-hydroxybenzoic acid which is known in the art, 4-hydroxy-3-methylbenzoic acid and 4-hydroxy-3,5-dimethylbenzoic acid which are prepared from 4-amino-3-methylbenzoic acid and 4-amino-3,5-dimethylbenzoic acid by diazotization and sulfuric acid hydrolysis, the 4-amino- acids being available from the abovementioned Aldrich Chemical Company.

Utilizing the abovementioned 4-substituted-benzoyl chloride and 4-hydroxybenzoic acids produces the 4-(4'-substituted-benzoyloxy)-benzoic acids. These latter acids are prepared by the reaction of the appropriate acid chloride with the appropriate hydroxybenzoic acid under reflux in toluene/pyridine solvent. After removal of the solvents on a rotary evaporator, the residue is dissolved in ether and washed with dilute HCl and water. Evaporation of the ether yields an oily or gummy residue which generally contains the desired product plus some anhydride by-product. The residue is allowed to reflux in aqueous dioxane for a sufficient period (about 4 hours) to hydrolyze any anhydride which is present. The dioxane is removed by azeotropic distillation and the product is isolated by filtration of the hot, aqueous mixture. Recrystallization from a suitable solvent produces the desired white, crystalline products in a 50 to 80% yield.

The 4-(4'-substituted-benzoyloxy)benzoic acids are permitted to reflux in an excess of thionyl chloride and the volatile materials are removed therefrom by vacuum distillation to produce 4-(4'-substituted-benzoyloxy)benzoyl chlorides.

In the synthesis, there are prepared 4-substituted-phenols. Examples of the latter are 4-n-butylphenol which is prepared from 4-n-butylaniline by diazotization and sulfuric acid hydrolysis, 4-hydroxy-3-methylbutyrophenone and 4-hydroxy-3,5-dimethylbutyrophenone which are prepared by Friedel-Crafts acylation of 2-cresol and 2,6-dimethylphenol, respectively, with butyryl chloride in nitrobenzene with aluminum chloride. The phenols 4-n-butyl-2-methylphenol and 4-n-butyl-2,6-dimethylphenol are prepared from the corresponding butyrophenones set forth above by the Huang-Minlon modification of the Wolf-Kishner reduction as set forth in the paper of Huang-Minlon, *J. Am. Chem. Soc.,* 68, 2487 (1946). The latter phenols, viz 4-n-butyl-2-methylphenol and 4-n-butyl-2,6-dimethylphenol are purified by distillation.

The esters set forth in Tables 1 and 2 are prepared by first dissolving an appropriate 4-substituted-phenol in pyridine. An equivalent amount of a 4-(4'-substituted-benzoyloxy)-benzoyl chloride is dissolved in toluene and added to the pyridine solution. After a substantial reflux period, such as an overnight period, the solvents are removed and the residue is taken up in ether. Following washes with water, 10% HCl, 10% NaOH and water, the ether layer is dried and evaporated. The residual product is chromatographed through a column of silica gel with chloroform. The product is then repeatedly crystallized from alcohol or methylcyclohexane and, in some cases, sublimed under high vacuum. The yields of the purified, white products range from 20 to 50%.

There are set forth below, esters included in the general structural formula of which many exhibit melting points below 70°C and whose mixtures are characterized by nematic phases in temperature ranges from room temperature to temperatures in excess of 100°C. In these esters, R is a radical selected from the group consisting of alkyl radicals containing from 1 to 12 carbons, alkanoyloxy radicals containing from 1 to 8 carbons, and alkoxycarbonyloxy radicals containing from 1 to 6 carbons, S is an alkyl radical containing from 1 to 12 carbons, A is methyl and X, Y and Z are all hydrogen. In Table 4, there are set forth examples of these latter nematic esters and their nematic ranges in °C. In this table, R is methyl to octyl and S is butyl to nonyl.

TABLE 4

	S					
R	BU	PE	HX	HP	OC	NO
ME	89–142	82–144	88–127	91–127	87–117	73–117
ET	86–129	92–132	68–117	68–116	71–106	62–107
PR	76–138	72–141	75–127	65–125	60–116	65–115
BU	72–124	54–128	79–116	52–115	62–106	53–107
PE	58–128	70–132	64–120	43–121	64–112	52–113
HX	50–117	54–122	66–110	53–112	61–105	51–105
HP	67–116	48–119	62–110	56–112	70–105	56–107
OC	68–108	58–113	66–105	59–106	70–102	64–102

Other examples of the latter esters are set forth in Table 5 below.

TABLE 5

R	S	Nematic range (° C.)
2-methylpentoxy	C_4H_9—	54-89
$C_6H_{13}O$—	C_5H_{11}—	70-152
$C_5H_{11}CO_2$—	C_5H_{11}—	61-176
$C_4H_9OCO_2$—	C_5H_{11}	77-161

A 1:1:1 mixture of the three esters where R is phenyl, heptyl and 2-methylpentoxy respectively, and S is butyl, A is methyl and X, Y, Z are hydrogen in all of these esters has a melting point of 27.5°C and a nematic clearing point of 111°C thereby providing a useful nematic range of 83°C. A 1:1 mixture of the following two esters exhibits a nematic range of 35° to 119°C.

R	S	A	X, Y, Z
Hexyl	Butyl	Methyl	Hydrogen.
Pentyl	Heptyl	do	Do.

A 1:1:1 mixture of the following three esters is characterized by a nematic range of 20° to 117°C.

R	S	A	X, Y, Z
Hexyl	Butyl	Methyl	Hydrogen.
Pentyl	Heptyl	do	Do.
Butyl	do	do	Do.

Alkyl-Stilbene Compounds

E.L. Steiger and H.J. Dietrich; U.S. Patent 3,742,067; June 26, 1973; assigned to Owens-Illinois, Inc. describe the preparation of p-methoxy-alkylene-oxy-p'-n-alkyl-stilbene liquid crystal compounds of the structure:

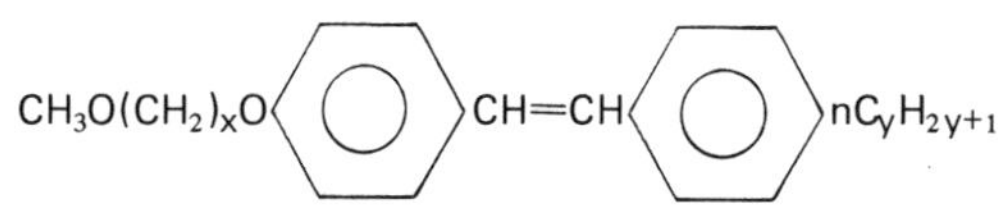

where x and y are the same or different integers of 1 to 10, typically 1 to 5. The liquid crystal compounds are prepared by reacting para-methoxy-alkyleneoxybenzaldehyde and para-alkylbenzyl-magnesium halide to form an intermediate product, and then heating the intermediate product so as to dehydrate it to the final p-methoxy-alkylene-oxy-p'-n-alkyl-stilbene product. The reaction is as follows:

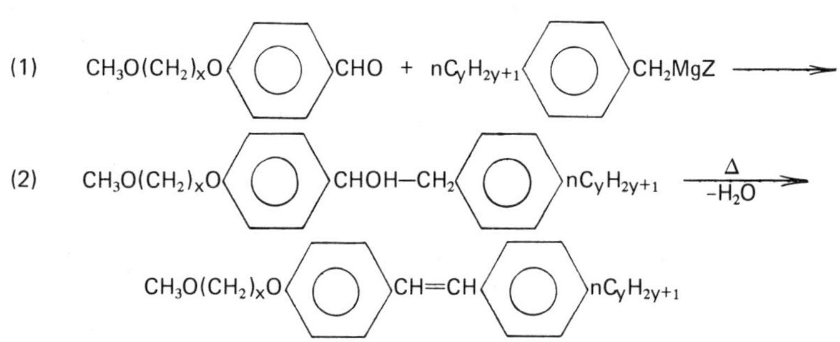

In the process x and y are as defined above. The halogen atom Z is typically selected from Br, I, or Cl. The dehydration reaction (2) may be in the absence or presence of dehydrating agents. Typical dehydrating agents include potassium acid sulfate, thionyl chloride, and hydrogen sulfate. The resulting substituted stilbene liquid crystal compound may be a mixture of cis and trans isomers. If desirable, these may be appropriately separated by chromatography or other suitable means.

MULTICOMPONENT MIXTURES

Low Voltage Crystal Display Mixture

P.Y. Hsieh; U.S. Patent 3,815,972; June 11, 1974; assigned to Ing. C. Olivetti & Co., SpA, Italy describes an improved orientation type liquid crystal display having a layer of a mixture of liquid crystal materials confined between a pair of substrates and electrodes for applying an electric field across the substrate. The mixture of liquid crystal materials includes between 5 and 60% by weight of an orientation-type liquid crystal material and between 95 and 45% by weight of a dynamic scattering type liquid crystal material. The following examples, all of which respond to electric fields in the orientation mode in their nematic range, are given to illustrate the process.

Example 1: A low voltage liquid crystal display was fabricated using a one-quarter to 1 mil thick layer of an orientation-type liquid crystal material mixture having the following composition:

Components	Parts by Weight
p-(p-ethoxybenzylidene)aminobenzonitrile	1
p-(p-n-butoxybenzylidene)aminobenzonitrile	1
4-methoxy-4'-n-butylbenzylideneaniline	3.5
4-ethoxy-4'-n-butylbenzylideneaniline	3.5

The mixture has nematic to liquid transition temperature of 59°C and is a fluid nematic at room temperature and gells upon cooling below 0°C. The mixture continues to respond to electric fields even in the gelled state although at a slower speed due to the increase in viscosity. This device operates with polarizing layers and requires switching voltage of between 6 and 8 volts.

Example 2: A low voltage liquid crystal display similar to that described in Example 1 was prepared using a mixture having the following composition:

Components	Parts by Weight
p-(p-n-butoxybenzylidene)aminobenzonitrile	1
p-(p-n-hexylbenzylidene)aminobenzonitrile	1
p-(p-octanoyloxybenzylidene)aminobenzonitrile	1
4-methoxy-4'-n-butylbenzylideneaniline	1.5
4-ethoxy-4'-n-butylbenzylideneaniline	1.5

The mixture has a nematic range of between 16° and 83°C and responds well to electric fields in the orientation mode in this temperature range.

Example 3: A low voltage liquid crystal display similar to that described in Example 1 was prepared using a mixture having the following composition:

Components	Parts by Weight
p-(p-ethoxybenzylidene)aminobenzonitrile	1
p-(p-n-butoxybenzylidene)aminobenzonitrile	1
n-(o-hydroxy-p-methoxybenzylidene)-p-butylaniline	4.16
n-(o-hydroxy-p-n-butyloxybenzylidene)-p-butylaniline	3.84

The mixture is a free-flowing fluid nematic at room temperature. It has a nematic liquid transition temperature of 72°C and gelled instead of crystallized upon cooling to 0°C. It responds to electric fields even at a gelled stage although at a slower speed due to the increase in viscosity.

Example 4: A low voltage liquid crystal display similar to that described in Example 1 was prepared using a mixture having the following composition:

Components	Parts by Weight
p-(p-n-hexylbenzylidene)aminobenzonitrile	1
4-butyl-4'-methoxyazoxybenzene	1

The mixture has a nematic range of between 5° and 85°C.

For Transparency Control

M. Tsukamoto and T. Ohtsuka; U.S. Patent 3,781,088; December 25, 1973; assigned to Matsushita Electric Industrial Co., Ltd., Japan describe a practically useful method for changing the transparency of a liquid crystal cell which can increase the decay time of the liquid crystal (i.e., maintain the light-transparent state for a longer time) so as to eliminate flickers. The decay time of a liquid crystal mixture is defined as a time period required for the liquid crystal to return from the light-transparent state to the initial opaque state (i.e., light-scattering state) from the time when the exciting electric field is removed.

This process provides a method for changing the transparency of a liquid crystal cell, comprising, in combination the steps of providing a cell including a liquid crystal mixture of the type which is normally opaque comprising: (1) a nematic liquid crystal having a positive dielectric anisotropy and (2) a steroidal compound, the cell having two electrodes attached to the respective opposite surfaces thereof; exciting the cell by applying a voltage higher than a threshold value above which the cell is transferred into a light-transparent state; and applying, across the cell, an AC bias voltage lower than the threshold value in order to maintain the light-transparent state for a longer time.

Figure 3.4a is a schematic drawing of the essential structure of a liquid crystal cell and the arrangement for applying voltages. Figure 3.4b is a graph showing the characteristic of the intensity of transmitted light versus peak value of the applied AC voltage for the liquid crystal cell. Figure 3.4c is a graph showing the characteristic of the intensity of transmitted light versus time upon application of an exciting voltage for the liquid crystal cell. Figure 3.4d is a graph showing the characteristic of the intensity of transmitted light versus time upon application of an exciting voltage during application of an AC bias voltage for the liquid crystal cell. Figure 3.4e is a schematic drawing showing a method for applying an AC bias voltage and an exciting voltage to the liquid crystal cell.

To illustrate the process simply, only one liquid crystal cell, and a driving circuit connected to the cell is schematically shown in Figure 3.4a. Referring to Figure 3.4a, the cell consists of a first plate **1** including a first electrode **2**, second plate **3** including a second electrode **4**, two spacers **5** for separating the plates to form a space between the first and the second electrodes and a liquid crystal mixture **6** used in this process, filling the space. Both the first plate and the second plate are light-transparent. The first electrode and the second electrode are opposed to each other so that they contact the liquid crystal mixture directly.

An AC bias voltage of rectangular wave-form generated by AC voltage generator **7** is applied to the liquid crystal mixture. The voltage of the AC bias voltage is lower than the threshold value which will be described in detail with reference to Figure 3.4b. A DC voltage generated by a DC voltage source **9** is added to the AC bias voltage when a switch **8** is connected to the DC voltage source. The exciting voltage which is the sum of the AC bias voltage and the DC exciting voltage must be large enough to bring the liquid crystal cell to an excited light-transparent state which will be described with reference to Figure 3.4b. An observer can see the cell modulating an incident light from a light source when he operates the switch **8**.

FIGURE 3.4: LIQUID CRYSTAL CELL WITH INCREASED DECAY TIME

a.

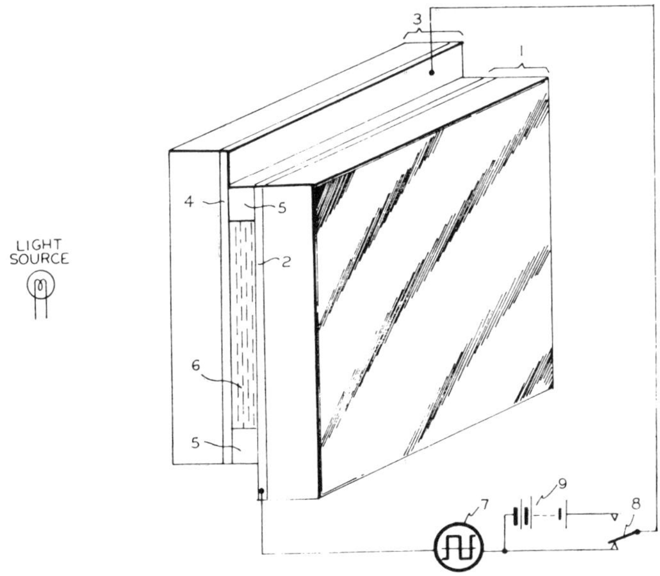

b.

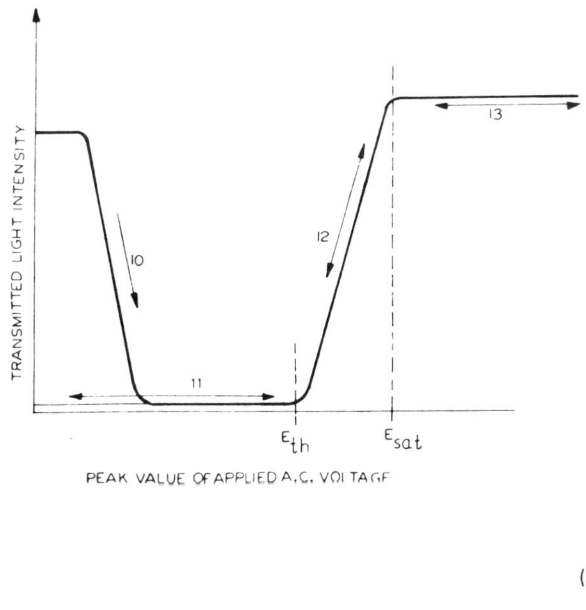

(continued)

FIGURE 3.4: (continued)

c.

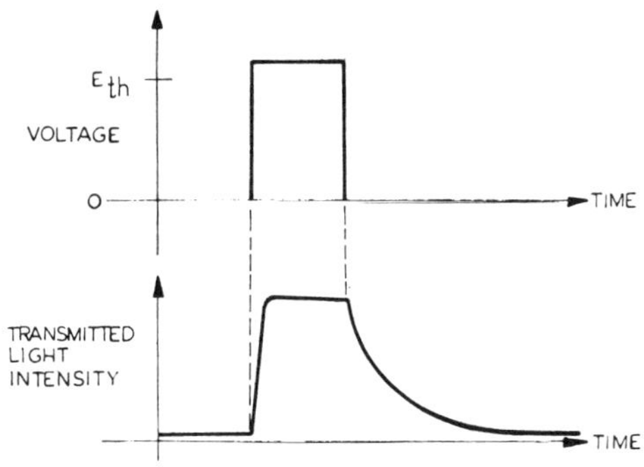

d.

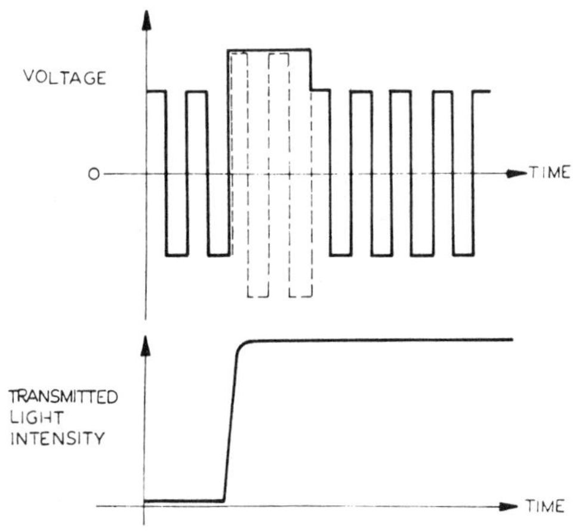

e.

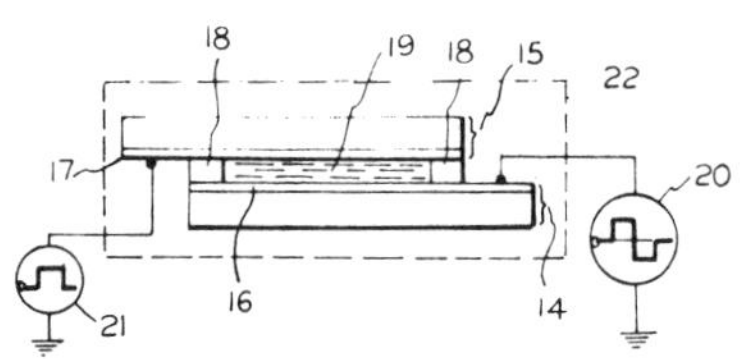

Source: M. Tsukamoto and T. Ohtsuka; U.S. Patent 3,781,088; December 25, 1973

Figure 3.4b shows the relationship between the intensity of light transmitted through the cell shown in Figure 3.4a and the peak value of the applied AC voltage. Details of the ingredients of the liquid crystal mixtures in the cell which have the electrooptical properties illustrated with reference to Figure 3.4b will be described later. Referring to Figure 3.4b, the curve is obtained by measuring the intensity of light transmitted through the cell shown in Figure 3.4a when it has applied thereto an AC voltage of rectangular waveform. During the measurement, the switch **8** is connected to the AC voltage generator **7**.

The AC voltage is varied within a range from 0 to a voltage beyond the threshold value of the liquid crystal mixture **6**. The liquid crystal cell is relatively transparent before application of an AC voltage. As the applied AC voltage is increased, the incident light from the light source is scattered and so the transmitted light decreases along a curve designated by numeral **10** and reaches a minimum intensity of transmitted light, or an unexcited light-scattering state **11** (i.e., an opaque state).

Further increasing the AC voltage produces hardly any change until the voltage reaches a threshold value E_{th}. Then the intensity of transmitted light begins to increase along a curve designated by numeral **12** after the peak value of the AC voltage exceeds the threshold value E_{th}, and, after that, continues to increase up to a saturation level **13** which occurs at a saturation voltage E_{sat}. The saturation state is called the excited light-transparent state. When the AC voltage decreases from a point above the saturation voltage E_{sat}, the intensity of transmitted light decreases again along the curves **12** and **11** and the light-scattering state as designated by the numeral **11** remains for several weeks even if the AC voltage is removed. If the AC voltage is increased again, the intensity of transmitted light follows the curves **11**, **12** and **13**. A curve similar to the curve shown in Figure 3.4b can be obtained by using DC voltage.

The threshold value is determined essentially by the amounts of ingredients of the liquid crystal mixture, the kinds of ingredients, and the frequency and waveform of the applied AC voltage. Therefore, in the following description, the threshold value E_{th} means the peak value of the AC voltage, above which a liquid crystal cell begins to be brought to the excited light-transparent state.

Figure 3.4c shows the characteristic of the intensity of transmitted light versus time for the liquid crystal cell upon application of an exciting voltage without any bias, while Figure 3.4d shows the characteristic of the intensity of transmitted light versus time for the liquid crystal cell upon application of an exciting voltage while the cell has applied thereto an AC bias voltage. Referring to Figure 3.4c, the voltage applied to the liquid crystal cell is high enough to bring the liquid crystal cell to the excited light-transparent state as illustrated in Figure 3.4b. The intensity of transmitted light reaches saturation or a maximum after a delay, and decays very fast after the exciting voltage has been removed.

On the other hand, when an AC bias voltage of rectangular waveform and having a peak value below the threshold value is applied to the cell, after the further application of the DC exciting voltage E_{th}, the decay time becomes very long as shown in Figure 3.4d. The decay time depends upon the voltage and the frequency of the AC bias voltage. The decay time rapidly increases when the peak value of the AC bias voltage comes close to the threshold value E_{th}. An AC bias voltage at a frequency within a range from 10 Hz to 10 kHz is effective to increase the decay time. A DC bias voltage increases the decay time, but the decay time obtained by the DC bias voltage is shorter by more than 2 orders of magnitude than the decay time obtained by an AC bias voltage at a frequency within a range from 10 Hz to 10 kHz.

More than a few minutes of decay time can be obtained with respect to most liquid crystal mixtures by using an AC bias voltage at a frequency within that range. An AC voltage shown by a broken line in Figure 3.4d may be used as an exciting voltage instead of a DC exciting voltage. The peak value of such an AC voltage should exceed E_{th}. Any exciting voltage may be used if its peak value exceeds E_{th}, such as a voltage generated by superposing a DC voltage on the AC bias voltage or a voltage generated by superposing an AC voltage field on the AC bias voltage. The liquid crystal mixture in the cell, which has the electrooptical properties as described above with reference to Figure 3.4b consists essentially of a nematic liquid crystal having a positive dielectric anisotropy and a steroidal compound.

Steroidal compounds, which are necessary in order to obtain the electrooptical properties described with reference to Figure 3.4b, are β-cholestanol, ergosterol, lanosterol, cholesterol and cholesterin derivatives. Above all, cholesterin derivatives are most desirable, because a cell containing a liquid crystal mixture consisting of a nematic liquid crystal having a positive dielectric anisotropy and cholesterin derivatives has a high contrast (i.e., a high ratio of the maximum intensity of transmitted light under excitation by an exciting voltage above the saturation voltage E_{sat} to the minimum intensity of transmitted light under excitation by an AC voltage below the threshold value E_{th}).

Decreasing of the concentration of the steroidal compound results in decreasing the contrast. For obtaining an operable contrast, at least more than 1 mol percent of a steroidal compound is required, because less than 1 mol percent of the steroidal compound results in an undesirably low contrast. Increasing of the concentration of the steroidal compound results in increasing the threshold value E_{th}. More than 15 mol percent of the steroidal compound results in an undesirably high threshold value.

For example, a liquid crystal mixture consisting of 99 mol percent of p-n-butoxybenzylidene-p-cyanoaniline and 1 mol percent of cholesteryl oleyl carbonate has a contrast of 3:1 which is substantially the lowest contrast suitable for display. A liquid crystal mixture consisting of 85 mol percent p-n-butoxybenzylidene-p-cyanoaniline and 15 mol percent of cholesteryl oleyl carbonate has a threshold of about 100 volts when the thickness of the liquid crystal layer (i.e., the distance between electrodes) is 25μ. A liquid crystal mixture having a threshold higher than such a high threshold is very difficult to use because of the breakdown of the liquid crystal layer, a requirement for expensive transistors, etc.

The following table shows half-decay times obtained by employing the method of the process with an exemplary liquid crystal mixture. The exemplary liquid crystal mixture consists of two nematic liquid crystals having a positive dielectric anisotropy (30 mol percent of p-n-butoxybenzylidene-p'-cyanoaniline and 18 mol percent of p-n-octoxybenzylidene-p'-cyanoaniline) one nematic liquid crystal having a negative dielectric anisotropy (47 mol percent of p-methoxybenzylidene-p'-n-butylaniline) and one cholesterin derivative (5 mol percent of cholesteryl oleyl carbonate). This liquid crystal mixture can be operated at a temperature of 2° to 75°C.

For measuring half-decay times of the liquid crystal mixture, a cell as shown in Figure 3.4a and containing the abovementioned liquid crystal mixture was used. The distance between the two electrodes was provided by Mylar spacers with a thickness of 25μ. The results in the table (half-decay times) were obtained by measuring a time required for the intensity of light transmitted through the cell to decrease to half the maximum value of the intensity of transmitted light in the excited light-transparent state of the liquid crystal mixture after removal of an exciting electric field. A DC electric field was superposed on a continuous AC bias voltage so that the liquid crystal mixture could be brought to the excited light-transparent state. An AC voltage of rectangular waveform was used for applying the AC bias voltage to the liquid crystal mixture in the cell.

Referring to the table, the first column gives the peak value of the AC bias voltage in units of volts. The second, the third and the fourth column give obtained half-decay times in a unit of seconds at frequencies of 500 Hz, 100 Hz and 20 Hz, respectively. Although not shown in the table, the operable range for the frequency of the applied AC bias voltage is 10 Hz to 10 kHz. Below 10 Hz, the half-decay time becomes too short (as in the case of DC bias as explained later) to be adapted to this process. The half-decay time at a frequency above 500 Hz up to 10 kHz is almost constant. Above 10 kHz, the half-decay time rapidly decreases. The lower limit (e.g., 10 Hz) and the upper limit (10 kHz) of frequency of the AC bias depend on the kinds of materials used for the liquid crystal mixture, and ambient temperature. The upper limit of the frequency is considered to be determined by the maximum frequency above which the dipole moment of the molecules of the liquid crystal cannot rotate following the frequency of the applied AC voltage. The lower limit of the frequency is considered to be determined by the frequency below which the ion flow in the liquid crystal mixture disturbs the orientation of the molecules remarkably. The lowest row in the table gives threshold values in units of volts at respective frequencies. The last column gives half-decay times and threshold value in units of seconds under DC bias voltage. The last column is provided for making clear the effectiveness of the AC bias voltage on the increase of decay time. In the table, very long means more than 10 minutes. Without any bias, the decay time of this liquid crystal mixture is about 50 msec. Then, if a fast turnoff is required, a method in which the AC bias voltage could be removed suddenly may be used. It is understood with reference to the table that an AC bias voltage below the threshold E_{th} increases the decay time effectively.

Figure 3.4e shows a method for applying an AC bias voltage and an exciting voltage to the liquid crystal mixture. Figure 3.4e is a cross section of a cell as shown in Figure 3.4a, with the cell designated **22**, and an AC voltage generator **20** is connected to the first electrode **16** for applying an AC bias voltage to the liquid crystal mixture and an exciting voltage pulse generator **21** connected to the second electrode **17** for generating an exciting voltage across the liquid crystal mixture cooperating with the AC bias voltage. The first plate **14** which is light-transparent and has the first electrode **16** attached thereon and the second plate **15** which is also light-transparent and has the second electrode **17** attached thereon have the space between them filled with the liquid crystal mixture **19**. The distance between the first electrode and the second electrode is determined by the thickness of spacers **18**. The AC voltage generator **20** generates a continuous AC voltage of rectangular waveform the amplitude of which is arranged such that the peak value of the AC bias voltage does not exceed the threshold value E_{th} of the liquid crystal mixture **19**. The exciting voltage pulse generator **21** generates an exciting voltage pulse of rectangular waveform.

The height of the exciting voltage pulse generated by the exciting voltage pulse generator **21** is such that the peak value of the exciting voltage, i.e., the sum of the AC bias voltage generated by the AC voltage generator **20** and the voltage of the DC voltage generated by the exciting voltage pulse generator **21**, exceeds the threshold value E_{th}. In practice, the exciting voltage pulse may be controlled by an external signal for controlling the transparency of the cell. If the excited light-transparent state illustrated with reference to Figure 3.4b is required to be held continuously, repeated exciting pulses should be used. The AC voltage generator **20** for the AC bias and the exciting voltage pulse generator **21** are not required to have low impedance because the cell **22** usually has a higher impedance than that of usual generators of such kind.

Thus, according to the method, the decay time of a liquid crystal mixture is increased by applying an AC bias voltage below the threshold value of the liquid crystal mixture. Therefore, flicker which is a deficiency in liquid crystal display devices using a time-shearing method can be removed, and temporary-memory display devices having a simple structure can be obtained by using the method.

Peak Value of AC Bias Voltage (v)	Half-Decay Time (sec) 500 Hz	100 Hz	20 Hz	DC
36	7	10	14	0.25
38	9	18	20	0.30
40	18	27	35	0.35
42	3,037		very long	0.39
44	>200	very long		0.43
46	very long			0.46
Threshold value E_{th} (v)	46	44	42	46

Ternary Nematic Mixtures

L.T. Creagh; U.S. Patent 3,655,270; April 11, 1972 describes a display device where a layer of a novel nematic mesomorphic composition that is normally transparent to light scatters light in response to a voltage applied across the composition. The composition consists essentially of a ternary mixture of 4-ethoxybenzylidene-4'-n-butylaniline; 4-methoxybenzylidene-4'-aminophenyl butyrate, and bis-(4'-n-octyloxybenzal)-2-chloro-1,4-phenylenediamine and exhibits the nematic mesophase over a broad temperature range that includes room temperature.

For Dynamic Scattering Mode at Room Temperature

A process described by *S.-Y. Wong; U.S. Patent 3,826,757; July 30, 1974; assigned to Hughes Aircraft Company* involves the formulation of room temperature nematic liquid crystals using compounds which are solid at room temperature, but when dissolved in certain solvents become true liquid crystals. There has been an increasing search for effective room temperature nematic liquid crystals. French Patent 1,537,000 has reported nematic liquid crystals for use in the dynamic scattering mode (DSM). Of the mixtures reported therein, only one exhibits solid-mesophase transition at room temperature, but the DSM effect is above room temperature.

This process is concerned with formulations which show DSM at room temperature. The compositions are also useful in ultraviolet-activated cells. The compounds used are solid at room temperature, but can be dissolved in certain solvents to produce solutions which are true room temperature nematic liquid crystals. The solvents useful in the process are: p-methylbenzylidene-p-n-butyl aniline and p'-n-butylphenyl-p-methyl benzoate. These solvents exhibit no nematic mesophase at room temperature by themselves, but when certain solid nematic liquid crystals are dissolved therein, they become effective room temperature nematic liquid crystals. The solid compounds used in the process correspond to the formula

(1)

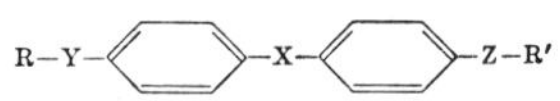

wherein

$$X \text{ is } -CH{=}N- \text{ or } -\overset{\overset{\displaystyle O}{\|}}{C}-O-;$$

R and R' are each alkyl having from 1 to about 10 carbon atoms;

$$Y \text{ is } -O-,\ -\overset{\overset{\displaystyle O}{\|}}{C}-O-, \text{ or } -O-\overset{\overset{\displaystyle O}{\|}}{C}-O-;$$

and

$$Z \text{ is } -O-,\ -O-\overset{\overset{\displaystyle O}{\|}}{C}-,\ -O-\overset{\overset{\displaystyle O}{\|}}{C}-O-$$

or a direct bond. The method for making these compounds has been reported in Vogel, *Practical Organic Chemistry,* Longmans pp 653-654. In preparing the solutions of the process, the compound of Formula 1 is dissolved in one of the named solvents in order to form a solution. Preferably, the composition will contain 20 to 50%, by weight, of the solvent. The compounds of Formula 1 may be used individually, or mixtures of them may be employed. They dissolve readily with stirring in the solvent. Specific examples of compounds which are useful within the scope of the process are set out in Table 1 which follows.

TABLE 1

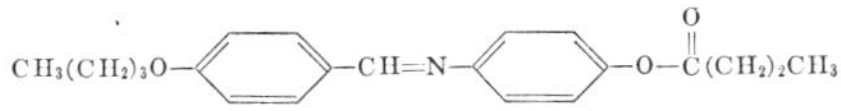

3

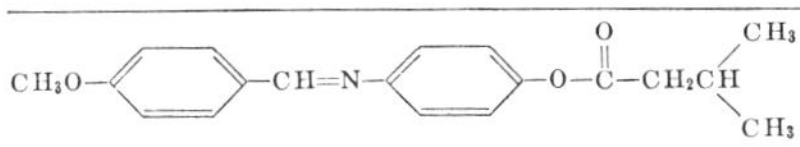

4

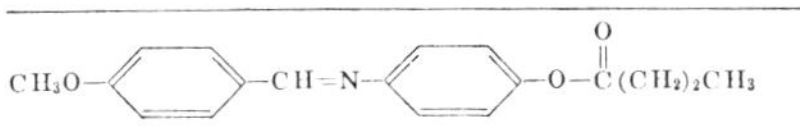

5

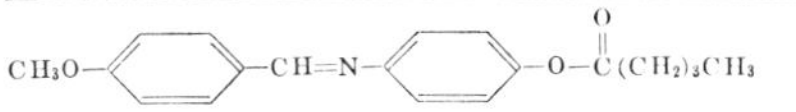

6

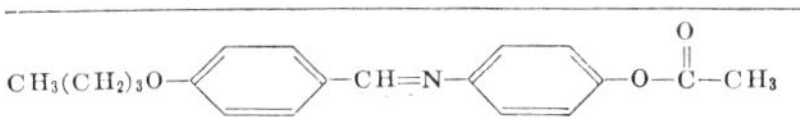

7

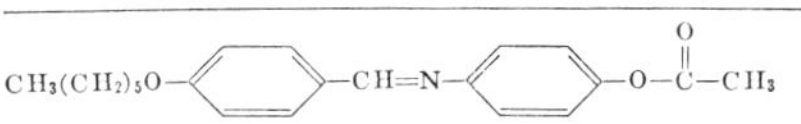

8

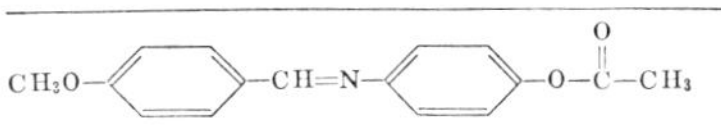

9

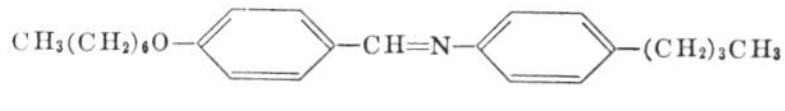

10

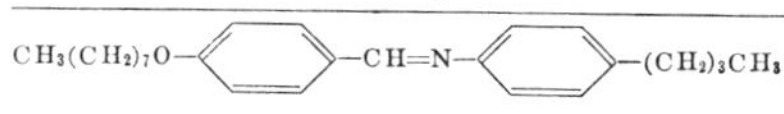

11

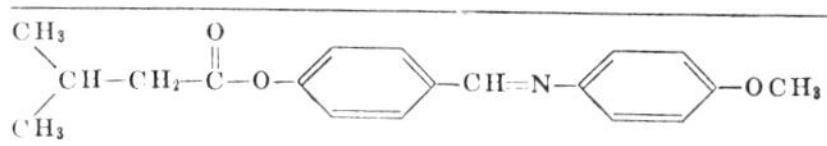

12

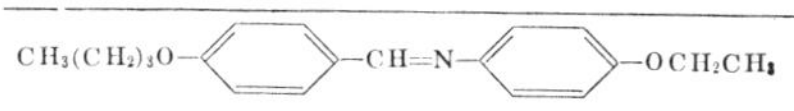

13

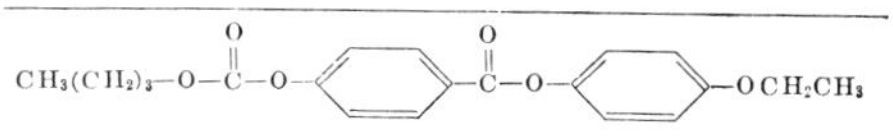

14

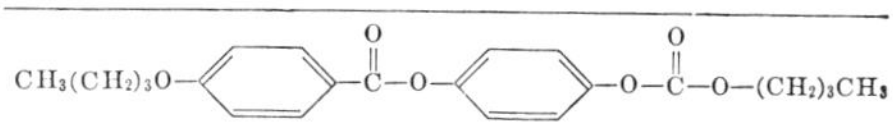

15

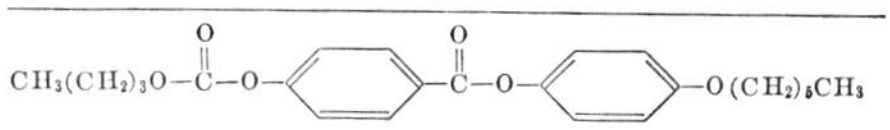

16

17 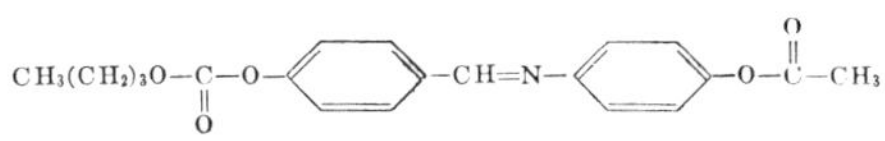

18 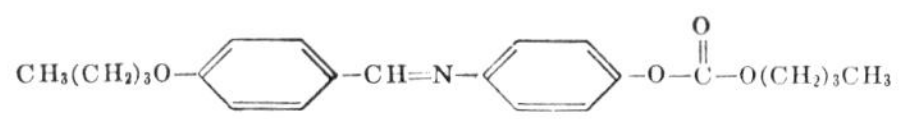

19 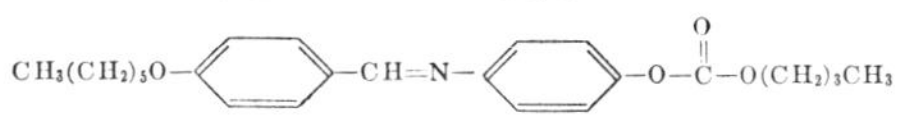

20 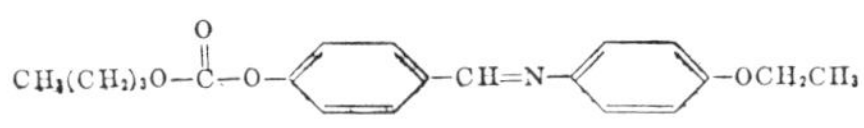

21 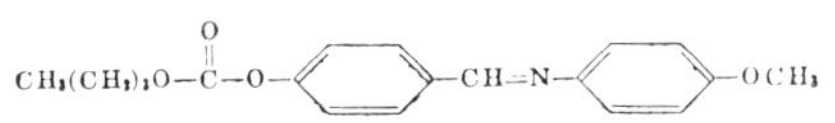

22 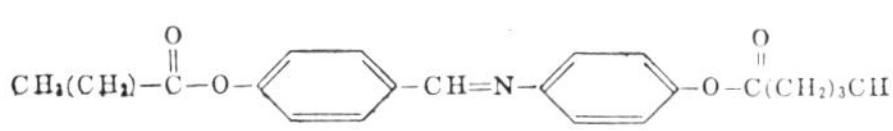

The following table sets forth 19 specific formulations made within the scope of the process. Compounds 1 and 2 are p-methylbenzylidene-p-n-butylaniline and p'-n-butylphenyl-p-methyl benzoate, respectively. While the solvent is nonmesomorphic at room temperature, the resulting solution with the nematic compounds is mesomorphic. In addition, a different scattering efficiency can be obtained from these formulations by varying the composition. For instance, the scattering effect of a solution of compound 11 in p-methylbenzylidene-p-n-butylaniline will be different at 10% solution as compared to 50% solution. This difference is due to the number of scattering centers created by the interaction of the liquid crystal compounds and the solvent. In addition, the compositions have wider and lower temperatures of nematic mesophase than prior art compositions. For example, composition 1 in Table 2 has a mesophase between -20° and 51°C. The lowest solid-mesophase transition previously recorded was 10°C. This is the first time that the art has been presented with a room temperature guest-host alignment liquid crystal. The formulations can be used in dynamic scattering mode between transparent electrodes for display system.

TABLE 2

Formulation number	Compound number	Percent by weight
1	1	40
	3	31
	4	29
2	1	50
	5	50
3	1	40
	10	30
	11	30
4	2	50
	14	25
	12	25
5	1	50
	12	25
	14	25
6	1	46.7
	10	6.5
	11	46.7
7	1	50
	10	25
	11	25
8	1	25
	11	25
	13	50
9	2	45
	20	45
	22	10
10	1	33
	18	33
	19	33
11	1	54
	18	23
	19	23
12	2	41
	14	44
	15	15
13	1	50
	4	26
	12	24
14	1	48
	18	26
	20	26
15	1	50
	18	25
	21	25
16	1	54
	17	25
	22	21
17	2	23
	16	77
18	1	54
	3	46
19	1	48
	3	27
	8	25

Nematic-Cholesteric Blend with Positive Anisotropy

W. Greubel, H. Krueger and U. Wolff; U.S. Patent 3,821,720; June 28, 1974; assigned to Siemens AG, Germany describe an improved image storage device of the type utilizing a liquid crystal composition comprising material having nematic liquid crystallinity and material having cholesteric liquid crystallinity. The liquid crystal composition has positive dielectric constant anisotropy and shorter extinction times compared to prior art.

Figure 3.5a is a schematic, fragmentary, perspective, vertical sectional view of a segment of a preferred image storage screen showing the individual layers utilized in this device. Figure 3.5b is a view similar to Figure 3.5a, but showing another example having no surface active layers. Figure 3.5c is a diagrammatic representation of a storage device with the liquid crystal materials in a quiescent condition. Figure 3.5d is a view similar to Figgure 3.5c with the liquid crystal molecules in a focal-conical configuration. Figure 3.5e is a view similar to Figure 3.5c with the liquid crystal molecules in a nematic-homeotropic configuration. Figure 3.5f is a view similar to Figure 3.5c with the liquid crystal molecules in a nematic-homeotropic configuration.

FIGURE 3.5: STORAGE DEVICE USING CHOLESTERIC–NEMATIC LIQUID CRYSTALS

a.

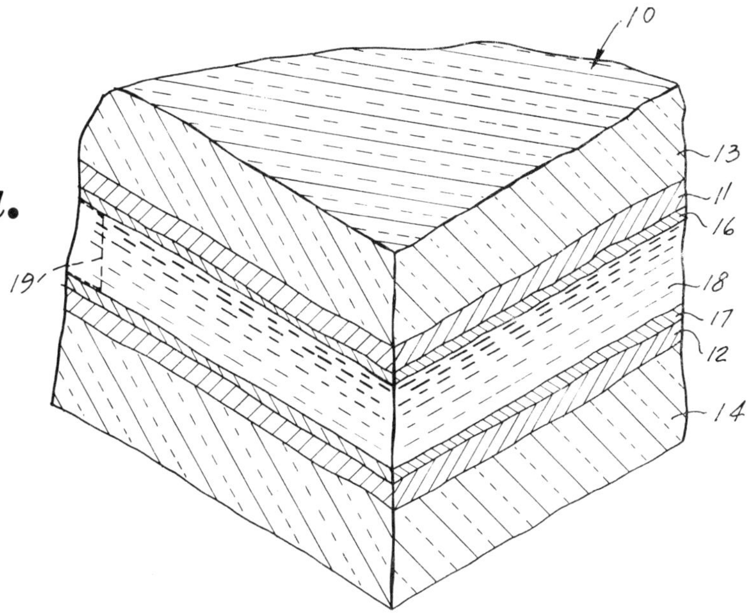

b.

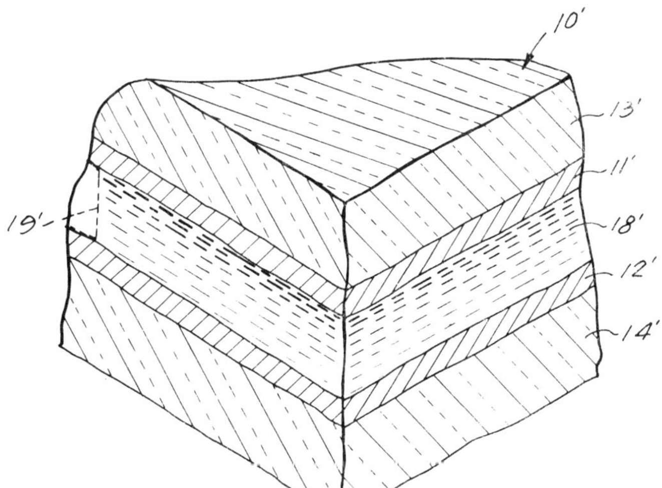

(continued)

FIGURE 3.5: (continued)

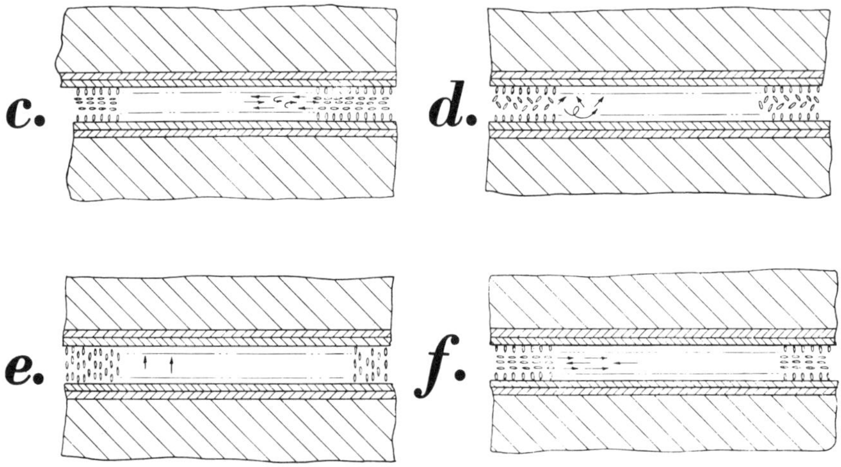

Source: W. Greubel, H. Krueger and U. Wolff; U.S. Patent 3,821,720; June 28, 1974

Referring to Figure 3.5a, there is seen illustrated a portion of an image storage device of this process, the device being designated in its entirety by the numeral **10**. Device **10** is seen to employ a pair of electrodes **11** and **12**. Each electrode **11** and **12** is a vacuum vapor-deposited substantially transparent film of conductive material and such film is deposited on glass plates **13** and **14** respectively. Those skilled in the art will recognize that the combination of electrode **11** with glass plate **13**, and of electrode **12** with glass plate **13** is known as conductive glass. Conductive glass is formed of plate glass having a thin coating of stannic oxide produced by spraying glass at about 900° to 1300°F with a solution of stannic chloride to produce a coating thickness of about 50 to 550 millimicrons.

The resulting stannic oxide layer will carry current densities of about 600 watts per square inch indefinitely. The coatings are hard and resistant to solvents. The light transmission is about 70 to 88% that of the original glass and the index of refraction is about 2.0 compared with 1.53 for glass. Examples of commercially available conductive glass include those available under trademarks such as Electrapane and Nesa.

Each conductor layer **11** and **12** is coated with a layer **16** and **17**, respectively, of a surface active material. Positioned between layers **16** and **17** is a layer **18** of a mixture composition of cholesteric and nematic liquid crystals. The layers **16** and **17** are thin so as not to interfere substanitally with the passage of an electric current therethrough, and the layers **16** and **17** are each adapted to align the longitudinal axes of the respective individual nematic cyrstals adjacent the layer in layer **18** in a direction substantially normal to the faces of conductors **11** and **12** when the mixture composition comprising layer **18** has a transparent planar structure relative to the face of either transparent electrode **11** and **12**. A preferred surface-acting material for use in layers **16** or **17** is lecithin.

Spacer means such as **19** (typically a thin strip of an inert, insulation polymeric material such as polytetrafluoroethylene film, or the like) maintain the spacing between electrodes **11** and **12** and serve to define a chamber for containing and defining the liquid layer **18**. Electrode leads permit electrical interconnection of each electrode **11** and **12** with appropriate sources of electricity of operating device **10**. In Figure 3.5b, is seen another form of a device of the process which is similar to that of Figure 3.5a, but here is numbered **10'**. Similar elements of device **10'** are similarly numbered to elements in device **10** but with the addition of prime marks thereto. As can be seen, device **10'** employs no surface-active material layers. While any convenient and conventional transparent electrode may be used, it will be appreciated that both electrodes **11** and **12** (or **11'** and **12'**) may be transparent, but one of these electrodes may be a specularly reflecting electrode which then serves as a back electrode in a device of this process. A device using such reflecting electrode preferably faces into a black background to maximize contrast between image and background (during use of such a device) and to direct the forward scattered photo radiation back through the liquid crystal layers to the viewer.

In such a device as **10** with layers **16** and **17**, the longitudinal axes of the liquid crystal molecules present in a liquid crystalline layer **18** adjacent to layers **16** and **17** are aligned in a direction normal to the faces of the bearing plates. A layer **16** or **17** is preferably as thin as practical and more preferable is about monomolecular in thickness. On the application of an extinction voltage pulse, a nematically homeotropic texture arises in the crystalline layer **18** in an optimally uniform fashion. This again promotes the formation of a uniform planar cholesteric texture. Furthermore, with this border orientation, the texture transition from focal-conical to nematically homeotropic takes place at lower field strengths.

It is preferred that a liquid crystal mixture possess a specific resistance of not greater than about 10^{11} Ω-cm, which is generally achievable when the liquid crystalline substances used possess a high degree of purity. Both ac and dc voltages may be employed as an extinction voltage. The thickness of a layer **18** or **18'** ranges from 6 to 12 microns in this process. A liquid crystal composition of this process substantially filling the space between electrodes comprises on a 100 weight percent total composition basis a mixture of from about 51 to 99 weight percent material having nematic liquid crystallinity and, correspondingly from about 1 to 49 weight percent material having cholesteric liquid crystallinity such composition in the chamber displaying a positive dielectric constant anisotropy such that

$$1.\ 20 \geqslant [\Sigma_{\Pi} - \Sigma_{\perp}]\ 0.2$$

where Σ_{π} is the dielectric constant in a direction parallel to the longitudinal axis of the molecules of the liquid composition in a storage device of this process (measured in the nematic homeotropic texture as shown in Figure 3.5e, and $\Sigma_{\perp}$ is the dielectric constant in a direction perpendicular to the helical axis of the liquid composition in a storage device of this process measured in the focal-conical texture shown in Figure 3.5d. The cholesteric material used in the composition has a helical pitch of from about 2 to 5 times smaller than the spacing between electrodes. One proposed liquid mixture system comprises from about 90 to 95 weight percent nematic p-methoxy-p'-n-butylazoxybenzene with the balance up to 100 weight percent thereof being cholesterylchloride.

In the storage device which displays positive dielectric constant anisotropy as indicated above, after a direct current or an alternating current voltage is applied across electrodes such as **11** and **12** in device **10**, the liquid crystal layer **18** possesses a light-scattering, focal-conical milky texture with a layer-wise helix formation and a screw axis parallel to the faces of the electrodes **11** and **12**. The light-scattering liquid crystal layer becomes transparent again if a short voltage pulse transforms the focal-conical texture into a transparent, nematically homoetropic texture, wherein all the molecular longitudinal axes are aligned in a direction normal to the surfaces of the bearing plates. After the extinction voltage pulse has died out, the likewise transparent, planar texture is spontaneously formed, when the molecular longitudinal axes run parallel to the faces of the carrier plates and a helical cholesteric crystalline structure is formed.

The extinction voltage must only be applied until the nematically homeotropic texture has formed. The greater the positive dielectric constant anisotropy and the greater the extinction voltage, the more rapidly the texture transition from focal-conical to nematically homeotropic takes place. Both long storage times and short extinction times may be achieved using the indicated helix pitch and indicated layer thickness for a cholesteric/nematic composition. In the process, a voltage in a pattern representation of information to be stored is selectively applied across opposed parallel faces of a 6 to 12 millimicron-thick layer of the composition described above. The voltage is either a direct current or an alternating current. The composition at the time of the application has a transparent, planar texture and the voltage used is applied for a time sufficient to generate in the layer a light-scattering, focal-conical, milky texture with a layer-wise helical nematic crystalline formation therein having a screw axis substantially parallel to the forces and in a pattern representative of the information to be stored. In this way, desired information is recorded in an imagewise pattern in such layer.

To electronically erase such image, one applies across the faces a dc or ac extinction voltage pulse for a total time of less than about 30 microseconds to 30 milliseconds, but which time with this interval is sufficient to transform the texture of the layer into a transparent, nematically homeotropic form wherein the nematic-cholesteric crystals are aligned in a direction substantially normal to the faces. Each of Figures 3.5c through 3.5f, shows, diagrammatically, a storage device in a different configuration. Each view is taken cross-sectionally through a storage device of this process. In the central portion is a liquid crystal composition. A lecithin layer is adjacent such composition on either face thereof. Behind the lecithin layer is an electrically conductive layer, and behind the electrically conductive layer is a backing layer. In Figure 3.5c, the liquid crystal mixture molecules directly containing each lecithin layer are vertical, while the molecules in the central region of such liquid crystal mixture are parallel to the lecithin layer and to the backing layer or carrier plate.

Figure 3.5d shows a device after a recording pulse has been applied across the electrode layers, for example, a pulse of 15 volts applied for about 100 microseconds, which leaves the liquid crystal mixture in a light-scattering condition. The drawing does not illustrate the helical axis in this focal-conical mixture. In Figure 3.5e is seen the appearance of the liquid crystal mixture of Figure 3.5d after the application of an erasure pulse thereto which leaves the nematic-homeotropic molecules in a transparent orientation which is perpendicular to the carrier plates. A suitable erasure pulse here would be, for example, about 30 volts applied for about 100 microseconds. In Figure 3.5f is shown the appearance of the device of Figure 3.5e after the same has undergone a spontaneous transition from the configuration shown in Figure 3.5e to a planar configuration, as similar to the nematic-homeotropic configuration of Figure 3.5c.

A large positive dielectric constant anisotropy is advantageous in that a low field strength is normally then sufficient for the erasing and the erasing time is also typically shortened. Cholesteric-nematic liquid crystal compositions with a large dielectric constant anisotropy can be found while choosing cholesteric and nematic liquid crystal materials which have large dipole moments and therefore, large dielectric constants parallel to their molecular longitudinal axes. The dielectric constants Σ_π and $\Sigma_\perp$ of cholesteric-nematic liquid crystal compositions can be measured exactly according to a nematic-homeotropic or focal-conical respectively, adjustment of the liquid crystals in a magnet field in a capacity gauging bridge. The following examples illustrate the process. All parts are parts by weight, unless others are indicated.

Example 1: An image storage device of this process is prepared and evaluated as follows: The liquid crystal composition used consists essentially of 94% by weight nematic p-methoxy-p'-n-butylazoxybenzene and 6% by weight cholesterylchloride. This composition is embedded in a layer of constant thickness of 10 microns between the two electrodes. Glass spacer elements are employed. The positive dielectric constant anisotropy of such layer is 0.2 measured as $[\Sigma_\pi - \Sigma_\perp]$.

The composition itself has a helical pitch of 2.6 microns. The inner opposed surfaces of the electrodes are each coated with an approximately monomolecular lecithin layer for the purpose of orienting the longitudinal axes of the respective individual liquid crystal molecules adjacent the lecithin layer in the device in a direction normal to the lecithin layer. Lecithin layers are applied by dipping an electrode surface into an ethanol solution of lecithin and evaporating. The composition in the liquid crystal layer has a specific resistance of about 10^{11} Ω-cm. On the application to the device of a direct current electric voltage of 10 volts and 1 microamp/cm^2 across the electrodes in an image-forming pattern with the composition having a transparent, planar texture, at the time of its application a visible pattern is produced in the layer corresponding to the image-forming pattern a light-scattering, focal-conical milky texture with a layer-wise helical nematic crystalline formation therein which has a screw axis substantially parallel to the layer faces. When the applied voltage is removed, the light-scattering state of the composition in the layer is retained for many hours.

Example 2: With a device as described in Example 1, the recording procedure of Example 1 is repeated except that, after the applied voltage is removed following image formation in the layer, there is applied across the electrodes a voltage pulse of 100 volts and 1 microamp/cm^2 of electrode surface area for a total time of about 30 milliseconds. During such application of voltage, the texture of the layer is transformed into a transparent, nematically homeotropic form wherein the nematic-cholesteric molecules are aligned in a direction substantially normal to the faces.

After such voltage is removed and the extinction voltage pulse has died out, the texture of the layer is transformed into the initial transparent, planar texture present in the composition before application of the image-recording signal. This procedure of recording and erasing may be repeated indefinitely without any observable electrochemical change in the device.

Example 3: Using a device as described in Example 1, there is applied an alternating voltage of 10 volts and 50 hertz, and a visible pattern similar to that obtained in Example 1 is produced. This pattern is retained for many hours.

Example 4: With a device as described in Example 1, the recording procedure of Example 3 is repeated except that thereafter the procedure for Example 2 is followed to erase the image so formed. Results equivalent to those in Example 2 are observed.

Alkoxybenzylidene Anils and Asymmetric Cholesteric Compounds

C.S. Oh and E.F. Pasierb; U.S. Patent 3,792,915; February 19, 1974; assigned to RCA Corporation have found that mixtures of certain asymmetric, optically active, p-alkoxybenzylidene-p'-aminobenzonitrile compounds with nematic liquid crystals which have a strong positive dielectric anisotropy, particularly one or more alkoxy- or acyloxybenzylidene anils form a cholesteric liquid crystal which, on application of an electric field, forms a nematic liquid crystal. When the electric field is removed, the mixture reverts to the cholesteric phase. The electric field-induced phase changes occur very rapidly and require only low fields to operate electro-optic devices incorporating them. The asymmetric optically active p-alkoxybenzylidene-p'-aminobenzonitrile compounds useful in the mixtures have the formula

(1)

where n is an integer from 1 to 5. The asymmetric carbon atom can have either R or S configuration. Mixtures of these compounds can also be employed, which generally will reduce the temperature of the transition points. The nematic liquid crystal compounds particularly useful in the mixtures are alkoxy- and acyloxybenzylidene anils having the formula

(2)

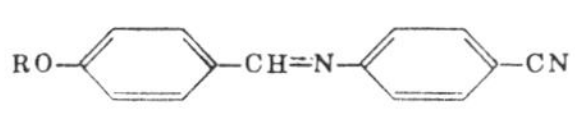

where R is a straight chain aliphatic radical selected from the group consisting of C_nH_{2n+1} and $C_nH_{2n+1}CO-$ where n is an integer from 1 to 7. These compounds and mixtures are described in U.S. Patent 3,597,044 issued August 3, 1971. The asymmetric compounds of Formula (1) can be prepared from an optically active isomer of 1-bromo-2-methylbutane by a Grignard reaction in known manner, as summarized in the following scheme.

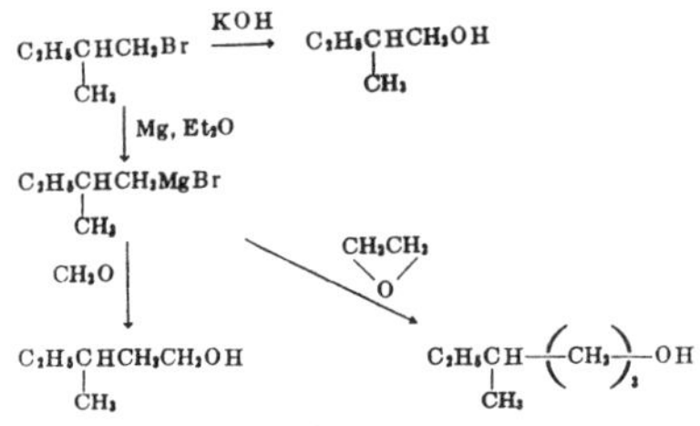

The desired alcohol is reacted with phosphorous tribromide to form the corresponding bromide; the bromide is reacted with p-hydroxybenzaldehyde in base to form the corresponding p-alkoxybenzaldehyde; and finally the p-alkoxybenzaldehyde is reacted with p-aminobenzonitrile to form the desired p-alkoxybenzylidene-p'-aminobenzonitrile. These reactions are summarized by the following scheme.

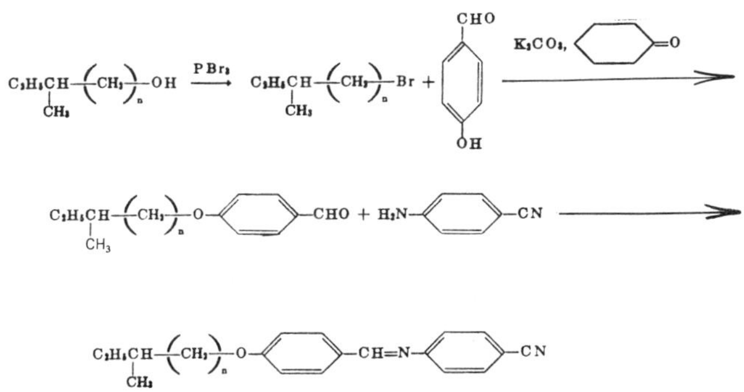

One or more asymmetric, optically active p-alkoxybenzylidene-p'-aminobenzonitrile compounds are then mixed with one or more alkoxy- or acyloxybenzylidene anil compounds to form the liquid crystal mixtures useful in the process. The addition of one or more nematic compounds to the cholesteric compound results in a reduction of smectic thermal stability and an increase in the mesomorphic temperature range. The mixture can contain from about 5% to about 80%, preferably about 20% to about 60% by weight of the mixture of the optically active component, which in general gives the most practical use temperature range for electrooptic devices. Mixtures containing from about 40% to about 50% by weight of the optically active compound are particularly preferred.

Lyotropic Nematic Liquid Crystals

D. Jones and S. Lu; U.S. Patent 3,690,745; September 12, 1972; assigned to Texas Instruments Incorporated describe a display device in which a thin layer of a lyotropic nematic mesomorphic composition is utilized to diffuse light from a source toward an observer by applying a suitable voltage, for example, 20 volts for a 1 mil layer. One form of the device transmits light through the layer. A second form of the device uses light diffused and reflected to the observer. A preferred lyotropc nematic mesomorphic composition includes three compounds of the structural formulas

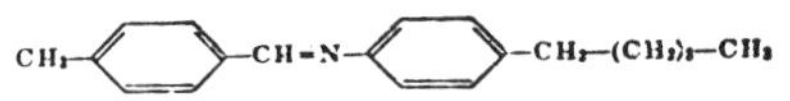

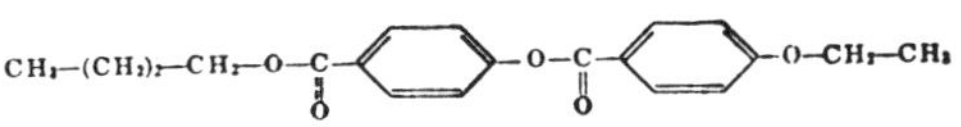

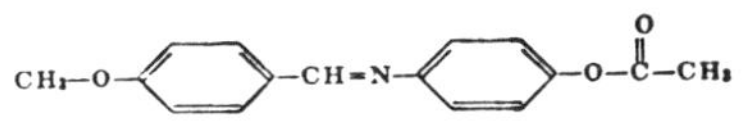

Complete details of the devices are given in U.S. Patent 3,716,289.

Ternary and Binary Mixtures

M. Lodolini; U.S. Patent 3,772,210; November 13, 1973; assigned to Eastman Kodak Company describes a nematic liquid crystal composition having a very broad mesophase. The family of nematic liquid crystal compositions provided by this process include at least one and preferably two compounds of the following Formula 1:

(1) 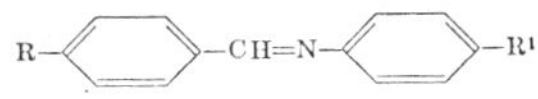

where R and R^1 are radicals selected from the group consisting of a saturated alkoxy radical of from 1 to 9 carbon atoms and an alkyl radical of from 1 to 9 carbon atoms. When R is a saturated alkoxy radical, R, R^1 is an alkyl radical and vice versa. Exemplary of compounds of this structure are N-(p-methoxybenzylidene)-p-ethylaniline; N-(p-methoxybenzylidene)-p-butylaniline; N-(p-butoxybenzylidene)-p-propylaniline and N-(p-ethoxybenzylidene)-p-butylaniline. Other compounds of the structure of Formula 1 and which may be used in this mixture are shown in the following table.

Compound:	R	R_1
1	C_2H_5O	$(CH_2)_3CH_3$
2	$CH_3(CH_2)_3O$	$CH_2)_3CH_3$
3	$CH_3(CH_2)_3O$	C_2H_5
4	$CH_3(CH_2)_4O$	C_2H_5
5	C_2H_5O	C_8H_{17}
6	$CH_3(CH_2)_6O$	C_2H_5
7	C_2H_5	OCH_3
8	C_2H_5	$O(CH_2)_3CH_3$
9	$(CH_2)_3CH_3$	$O(CH_2)_3CH_3$
10	C_8H_{17}	OC_2H_5
11	C_2H_5	OC_8H_{17}

The nematic liquid crystal compositions provided by this process include at least one and preferably two compounds of the following Formula 2:

(2) 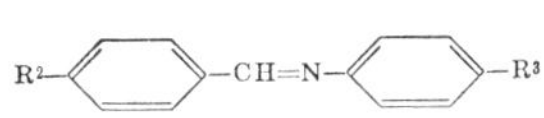

where R^2 and R^3 are radicals selected from the group consisting of a saturated alkoxy radical having from 1 to 9 carbon atoms and a saturated acyloxy radical having from 2 to 5 carbon atoms. When R^2 is a saturated alkoxy radical, R^3 is a saturated acyloxy radical and vice versa. Exemplary of compounds of Formula 2 are p-methoxybenzylidene-p-aminophenyl butyrate; p-butoxybenzylidene-p-aminophenyl propionate; and p-methoxybenzylidene-p-aminophenyl valerate. Other examples of compounds of Formula 2 which may be used in this nematic liquid crystal are shown in the following table.

Compound:	R^2	R^3
12	$CH_3CH_2\overset{O}{\overset{\|}{C}}O$	OCH_3
13	$CH_3(CH_2)_2\overset{O}{\overset{\|}{C}}O$	OCH_3
14	Same as above	OC_6H_{13}
15	C_4H_9O	$O\overset{O}{C}CH_3$
16	$C_6H_{13}O$	Same as above.
17	$C_8H_{17}O$	Do.
18	$C_9H_{19}O$	Do.

Compounds of this type are prepared by the methods described in U.S. Patent 3,540,796. The liquid crystal compositions include at least one and preferably only one compound of the following Formula (3).

(3) 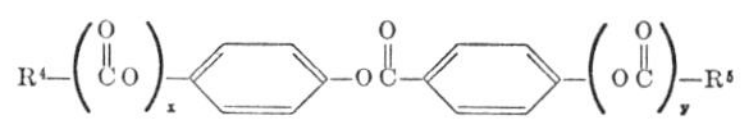

where R^4 and R^5 are saturated alkoxy radicals having from 1 to 9 carbons, x and y differ and are selected from the group consisting of 0 and 1. Typical examples of useful alkoxy radicals are set out above [in the description of substituents in Formula (1)], and preferably having 1 to 5 carbons. The first five of the following examples will illustrate the preparation of the compounds comprising a preferred nematic liquid crystal mixture.

Example 1: p-Methoxybenzylidene-p-aminophenyl butyrate is prepared by the esterification of p-methoxybenzylidene aminophenol with butyric anhydride. The reaction is carried out in a refluxing benzene solution in the presence of pyridine. The resulting compound is recrystallized from an isopropyl alcohol solution.

Example 2: p-Butoxybenzylidene-p-aminophenyl propionate is prepared by refluxing a solution of p-butoxybenzylidene aminophenol with propionic anhydride in the presence of pyridine. The resulting ester is purified by recrystallizing from hot isopropanol.

Example 3: Butyl-p-(p-ethoxyphenoxycarbonyl)phenyl carbonate is prepared by the esterification of p-hydroxybenzoic acid with p-ethoxyphenol to form p-ethoxyphenyl-p-hydroxybenzoate. This intermediate is treated with butyl chloroformate in the presence of pyridine or triethylamine to yield the desired product.

Example 4: N-(p-ethoxybenzylidene)-p-butylaniline is prepared by treating p-butylaniline with p-ethoxybenzaldehyde in alcohol. The pure product is obtained by recrystallization from alcohol.

Example 5: N-(p-butoxybenzylidene)-p-propylaniline is prepared by treating p-propylaniline with p-butoxybenzaldehyde in alcohol. The product is recrystallized from alcohol.

Example 6: A nematic liquid crystal composition is prepared by mixing 488 parts of the product of Example 1, 319 parts of the product of Example 2, 75 parts of the product of Example 3, 75 parts of the product of Example 4 and 75 parts of the product of Example 5. The mixture was melted and stirred to obtain a homogeneous nematic melt. The resulting melt was filtered through a Super-Cel mat to remove trace insoluble materials and then stirred over a Linde 4A molecular sieve to remove traces of moisture. The resulting composition had the following physical and electrooptical properties. All measurements were made in a 0.5 x 0.5 cm cell of NESA coated glass with a 0.5 mil Teflon spacer.

Mesomorphic range (by differential scanning calorimetry)	15° to 97°C
Rise time	10 milliseconds
Decay time	350 milliseconds
Resistivity	6.67×10^9 ohm cm
Transmittance in the clear state	78%
Transmittance in the saturated state	0.19 in 8 milliseconds
Optimum voltage	40 volts DC; 50 to 60 volts AC
Threshold voltage	4 volts
Contrast ratio at 38 volts DC	39:1
Contrast ratio at 24 volts DC	26:1
Readily supercools below 0°C	

By rise time, it is meant the time necessary to align the domains. By decay time, it is meant the time it takes after removal of the applied field for the liquid crystals to return to its random state. It can be seen from the foregoing that liquid crystal compositions of extremely wide mesomorphic range can be obtained by mixtures of the various compounds set forth above.

In related work *D.C. Batesky; U.S. Patent 3,772,209; November 13, 1973; assigned to Eastman Kodak Company* describes a nematic liquid crystal composition having a very broad mesophase comprising a mixture of at least two compounds of the formula:

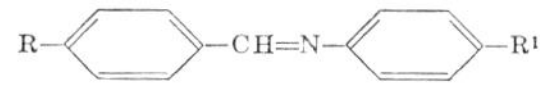

where each R and R^1 are radicals selected from the group consisting of a saturated alkoxy radical and an alkyl radical each having from 1 to 9 carbons and only one of each R and R^1 is alkyl, and at least one compound of the formula:

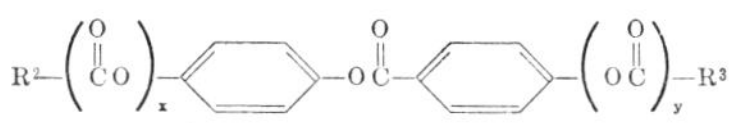

where R^2 and R^3 are saturated alkoxy radicals of from 1 to 9 carbons, one of x and y is 0 and the other is 1.

p-Alkoxyphenyl Esters of Capronyloxybenzoic Acid

B. Scheurle and K. Kelker; U.S. Patent 3,689,525; September 5, 1972; assigned to Farbwerke Hoechst AG, Germany relates to nematogenic compounds of the general formula:

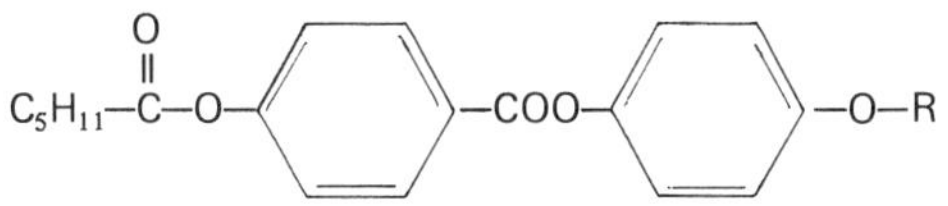

in which R represents an ethyl or butyl radical. The process further involves nematogenic mixtures which consist of one of the above compounds, N-(4-methoxybenzylidene)-4'-o-n-butyrylaminophenol and 4-methoxy- or 4-ethoxybenzylidene-4'-n-butylaniline. Azomethine type compounds 4-methoxybenzylidene-4'-n-butylaniline

(1) 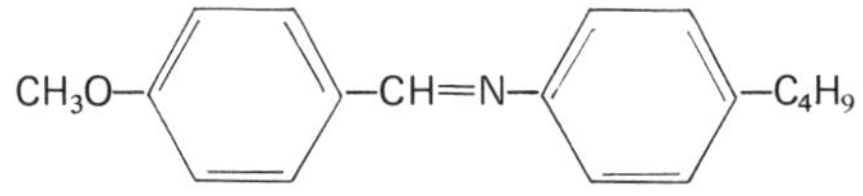

and 4-ethoxybenzylidene-4'-n-butylaniline

(2) 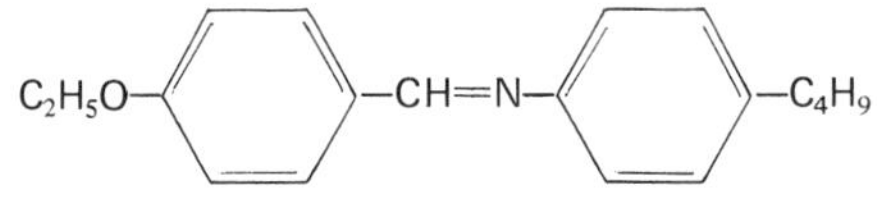

which form enantiotropic nematic phases in the range of room temperature or slightly above and the nematic phases of which lie within the range of persistency from 20° to 30°C are of interest in this process. In particular, 4-methoxybenzylidene-4'-n-butylaniline having a melting point of 20°C and a transition point of approximately 42°C represents a compound which persists working with nematic liquids without complicated thermostatization slightly above normal room temperature.

However, it is desirable to find nematic liquids having a melting point that is still lower and having a range of persistency that is still wider so that thermostatization may not be necessary at all. It has been found that the compound N-(4-methoxybenzylidene)-4'-O-n-butyrylaminophenol of the formula:

(3) 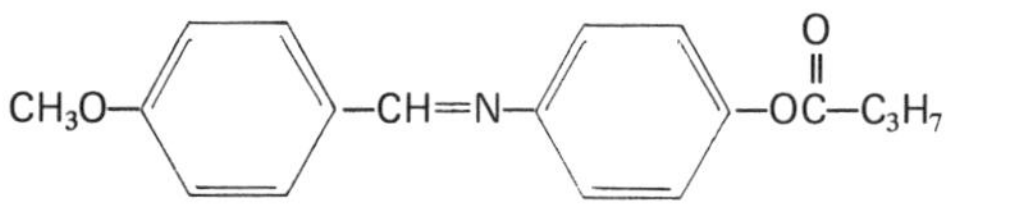

is a mixing component the addition of which leads to a higher nematic isotropic (n/i) transition point of the above compounds (1) and (2). A mixture of 2 mols of compound (1) and 1 mol of compound (3) has a n/i transition point of 70°C, which is, thus, by about 30°C higher than that of pure compound (1). However, the melting point of the mixture lies hardly below that of pure compound (1). It has also been found that nematic mixed phases having a melting point below 0°C can be obtained if 4-capronylhydroxybenzoic acid-4'-ethoxyphenyl ester of the formula:

(4) 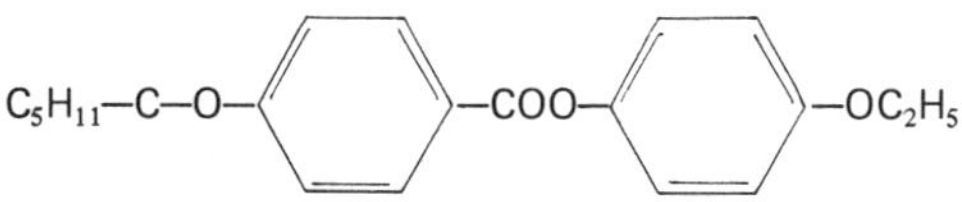

or 4-capronylhydroxybenzoic acid-4'-butoxyphenyl ester of the formula:

(5) 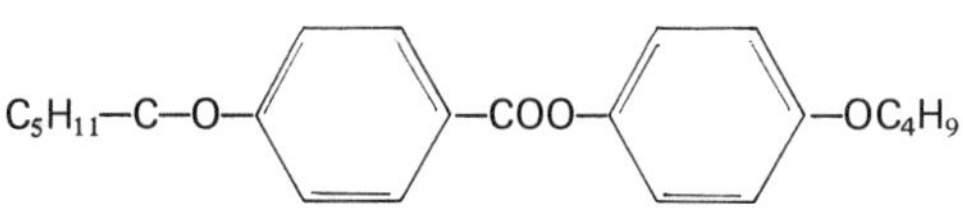

is added as third component. Compound (4) was prepared as follows. A solution of 21 grams of 4-hydroxyphenetole in 300 ml of benzene was added to a solution containing 49.3 grams of 4-capronylhydroxybenzoyl chloride in 250 ml of benzene. 15 ml of pyridine in 100 ml of benzene were added dropwise to this solution. The reaction mixture was then boiled under reflux for 3 hours. After cooling the pyridine hydrochloride precipitated was filtered off. Then the benzenic solution was washed with diluted sodium hydroxide solution and subsequently with water, dried over sodium sulfate and finally the benzene was distilled off. The distillation residue was recrystallized from hexane. The 4-capronylhydroxybenzoic acid-4'-ethoxyphenyl ester melts at 64° to 65°C to yield a nematic melt having a n/i-transition point between 100° to 101°C.

Compound (5) is prepared in analogous manner using 4-capronylhydroxybenzoyl chloride and 4-hydroxyphenylbutyl ether; it has a melting point of from 59° to 60°C and a n/i-transition point between 89° to 90°C (from hexane). Mixtures having the lowest melting points and the widest range of persistency of the nematic liquid have the following composition:

	Melting Point	n/i-Transition Point
2 mols of compound (1) 2 mols of compound (3) 1 mol of compound (4)	-3°C	+80°C
2 mols of compound (1) 2 mols of compound (3) 1 mol of compound (5)	-2°C	+75°C
3 mols of compound (1) 2 mols of compound (3) 1 mol of compound (4)	-3°C	+62°C

The last has a melting point below 0°C but a less wide range of persistency. Of course, mixtures containing compound (2) have higher melting points.

An important advantage of these nematogenic mixtures is their stability. They have no tendency to crystallize out or to dissociate. They may be mixed with relatively large amounts of dissolved foreign substances without losing their nematic character, for example, with up to 9% of benzene. They show a pronounced dynamic scattering effect and are well suitable as solvents for nuclear magnetic resonance spectroscopy.

ALIGNING AND DOPING TECHNIQUES

Hexadecyltrimethyl-Ammonium Bromide

A process described by *I. Haller and H.A. Huggins; U.S. Patent 3,656,834; April 18, 1972; assigned to International Business Machines Corporation* relates to liquid crystal compositions where the orientation of films can be controlled.

In the absence of external aligning influences, the direction of the orientational order in a nematic liquid, i.e., the direction of the optic axis varies randomly from place to place. The variation is generally both continuous and abrupt, the location of abrupt changes in orientation being termed disinclination lines. In the vicinity of a boundary surface, the orientation direction of the liquid crystal is determined by the nature and past history of the surface. For example, most nematic liquids in contact with an untreated glass surface orient in the surface layer with the long axis of the molecules lying in the plane of the surface. The orientation direction in the plane varies, however, from point to point, subdividing the surface into random regions or domains of different orientation.

When a nematic liquid is placed between two surfaces, spaced apart about 10 to 1,000μ, it exhibits one of several kinds of characteristic birefringence patterns, termed textures. These textures can be observed visually at low magnification between crossed polarizers. They result from the variation of the optic axis direction in the nematic layer which in turn is determined by the orientational domain pattern of the bounding surface as well as the deformational characteristics of the liquid itself. For example, most nematic liquid crystals, when enclosed between two untreated glass surfaces, exhibit a threaded texture. The threads in this texture have their origin in the disinclination lines while the remainder of the texture originates in the continuous variation in the optic axis direction in the bulk to accommodate the optic axis directions in the boundary layer imposed by the glass.

Another example is the homogeneous, or aligned, texture, which is characterized by the optic axis pointing uniformly in a predetermined direction but lying in the plane of the boundary surface. This texture is prepared by surface treatments such as rubbing prior to the enclosure of the liquid crystal.

It is often desirable for device applications and in the measuring of certain physical properties to provide a nematic layer in a homeotropic texture, i.e., large areas in which the optic axis is oriented perpendicularly to the boundary. Commonly, homeotropic texture layers have been prepared by pretreating the surfaces which subsequently enclose the nematic layer with a strongly oxidizing acid such as nitric or chromic-sulfuric acid. This technique has been found to be undesirable in many situations because it cannot be used with oxidation-sensitive (transparent conductive coated) surfaces. Particular care is required to avoid the touching of a surface so treated, and homeotropic nematic layers produced by this technique tend to be unstable, i.e., their homeotropic property tends to decrease in an undesirably short period.

According to this process, there is provided a nematic liquid crystal material which assumes a homeotropic texture comprising a nematic material having dissolved therein a material having the formula: $RR_3'N^+X^-$ where R is selected from group consisting of alkyl radicals having 10 to 24 carbons, R' is selected from the group consisting of methyl and ethyl and X^- is selected from the group consisting of anions derived from simple acids.

Typical examples of nematic materials may be such compounds as p-methoxybenzylidene-p-butylaniline, p-methoxybenzylidene-p-aminophenyl acetate, p-azoxyanisol and butyl-p-(p-ethoxyphenoxycarbonyl)phenyl carbonate. A suitable example of R may be a hexadecyl radical. The anion may be a halide ion such as Cl^-, Br^-, I^-, or an anion derived from oxyacids such as NO_3^-, and the like. A typical example of the material symbolized by the structural formula may be hexadecyltrimethyl-ammonium bromide which may be present in its solution in the nematic material in a range of 0.25 to 2.5% depending, of course, on its solubility characteristics relative to the spectific nematic material which is employed.

Example 1: An 0.49% solution of hexadecyltrimethyl-ammonium bromide in p-methoxybenzylidene-p-butylaniline was prepared by mixing 0.0024 and 0.495 grams, respectively, of the materials in a vial, heating to a temperature where the nematic liquid first turned isotropic (approximately 46°C) and shaking until the last specks of the solid mass dissolved. A drop of the solution at room temperature was placed between untreated glass plates whose separation was effected by a 2-mil thick Teflon gasket. The sample was then examined between crossed polars on a polarizing microscope over its entire nematic temperature range and extinction was observed independently of the direction of the sample relative to the polarization direction. Upon the shearing of the liquid crystal by selective motion of the two plates, light was observed to be transmitted. These observations indicated that the optic axis of the nematic fluid, at rest, is perpendicular to the glass plates, i.e., because of the additive, it assumed a homeotropic texture.

In a second experiment, the same procedure as described above was followed with the exception that the glass slides which were employed were first repeatedly rubbed on a cheese cloth in a given direction. Such rubbing treatment is known to normally cause pure p-methoxybenzylidene-p-butylaniline, which is subsequently enclosed, to assume a texture in which the optic axis lies in the plane of the glass and in the rubbing direction. The inventive solution, as described hereinabove, however, was again observed to assume the homeotropic texture.

In a third experiment, the same solution was enclosed between electrically conducting tin oxide-coated glass plates where it again exhibited a homeotropic texture. A DC voltage was then applied to the plates, while the appearance of the sample was monitored on the polarizing microscope and the intensity of the transmitted light (between crossed polarizers) was measured with a CdS photocell replacing one of the eyepieces of the binocular microscope. With an applied voltage of 0 to 5 volts, the structure appeared uniformly dark. At 7.5 volts, however, the liquid crystal material became transparent and, under magnification, it appeared stationary.

The contrast ratio, i.e., transmitted light intensity at 7.5 volts relative to that at zero volts was found to be 16. With applied voltages of 10 volts or higher, flow patterns and turbulence appeared similar to those patterns and turbulence observed in nematic materials of any texture. The maximum contrast which was achieved was 31 at an applied voltage of 24 volts. In a comparison, the maximum contrast ratio measured under similar conditions, of elements containing nematic liquid crystals with no additive, and aligned in the plane of the boundary surfaces in the polarization direction was found to be only 11 to 17.

Example 2: Hexadecyltrimethyl-ammonium bromide was dissolved in butyl-p-(p-ethoxyphenoxycarbonyl)phenyl carbonate in a concentration of 0.4 to 1.1% by weight, to provide the inventive nematic liquid crystal material.

Example 3: Hexadecyltrimethyl-ammonium bromide was dissolved in p-methoxybenzylidene-p-aminophenyl acetate in a concentration of 0.4 to 1.1% by weight to provide the inventive nematic liquid crystal material. A sample containing 0.95% of the additive showed particularly good results when tested.

Hexadecylpyridinium Bromide

According to a process described by *H. Sorkin and R.I. Klein; U.S. Patent 3,698,449; October 17, 1972; assigned to RCA Corporation* to provide a uniform composition of the liquid crystal material throughout the device, the interior surfaces of the device are coated or contacted, prior to filling with the liquid crystal material, with certain ionizing additive materials normally contained in the liquid crystal material to improve the device performance. Various methods are possible to achieve the precontacting operation. In one method, a preliminary quantity of the liquid crystal material itself can be flushed through the container with the result that various ingredients of the liquid crystal material are stripped therefrom to desirably precontact or dope the container surfaces.

A sufficiently large quantity of the liquid crystal material is used whereby a thorough flushing of the container occurs. Thereafter, a new or fresh batch of the liquid crystal material is used to fill the container. Since the container surfaces have been precontacted or predoped with the various liquid crystal material ingredients, little or no additional stripping of these ingredients from the fresh batch of liquid crystal material used in the final filling of the container occurs, whereby the composition of the liquid crystal material throughout the container is substantially uniform.

In another method, a gaseous medium carrying various ones of the additive materials which tend to be stripped from the liquid crystal material, or a vapor of these additive materials alone, is flushed through the cell prior to the filling of the cell with the liquid crystal material. The additive materials are thus deposited on the interior surfaces of the liquid crystal cell.

One material having utility as an ionizing additive in various liquid crystal compositions is hexadecylpyridinium bromide. According to one example of this process, a solution of this material (HDPB), comprising 0.01 to 0.2%, by weight, of the HDPB, balance isopropanol, is flushed through the cell via the two tubulations. The extent of flushing is not critical, three to five times the cell volume of solution can be used. The excess solution is then pumped out of the cell by drawing a vacuum through one of the tubulations. The cell is then heated in a vacuum oven at a moderate temperature, in the range 30° to 60°C, so as not to decompose the additive material, for 3 to 4 hours. The dried cell is then filled with the liquid crystal material, and the cell sealed off.

While the liquid crystal material used in the above example can contain some of the HDPB, in the order or 0.001%, by weight, in some instances, the liquid crystal material can contain no HDPB at all. In the past, obtaining adequate device performance required the use of the HDPB in the liquid crystal material at a concentration in the order of 0.05%. Using a liquid crystal material containing no HDPB, in accordance with the method, the electrical resistance of the cells is increased by a factor in the order of 100 in comparison with the prior art devices.

In another example of the process, a mass of the HDPB is gently heated in a closed container to a temperature in the order of 65°C, and the vapors so produced are passed through the liquid crystal cell maintained at a temperature of about 95°C. Preferably, to obtain more uniform coating of the cell interior surfaces, the vapor is caused to flow through the cell in one direction, for a period of time in the order of 15 minutes, and then caused to flow through the cell in the other direction, as by admitting the vapor into the cell through the other tubulation, for about the same period of time.

Comparing the above-described two methods, somewhat better results with respect to the uniformity of coating of the cell surfaces, are obtained using the liquid carrier method.

Long Chain Acids, Ureas and Silanes

J.E. Goldmacher and M.G. Tayag; U.S. Patent 3,809,456; May 7, 1974; assigned to Optel Corporation describe an electrooptical display device which comprises a liquid crystal display cell comprising spaced, parallel support plates having electrodes thereon, at least one liquid crystal compound between the electrodes and at least one aligning agent between the electrodes, the aligning agent selected from the group consisting of long-chain carboxylic acids, long-chain alcohols, ureas and silanes. The aligning agent may be present in a concentration of from as little as a monolayer on the surface of an electrode to up to 2% in the liquid crystal material.

Examples of specific useful aligning agents include p-n-octyloxybenzoic acid, dodecanol, n-hexadecyl urea, and dodecyltriethoxy silane. Generally, preferred chain lengths are from 4 to 18 carbons. Wetting and adhesion to the surface is generally hampered with increasing chain length, thereby causing the liquid crystal to align homeotropically.

The addition of the aligning agent to the device increases the attainable contrast ratio as compared with similar devices which do not incorporate the aligning agents. In a typical light-scattering type of electrooptical cell, in the absence of an applied signal to the electrodes, the liquid crystal cell appears highly transparent and upon application of a voltage the material appears milky white in the region of activated segments. Thus, the contrast ratio of the device is improved. A similar increase in contrast ratio is observed in devices employing liquid crystal which affect a change in transmission or reflection of polarized light.

A typical device would comprise a 12μ thick layer of a nematic liquid crystal mixture of for example, p-methoxybenzylidene-p'-butylaniline and p-ethoxybenzylidene-p'-butylaniline containing 2 weight percent p-octyloxybenzoic acid as an aligning agent and 1×10^{-3} weight percent of tetraphenyl arsonium chloride or triphenylheptylphosphonium bromide. At 15 volts, 32 Hz this device shows uniform dynamic scattering while utilizing only 2μ amperes of current per square centimeter.

Surface Treatment for Orientation Purposes

L.H. Hutfles; U.S. Patent 3,459,839; August 5, 1969 describes the making and reproduction of light polarizing patterns on surfaces which transmit or reflect light. It has been found that one can easily transmit to successive receptor surfaces polarization patterns of a master such as develop naturally, e.g., by crystallization or by rubbing a surface of the master along lines of desired polarization, e.g., as an artist might apply to prints with oriented brush strokes in different directions to emphasize or develop depth effects, etc., whether uniformly over an entire area on the master or in different directions on different smaller areas, for more complex patterns.

This is accomplished according to this process, if the two are in intimate surface contact while the material of the receptor is softened so that its molecules can respond to orienting forces, and if the materials of both have polar molecules of such geometry that the attractive forces at the surface act to orient such molecules of the receptor into parallelism. This is achieved by exposing to heat or a solvent the receptor surface in which orientation is desired, while the contacting master surface is maintained under conditions such that its molecules are held against reorienting.

In the simplest case, a sheet is parallel-oriented throughout, e.g., by rubbing always parallel to a single directrix. When such matrix is pressed against a softened body of a polar plastic, the molecules of which have substantial freedom of orientation, the same pattern of orientation is reproduced on the softened body. There need be no visible indication on the receptor that it has been thus oriented, although if there are surface irregularities or patterns on the matrix, they can be expected to appear molded into the contacting surface of the receptor. However, when the surface is treated to make it light-polarizing, e.g., with a nematic dichroic dye, the dichroic material will be oriented accordingly and will thus polarize light.

It is possible to reproduce a light-polarizing pattern or other anisotropic surface characteristic in multiple copies from a master, or matrix, by a technique analogous to ordinary printing. Thus, a first copy is made by rubbing the surface of the body (ordinarily a plate or sheet or film) which is to form the master, with the direction of rubbing in each area uniform and corresponding to the polarization desired in that area. This handmade master can be put in a press with a sheet or plate of polar thermoplastic which is to receive the oriented pattern, its surface being softened by heat or solvent, and then allowed to set while the matrix is in intimate contact. Surprisingly the only copy made in this way can be used as the matrix for producing other copies for ultimate use, e.g., as light polarizers, etc., or for producing other matrices for use on other presses to increase production.

The oriented pattern can be transmitted, not ohly successive materials applied onto the matrix, but successively from one matrix to a responsive copy material and then, when that material is set with the oriented pattern, it can in turn transmit the pattern to another soft material, and so on through an indefinite number of generations. This has the practical advantage that a matrix, which can be used like a printing plate to form an indefinite number of copies, can be formed of a plastic film and such film can then be used as the matrix. A single matrix can thus make any number of reproduced matrices, each of which can turn out multiple copies of the original. For example, the films with polarized images now need no release additives and can have good bond to its substrate. The strip-off plastic film can be much stronger than the dichroic film and is less apt to break during the strip-off. The surface of the matrix has not been found to be critical, nor does normal handling substantially impair the matrix. The repeated application of the sheet-forming material and its stripping from the matrix keeps the surface of the matrix clean.

It is desirable that the plastic material of which copies are being made, should be softened adjacent the contacting surface to facilitate reorientation of the molecules. To this end, it is advantageous to heat the plastic material, or less advantageously, to soften with a solvent, at, or close ahead of, the position at which pressure is applied to bring the oriented surface and the plastic copy material into intimate contact; and, beyond the point of pressing these together, the plastic should be cooled so that it is set on the matrix before being stripped. In order to do this efficiently, the matrix or the supporting cylinder, or other carrier, is made thin, with as little heat capacity as possible.

Thus, the cylinder, or other loop of the matrix, as it passes beyond the heated press roll, is exposed to the atmosphere, or to a blast of cooling fluid, so that it is quickly set, fixing the latent image induced in the copy by the contiguity of the oriented image on the matrix. The oriented copy is then stripped off from the matrix, which moves on toward its initial position adjacent the pressing rolls, where additional plastic material is supplied, and is then passed under the pressure roll to form another oriented copy by repetition of the process. This can be done at a speed which is entirely economical.

The plastics used, and the dichroic materials used, may be the same as those used in the prior art for making such latent images and for developing them into light-polarizing plates or films, see for example, U.S. Patent 2,544,659. For best results the plastic of the film should be one having polar linear molecules and high tensile strength, relatively high melting point, and toughness (in the strict sense of energy absorbed in its deformation). For the plastic film, some of the newer materials have been found particularly advantageous, for example, Videne, Mylar, Terafilm and other commercially available terephthalate polyester films. It has been found that the para-terephthalates are better than those in which the two active groups are not in the para arrangement. Also, it is better to use films free from molecular orientation produced in manufacture of the film. For additional detail on these materials see "Polymers and Resins" by Brage Golding (Van Nostrand), page 285, U.S. Patent 2,965,613, and the *Soc. Photograph. Eng. Jour.* (1961) vol 17, No. 10, pages 1-5.

Polyamide Resins

W.E.L. Haas, J.E. Adams and B. Mechlowitz; U.S. Patent 3,803,050; April 9, 1974 and W.E.L. Haas, J.E. Adams and J.B. Flannery, Jr.; U.S. Patent 3,687,515; August 29, 1972; both assigned to Xerox Corporation describe liquid crystalline compositions containing an advantageous aligning agent resin, which causes the compositions, upon application to any suitable substrate to spontaneously align or to have a greater tendency to become aligned in an orientation where major axes of the molecules comprising liquid crystalline material are substantially perpendicular to the plane of the substrate. Such liquid crystalline compositions are particularly suited for use in phase transition systems where a film comprising cholesteric liquid crystalline material having the tendency to align is transformed to the nematic liquid crystalline state by the application of electrical fields of sufficient field strengths. Such compositions are also used in other liquid crystalline imaging systems making use of differentially aligned portions of liquid crystalline films.

The aligning agents suitable for use in the process include a wide variety of chemical compounds including a variety of resins and surfactants (surface active agents). Resins suitable for use as aligning agents include polyamide resins, for example, Versamid 100, a polyamide resin available from General Mills, Inc.; and other Versamid resins such as Versamid 115, 125 and 140; epoxy resins such as Easypoxy (part 2), available from Conap, Inc.; Dow's Novolac phenol-formaldehyde resins; Polyscience's Resin DER 334; Catalyst "9" available from Emerson & Cuming; Marson's Fiberglass Epoxy; Dupont's Elvacite 2046, an n-butylisobutyl methacrylate copolymer; and other various types of resins including resins and resinous derivatives of fatty oils and acids, phenoplasts, aminoplasts, aliphatic amines, glycidyl ethers, polyglycols including amine-terminated and epoxy-terminated polyglycols, polyols, polysulfides, and others.

Other materials which are sometimes identified as surfactants (surface active agents) are also suitable as the advantageous aligning agent in the process. Such surfactants include esters of polyhydric alcohols, alkoxylated amides, esters of polyoxypropylene glycols, ethers of polyoxyalkylene glycols, alkylphenyl ethers, tertiary acetylenic glycols, polyoxyethylated alkyl phosphates, carboxylic acids, soaps and sulfonated hydrocarbons and their salts. The aligning agents are typically added to the liquid crystalline material in amounts in the range between 0.1 and 10% by weight, which amounts are typically sufficient to facilitate the advantageous spontaneous alignment. The following examples illustrate the process.

Example 1: A nematic liquid crystalline mixture comprising p-methoxybenzylidene-p'-n-butylaniline (ABUTA) is doped with 0.5% Versamid 100, a polyamide resin, and this composition is spread on a thin film on a standard microscope slide. The nematic liquid crystalline composition containing the aligning agent spontaneously aligns in a nematic homeotropic texture where the major axis of the molecules comprising the nematic liquid crystalline material are substantially perpendicular to the plane of the substrate.

Example 2: A thin line of a specific aligning agent such as Novolac (a phenyl-formaldehyde resin), Elvacite 2046 (an n-butyl/isobutyl methacrylate copolymer), diethylenetriamine, T4573 and triethylenetetraamine, P4572 (Eastman Chemicals) is placed on a microscope slide, and a drop of ABUTA is placed over the thin line of the aligning agent thereby allowing the agent and the nematic liquid crystalline material to mix at their interface to an extent not greater than about 10% by weight of the aligning agent.

In each case the liquid crystalline material doped by the aligning agent assumes the aligned orientation where major axis of the molecules comprising nematic liquid crystalline material are substantially perpendicular to the plane of the substrate.

Example 3: A thin line of each one of a surfactant-type aligning agent such as sulfated nonylphenyl ether tetraethylene glycol ammonium salt, polyoxyethylated nonylphenol (alkylphenyl ether), zinc naphthenate, cobalt naphthenate, etc., is placed on a standard microscope slide and a drop of ABUTA placed on the line of the aligning agent.

Example 4: A liquid crystalline electrooptic cell is prepared by first preparing a mixture of liquid crystalline materials comprising about 35% cholesteryl chloride and about 65% of a mixture of p-ethoxybenzylidene-p'-n-butylaniline (EOBUTA) and p-butoxybenzylidene-p'-n-butylaniline (BOBUTA) which is a mixture of materials which have cholesteric liquid crystalline characteristics. Four parts are mixed with one part of a composition comprising ABUTA and about 2½% of the aligning agent Versamid 100. A layer of the aligning agent, doped cholesteric liquid crystalline mixture, is placed between and in contact with a pair of substantially transparent electrodes of tin oxide-coated glass which are separated by an about ½-mil thick Tedlar (polyvinyl fluoride) spacer. The electrodes are connected in circuit with an electrical generator. When an electrical field provided by placing a voltage of about 275 volts DC is placed across this sample, the electrical field-induced cholesteric-nematic phase transition occurs.

As a control cell, a second cell having the same component mixture without the Versamid 100 aligning agent is used, and it is found that the electrical field-induced cholesteric-nematic phase transition threshold voltage is about 400 volts DC. It is therefore clear that the addition of the aligning agent of the process significantly reduced the total electrical energy necessary to achieve the cholesteric-nematic phase transition.

It has also been found that the liquid crystal layers or films of the process are more readily spontaneously aligned when the liquid crystalline composition is doped with a small amount of a surfactant additive. For example, polymeric resin dopants, such as Versamid 100, a polyamide resin and other Versamid resins such as Versamid 115, 125 and 140, added to the liquid crystalline composition in amounts in the range between 0.1 and 10%, facilitate the spontaneous alignment which is desired in the process.

Example 1: A nematic liquid crystalline electrooptic cell is provided by placing a pair of aluminum strips having square cut ends on a standard microscope slide and adjusting the square ends of the strips so that they are about 4 mm apart. The aluminum strips are of thickness of about 15μ. In the space between the aluminum electrode strips on the substrate slide, a layer of p-methoxybenzylidene-p'-n-butylaniline (hereafter ABUTA) is provided, and is leveled to about the thickness of the aluminum electrodes by placing a standard microscope cover slide over the layer of nematic liquid crystalline electrooptic composition. The nematic composition assumes the nematic homeotropic texture where the optic axis of the composition layer is aligned normal to the plane of the layer. This imaging cell is then placed under a Leitz Dialux microscope between crossed polarizers. In parallel or collimated transmitted light, the cell appears dark in all stage positions, and in convergent light the conoscopic figure is a uniaxial cross. The optic sign of the composition is positive, i.e., light travels fastest in the direction of the optic axis.

A DC field of about 200 volts/centimeter is applied to the sample, and during the application of the field the transmitted light is extinguished only if the direction of the field, which is normal to the line of transmission and parallel to optic axis of the layer of electrooptic composition, is parallel to the direction of vibration of either the polarizers or the analyzer. It is observed that during rotation of the microscope stage through 360°, the extinction occurs 4 times. Maximum transition is observed to occur when the field direction is at an angle of about 45° from the direction of vibration of the polarizer or analyzer, which are crossed at about 90°.

The sample is observed to transmit light uniformly and transmission increased with increasing electrical fields. Initially, the color of the transmitted light is white, but as the field strength is increased uniform interference hues become visible. Field reversal does not affect the result. When the electric field strength exceeds about 1,500 volts/centimeter of sample width, the sample uniformity decreases. At high fields of about 3,500 volts per centimeter of sample width, dynamic scattering is observed to occur.

In convergent light the conoscopic figure is clearly uniaxial before the application of the field. The interference figure is exceptionally sharp. The application of the DC field results in the breaking of the uniaxial figure, and when the film is observed in the 45° position (electrical field direction of about 45° to the direction of vibration of the polarizer or analyzer) two hyperbolic isogyres are seen, the separation of which increases with field strength.

Example 2: A nematic liquid crystalline electrooptic cell is provided as described in Example 1, however, in this cell the nematic composition is doped with about 0.5% Versamid 100, a polyamide resin. The dopant enhances the spontaneous alignment of the composition to the nematic homeotropic texture.

Example 3: A nematic liquid crystalline electrooptic cell is provided as described in Example 1, however in this cell the nematic composition is methoxybenzylidene-aminobutyrophenone (MBAB). This cell exhibits electrooptic effects as described in Example 1.

Example 4: A nematic liquid crystalline electrooptic cell is provided as described in Example 3, however, the nematic composition is doped with about 2.0% Versamid 115, a polyamide resin. The dopant enhances the spontaneous alignment of the composition to the nematic homeotropic texture.

Oriented Light Scattering

J.V. Cartmell; U.S. Patent 3,749,474; July 31, 1973; assigned to The National Cash Register Company describes an electrooptical, light-scattering display device which employs nematic liquid crystal material. The faces of the device bearing the conductive electrodes are unidirectionally oriented by rubbing. The liquid crystals align with their molecular axes parallel to the surface of the faces and along the rubbing direction with the field off. When the field is on, dynamic light scattering occurs in the electrode area. Due to this rubbing alignment, a rapid response between the on-off states is evident.

Example 1: A test cell comprising a seven-bar figure eight design was constructed using 2" x 2" glass substrates. These glass plates were initially coated with transparent electrode (indium oxide) and etched on the HCl to the seven-bar design. The interior surfaces of each side of the cell were rubbed unidirectionally with optical cloth such as to cause alignment of the nematic liquid crystal on the interior cell surface along the rubbing direction. Upon assembly of the cell the substrates were positioned such that the rubbing direction of the aligned interior surfaces were 90° to one another. Thus when the liquid crystal layer was placed between the glass substrates a 90° helix of the molecular axes was imposed in the thin layer, i.e., about 0.15 mil.

The liquid crystal mixture utilized in this test cell consisted of a 1 mol to 1 mol mixture of p-methoxybenzylidene-p-n-butylaniline (MBBA), p-ethoxybenzylidene-p-n-butylaniline (EBBA) with a liquid crystal range of 0° to 59°C. The test cell was sealed around the exterior edges with ethylene acrylic acid polymer. Upon placing uncrossed polarizers in front of, and behind the test cell a dark state (cross polar effect) was observed due to the rotation of the polarized light by the liquid crystal 90° helix.

Upon subjection of an electric field to the test cell and the resulting turbulence caused by dynamic scattering, those areas with the electrode seven-bar pattern exhibit and transmit light. Contrast ratios of 50:1 are typical. This device also functions without polarizers with the viewer observing scattered light in the field imposed areas. Typical response times for this cell of 30 volts, 60 cycle AC drive were 10 milliseconds on and 40 milliseconds off.

Example 2: A second cell was constructed identical to the above except elimination of the rubbing of the interior cell surfaces. In this case, placing polarizer behind, and in front of the cell does not result in a dark background in that random orientation of the liquid crystal layer on the interior cell surfaces causes some light to be transmitted through the polarizers regardless of their orientation. The observed scattering of light for this test cell was markedly inferior to that of the rubbed cell. Typical response times for this cell were 40 milliseconds on and 100 milliseconds off.

Parallel Grooves in Contact Surfaces

D.W. Berreman, S. Meiboom, D.L. White and F.J. Kahn; U.S. Patent 3,787,110; January 22, 1974; assigned to Bell Telephone Laboratories, Inc. describe a device for the alignment of liquid crystals. In general, such alignment results from minimization of elastic energy in the body of the liquid crystal. Alignment results when use is made of a substrate having ridges or grooves of dimensions and spacings of the order of tens of angstrom units or greater. Such grooves or ridges may be produced on an inorganic or organic media totally lacking in the characteristics required for the traditional mechanism utilizing Van der Waals forces. Such substrate surfaces may be produced in a variety of manners as by mechanical scoring, casting, etching, photolithography, ion bombardment, etc. Substrate may be totally amorphous, may be polycrystalline, or even crystalline. In the latter, surfaces may be modified with ridges running in a direction different from that which would otherwise be induced by virtue of any influence resulting from the crystallographic lattice itself.

Devices constructed in accordance with the process may exhibit certain operational advantages over prior art devices. Improved lifetime, for example, may arise simply by reason of the broader class of substrate material made available. Similarly, longer lifetime may result by elimination or at least lack of dependence on orienting linear long-chain molecules which may react with liquid crystals or other media or which may deteriorate due to any of a variety of mechanisms (e.g., reaction between liquid crystal and substrate oxidation of substrate, etc.).

Figure 3.6a is a cross-sectional view of a fundamental type of device. Figure 3.6b is a perspective view of a specific device which may operate in the manner of the generic device shown in Figure 3.6a. Figure 3.6c is a cross-sectional view of a different device. Figures 3.6d, 3.6e and 3.6f are a composite view and partial electrode detail view of a different device utilizing grooved material.

FIGURE 3.6: LIQUID CRYSTAL DEVICE

a.

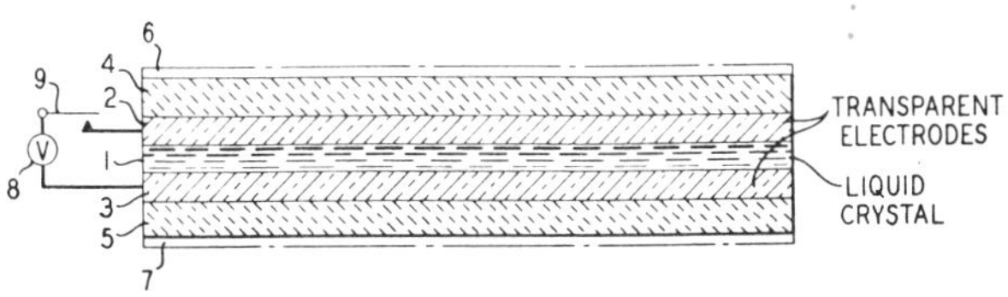

b.

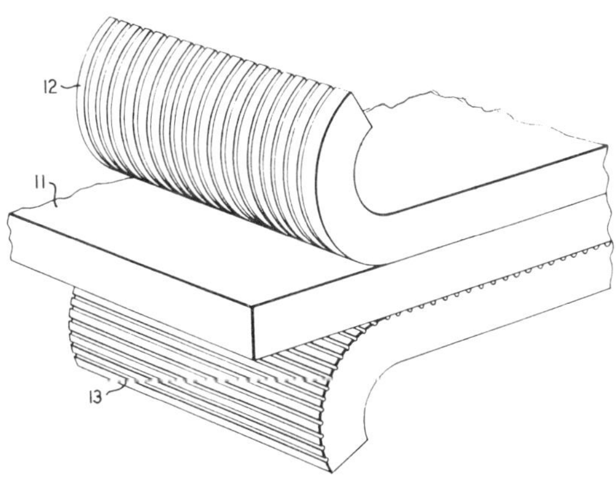

c.

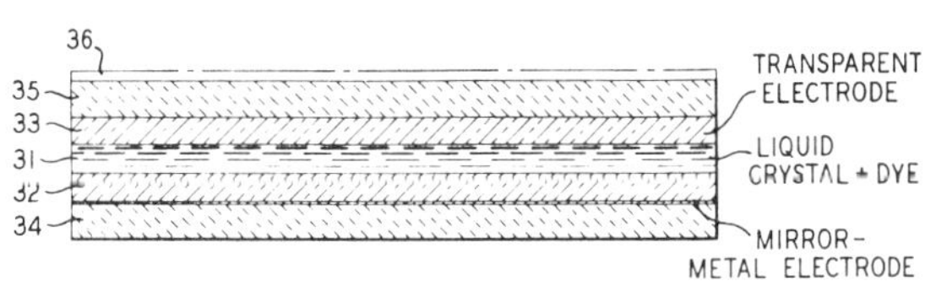

(continued)

FIGURE 3.6: (continued)

d.

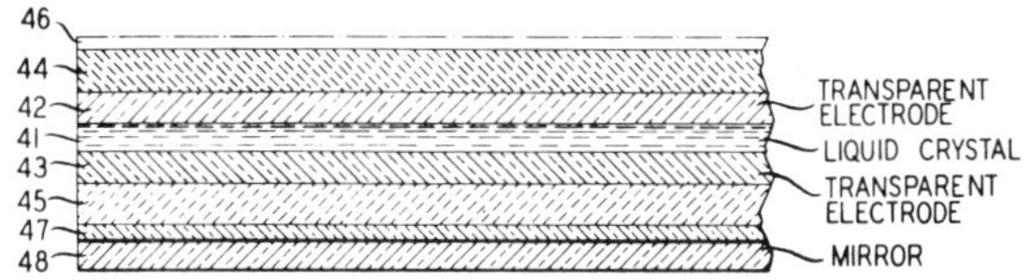

e.

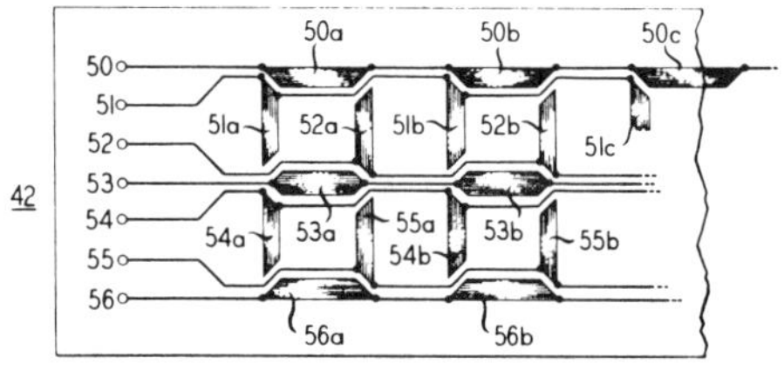

f.

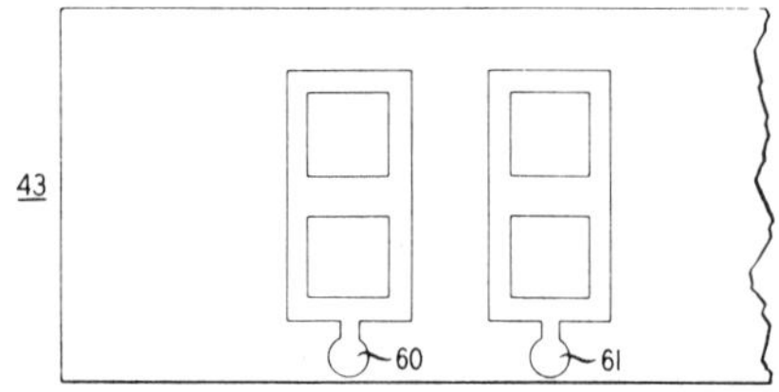

Source: D.W. Berreman, S. Meiboom, D.L. White and F.J. Kahn; U.S. Patent 3,787,110; January 22, 1974

The arrangement of Figure 3.6a includes a liquid crystal layer **1** sandwiched between two electrodes **2** and **3**, at least one of which is grooved so as to orient the liquid crystalline material of the layer. One or both electrodes may be transparent depending upon whether the device is designed to operate by reflection or transmission. The electrodes are optionally supported by layers **4** and **5** which are transparent or reflective in accordance with the intended operation of the device shown. Either or both of layers **4** and **5** may be provided with grooves which are replicated by electrodes **2** and **3** so as to finally influence alignment of layer **1** or layers **4** and **5** may be grooved. Elements **6** and **7**, shown in phantom, are polarizers which may be crossed or not and which function to introduce or enhance contrast as between two operating states of Figure 3.6a. Electrodes **2** and **3** are energized by electrical source **8** when switch **9** is closed. Electrical source **8** may function as a current source, in the instance of dynamic scattering devices, or, primarily, as a voltage source in the instance of field effect devices.

The device of Figure 3.6a may also be utilized as a simple polarizer by appropriate choice of composition of material **1** and groove orientation of interface/s formed between layer **1** and **2** and/or **3**. Polarization may result in the absence of energizing, in which event, electrical source **8** may be eliminated.

Figure 3.6b depicted in the form of a device of Figure 3.6a, represents a species of that device in that electrodes **12** and **13** are grooved orthogonally with respect one to the other on the surfaces forming interfaces with liquid crystalline material **11**. The function of the grooving arrangement shown is to impose a twist in the liquid crystal alignment proceeding from one electrode to the other. Such a device is exemplary of the family of devices herein which produces rotation of the plane of polarization of light. Energization by electrical source **18** upsets the order or, alternatively, reorders the crystals with their major axes perpendicular to the plane surface. Unenergized, the 90° twist imposed on the alignment of the liquid crystal from electrode **11** to electrode **12** results in transmission through crossed polarizers **16** and **17**. With a field imposed or with the passage of current, the effect of the crossed polarizers is to block transmission.

The device of Figure 3.6c is a dye cell which may operate by reflection or transmission. It consists of layer **31** of a liquid crystal and dyestuff combination with the dyestuff being of such nature as to align with the liquid crystal. Layer **31** is sandwiched by layers **32** and **33** which may both function as electrodes with current or voltage supplied by source not shown. Both electrodes **32** and **33** may be transparent or, for reflective operation, one, for example, **32**, may be mirrored. In common with all devices depicted and described, alignment of the liquid crystal layer **31** is influenced by grooves or ridges which are parallel over at least a portion of one or both of the electrode surfaces forming interfaces with layer **31**. Layers **34** and **35**, which may serve as the actual substrate bodies, may be constructed of any of the materials described herein.

Detailed requirements for the nature of materials contacting the liquid crystalline medium are described in the following section. Layer **36** is so designed as to selectively transmit some portion of light, depending upon the state of liquid crystal layer **31**. For example, for a noncholesteric nematic crystalline layer oriented parallel to a major electrode surface in the nonenergized state, polarizer **36** may be crossed with respect to the alignment direction of the liquid crystalline layer. Under such circumstances, light is not transmitted (or reflected) in the unenergized state. With passage of current or application of voltage, layer **31** is disordered or reordered so as to permit transmission (or reflection) of light energy of the appropriate polarization. The purpose of the dyestuff in such a cell is to increase absorption of the liquid crystal for the incident light so as to increase the polarization extinction ratio.

Figures 3.6d, 3.6e and 3.6f are an elevational view and electrode details, respectively, of a form of liquid crystal device of a coincident array design designed to operate as a numerical display. The device depicted consists of liquid crystal **41** sandwiching electrodes **42** and **43**, supporting layers **44** and **45**, and polarizers **46** and **47**. Layer **48**, shown in phantom, is a mirror which permits use of the depicted device in a reflective mode. Orientation of layer **41**, and functioning of all other elements depicted, is, as described in any of the figures above except for the detailed operation of electrodes **42** and **43**.

Detail Figure 3.6e depicts the pattern of electrode **42**, while detail Figure 3.6f depicts the pattern of electrode **43**. Electrode array **42** consists of a seven-segment electrode arrangement found sufficient for numerical display. Electrode **50** results in orientation within the region of crystalline material corresponding with segments **50a, 50b, 50c**, etc. Electrode **51** similarly influences regions corresponding with segments **51a, 52a,** etc., and similarly denoted segments are influenced by remaining electrodes **52** through **56**. Electrodes of electrode layer **43**, shown in detail 3.6f, two of which are designated **60** and **61**, define figure eights and are addressed sequentially at times corresponding with desired portions of electrode layer **42**. Such a device may operate, for example, in the manner of the device of Figure 3.6b with operation as described.

Disalignment or realignment results only in those portions of liquid crystal layer **41** sandwiched between energized segments of electrode layers **42** and **43**. The patterns depicted are in commercial use for numerical displays.

The device of Figure 3.6d is representative of a family of coincident energy devices which may serve a variety of functions, as displays, either transmissive or reflective. A variety of electrode patterns operating coincidentally, either by passage of current or imposition of field, may result in a variety of symbols or designs of which the patterns of details 3.6e and 3.6f are exemplary.

Example 1: Scored surfaces were produced on silicate glass microscope slides by use of diamond paste with average particle size of one micrometer imbedded in leather. Slides were rubbed unidirectionally several times in order to result in grooves of depth of approximately 500 angstrom units separated by approximately 1000 angstrom units. The slides were cleaned first with organic solvent, then with chromic and sulfuric acid, rinsed with distilled water, and followed by baking at 500°C for a period of about 1 hour. A sandwich of para-azoxydianisole of a thickness of approximately 10 micrometers was formed between the two scored surfaces with the grooving parallel. The cell was rotated between cross polarizers with a maximum extinction ratio of 20 or greater.

Examples 2 through 5: Similar cells were constructed, however, utilizing fused silica in lieu of silicate glass:

Example	Liquid Crystal Composition
2	para-azoxydianisole
3	MBBA
4	heptyloxyazoxybenzene
5	octyloxyazoxybenzene

Extinction ratios are as reported in Example 1.

Examples 6 and 7: Whereas the above examples are utilized nematic phase material, these examples were operated at reduced temperature so that the liquid crystal media, the compositions utilized in Examples 5 and 6, were in the smectic phase. The observed extinction ratio was somewhat less than for the nematic phase due to a slight angular displacement as between the two permitted domains. For this smectic C phase, the overall extinction ratio was about 5:1.

Inert Coatings on Substrates

H.S. Cole, Jr.; U.S. Patent 3,836,231; September 17, 1974; assigned to General Electric Company has found that the application of a thin inert coating to the surface of the substrates which may include conductive or semiconductive regions thereon unexpectedly produce a uniformly clear liquid crystal display device and one exhibiting substantially longer lifetimes than those without such a coating. The primary requisites of the inert materials are insolubility in the liquid crystal material, adhesion to the substrate material with its conductive coatings thereon and a degree of transparency. Inert materials useful for the process include such materials as vinyl acetate copolymers, ethylene-based resins, epoxies, anaerobic adhesives, polyimides and linear saturated polyester polymers, for example.

Typical materials which meet these requirements and produce liquid crystal display devices with the aforementioned desirable characteristics include: (1) The 200-series ELVAX resins, ELVAX 250, having a vinyl acetate concentration of between 27 and 29% of the total composition, is particularly useful. (2) VITEL resins, identified by the #PE-200. These resins are linear, saturated polyester polymers having molecular weights in the range of 20,000 to 30,000. (3) LOCTITE anaerobic adhesives. These adhesives are single-component acrylic diesters which cure in the absence of air. LOCTITE adhesive #404 is particularly useful. (4) Epoxies, such as Crystal Clear Epoxy; and (5) Polyimides, such as those described in U.S. Patent 3,448,068.

Example 1: A pair of transparent glass substrates having a conductive coating of tin oxide thereon are provided with an ELVAX-250 coating having a thickness of approximately 200 A. This coating is applied by spin-coating from a xylene solvent containing 3% ELVAX-250. The substrate with its coating is baked in an oven for approximately 15 minutes at a temperature of approximately 140°C to remove residual solvent. The plates are then placed in parallel relation, about 15μ spacing, and a suitable gasket provided along the periphery thereof and the space between the parallel plates is filled with a nematic liquid crystal, such as N-(p-methoxybenzylidene)-p-butylaniline, having a nematic temperature range of 20° to 46°C.

The fabricated display device is substantially transparent in the unactivated condition and exhibits negligible light scattering as a result of ordering influence of the polymer coating. The application of a DC voltage of approximately 30 volts between the conductive coatings produces light scattering in the activated areas of the display device and produces a contrasting image with respect to the background or unactivated areas of the display device.

Example 2: A display device similar to that described in Example 1 is fabricated in substantially the same manner but the inert polymeric material is VITEL PE-200. The thickness of this linear saturated polyester is 1,000 A and the resultant display device exhibits substantially the same characteristics as that described in Example 1.

Example 3: A display device similar to that described in Example 1 is fabricated with an inert coating of LOCTITE adhesive 404, an anaerobic adhesive. The thickness of this coating is approximately 500 A. The coated substrates are baked under vacuum for one-half hour at 140°C. The resulting display device functions in substantially the same manner described above in Example 1.

Silane Derivative Chemisorbed on Substrate

According to a process described by *F.V. Allan and P.Y. Hsieh; U.S. Patent 3,728,008; April 17, 1973; assigned to Ing. C. Olivetti and Co., SpA, Italy* a silane derivative is chemisorbed on the interior surfaces of the substrates of a liquid crystal display to eliminate cloudiness of the display and to increase contrast.

According to the process, the substrate and the electrodes deposited on the surface thereof are rendered hydrophobic by chemisorbing a very thin layer of a cationic silane derivative on the surface. The silane derivative may be generally described by the formula R_nSiX_{4-n} where R is an organofunctional group which includes an alkyl and X designates a hydrolyzable group attached to the silicon; for example, an alkoxy group (usually methoxy) or a halogen.

The chemisorbed coating of the silane derivative may be deposited by immersing the electrode coated substrate in a solution of the silane derivative in water or an organic solvent or in a silane derivative vapor. In the case of glass substrates and most electrode materials, the X group reacts with OH groups on the substrate or electrodes eliminating a volatile compound, for instance, an alcohol or HCl, so that the silicon forms a strong chemical bond with the substrate or metal coated thereon, and the alkyl groups, being nonpolar, cause the surface to be hydrophobic. The silane derivatives are also chemisorbed by plastic substrates.

A very important effect of having the exposed surfaces of the substrate and electrodes hydrophobic is a great improvement in contrast. This occurs not only because the unenergized (background) portion of the liquid crystal is much clearer, but also because liquid crystal on the hydrophobic surface forms a close-knit honeycomb domain structure at a voltage slightly above the threshold voltage, instead of the usual sausage-like structure. This honeycomb structure gives much more effective light scattering. At higher voltages, the liquid crystal domains appear to be very small spheres, which give very efficient light scattering.

Some commercially available water-soluble cationic silane derivatives, which have been found to be desirable for use in accordance with the process, are marketed as XZ-2-2300 and Z-4141, and Siliclad. These materials may be used in an aqueous solution in any concentration ranging from 0.02 to 5%, depending on the application. When coating a glass substrate, it is preferred to use solutions having concentrations of between 0.02 to 0.5%, while the higher concentration solutions are preferred for coating metals. Also useful are Owens Illinois glass resins #100 and #650 and the chlorosilanes, which are soluble in organic solvents. The structure of the Dow Corning XZ-2-2300 compound is:

$$(RO)_3Si{-}(CH_2)_3{-}\overset{+}{N}(R')_2{-}R''$$

where R and R' are small alkyl groups (CH_3 or C_2H_5) and R" is a long alkyl chain (C_{16} to C_{18}). The other materials also all have the general formula R_nSiX_{4-n} given above. The following examples give more detailed descriptions of effective methods of preparing displays according to the process.

Example 1: Two pieces of Nesa glass are cleaned in a chromic acid solution or a caustic soda solution at pH 8 to 9 and rinsed thoroughly, first with clean tap water and then distilled water. The cleaned glass is immersed in a 1% aqueous soltuion of Siliclad for about 5 seconds. After complete immersion, they are removed and rinsed thoroughly with distilled water to remove any excess Siliclad from the surface, and the coated surfaces are air dried. The length of immersion time is not critical. A longer immersion may give a thicker layer of the coating; otherwise, there is no detrimental effect on the surface properties.

Air drying is adequate to get a fully cured treatment, but heat is often employed to speed up the curing process. The treated surfaces may be heated to a temperature of 100°C for 10 minutes. Any temperature up to 260°C may be used. The treated surfaces have a smooth hard film which is hydrophobic and gives a high contact angle with liquid crystal. The Nesa glass is used to prepare a liquid crystal display panel containing liquid crystal film of 1 mil thickness which is perfectly clear and which has high contrast when energized at 30 volts DC or AC. The panel shows no spotty appearance after a series of heating and cooling cycles.

Example 2: A clean glass plate on which a reflective layer of aluminum is evaporated by vacuum deposition is immersed in a 5% aqueous solution of Siliclad for 10 minutes. After the immersion, the plate is removed and rinsed thoroughly in distilled water to remove any excess Siliclad from the surface. The plate is air dried for 24 hours and fabricated into a display panel by sandwiching liquid crystal between it and a piece of Nesa glass treated as in Example 1 with an electrode spacing of 1 mil. The panel shows a reflectivity nearly equal to a bare aluminum portion which is not covered by the liquid crystal. The contrast is much higher than a panel which is prepared from an untreated aluminum coated glass plate.

Example 3: A solution of Dow Corning XZ-2-2300 is used in place of the Siliclad in Examples 1 and 2. The quaternary nitrogen atom in the XZ-2-2300 is believed to increase further the strength of the chemisorption of this compound on the surface of the substrate. This material is supplied in a 50% solution in an inert water-miscible solvent. It is diluted with distilled water to a concentration in the range of 1 to 0.1%. The immersion time is 1 minute or longer. Cells made in this way have the same excellent properties as in the preceding examples.

Squarylium Dye or Tetracyano-p-Benzoquinone

According to a process described by *S-Y. Wong; U.S. Patent 3,838,059; September 24, 1974; assigned to Hughes Aircraft Company* the operational characteristics (fast responding) are improved for the electrical field induced dynamic scattering mode (DSM) in nematics, and for emulsion storage scattering (ESS) in nematic-cholesteric liquid crystals. For example, small amounts of Squarylium dye material or of tetracyano-p-benzoquinone are particularly favorable dopants. The improved longevity and fast rise and decay characteristics also pertain to application of the doped liquid crystal compositions in photoactivated liquid crystal cells as real time projection displays, stored-image projection displays, and real-time light values for noncoherent modulation of coherent light, including application to electro-optical projection, storage, and memory devices.

Complex Ions Added for Improved Dynamic Scattering

A. Aviram, I. Haller, R.D. Miller and W.R. Young; U.S. Patent 3,814,700; June 4, 1974; assigned to International Business Machines Corporation describe a method for controllably varying the electrical properties of nematic liquids over a relatively large range by using as dopants for the nematic liquids organic materials that provide ions in which the charge is distributed over several atoms and is shielded from the charge of the counterions by a nonpolar shell. The dopant materials can be conveniently divided into two classes: (1) compounds consisting of large organic ions or precursors for large organic ions; and (2) complex ions of an alkali metal and a chelating agent. The dopant materials result in ions in which the charge is distributed over several atoms and have sufficient nonpolar groups to insure their solubility in a low-dielectric constant liquid (the nematic liquids), i.e., a dielectric constant of 1 to 20.

Thus, for example, a 60:40 mixture of 4-butyl-4'-ethoxy-α-chlorostilbene and 4-octyl-4'-ethoxy-α-chlorostilbene which is a nematic liquid between 8° and 59°C and has a resistivity ρ of approximately 10^{12} ohm-cm, does not, in itself, exhibit continuous dynamic scattering in 1 mil layers between transparent electrodes either under DC or AC excitation. It has been found that, when a material of the two abovementioned classes (examples being enumerated under compound numbers 1 to 22 below) is added to this mixture under the conditions set forth, dynamic scattering occurs. Typical examples of materials of the first class are those set forth in the table legended Class 1.

Class 1

Compound	Name
1	N,N'-dimethyldocosanylamine-n-tetracyanoquinodimethane complex
2	N-butylpyridinium p-toluenesulfonate
3	N-dodecylpyridinium p-toluenesulfonate
4	Hexaphenylethane
5	p-Methoxybenzylidene-p-butylaniline
6	Quinhydrone
7	Tetrabutylammonium tetraphenylboride
8	Tetramethyl-2,2,6,6-piperidinol-4-oxyle-1

Examples of materials in Class 2 are subdivided into three groups. A first subgroup of the second class of compounds have the following formula: Dicyclohexyl-18-crown-$6M^+X^-$ where M^+ is a metal, suitably sodium or potassium, and X^- is an anion. The structural formula for this first subgroup is as follows:

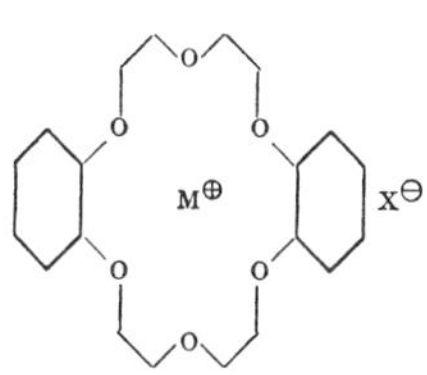

(Dicyclohexyl-18-crown-$6M^+$)

The anion X^- may suitably be selected from the following group: (1) An aliphatic carboxylate having the general structural formula:

$$C_nH_{2n+1}-\underset{\displaystyle R}{\underset{|}{C}}H-\overset{\displaystyle O}{\overset{\|}{C}}-O^{\ominus}$$

where n may have a value of 4, 5, 6 or 7, and R is H, CH_3 or C_2H_5. A typical example of a first subgroup compound having this ion is dicyclohexyl-18-crown-6 potassium 2-ethylhexanoate (Compound #9) where the anion has the structure:

$$C_4H_9{-}\underset{\displaystyle C_2H_5}{\underset{|}{CH}}{-}COO^{\ominus}$$

(2) An aromatic carboxylate having the general structural formula:

$$C_nH_{2n+1}{-}\underset{\displaystyle R}{\underset{|}{CH}}{-}CH_2O{-}\phi{-}COO^{\ominus}$$

where n may be 4, 5, 6 or 7, and R is H, CH_3 or C_2H_5. A typical example of a first subgroup compound having this ion is dicyclohexyl-18-crown-6 potassium 2-ethylhexyloxybenzoate (Compound #10) where the anion has the structure:

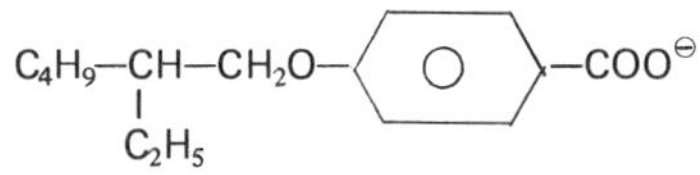

(3) A phosphate having the general structural formula:

$$(C_nH_{2n+1}{-}\underset{\displaystyle R}{\underset{|}{CH}}{-}CH_2O)_2POO^{\ominus}$$

where n = 4 through 7 and R = H, CH_3 or C_2H_5. A typical example of a first subgroup compound having this ion is dicyclohexyl-18-crown-6 potassium di(2-ethylhexyl)phosphate (Compound #11) where the anion has the structure:

$$(C_4H_9{-}\underset{\displaystyle C_2H_5}{\underset{|}{CH}}{-}CH_2O)_2\overset{\displaystyle O}{\overset{\|}{P}}{-}O^{\ominus}$$

(4) A borate having the general structural formula: $(C_nH_{2n+1}{-}\phi)_4B^{\ominus}$, where n = 0 to 3. A typical example of a first subgroup compound having this ion is dicyclohexyl-18-crown-6 sodium tetraphenylborate (Compound #12) where the anion has the structure: $\phi_4B^{\ominus}$.

(5) A sulfate having the general structural formula: $C_nH_{2n+1}O{-}SO_2{-}O^{\ominus}$ where n = 10 to 18. A typical example is the dicyclohexyl-18-crown-6 potassium salt (Compound #13) of the anion which has the structure: $C_{12}H_{25}O{-}SO_2{-}O^{\ominus}$.

(6) A xanthate having the general structural formula:

$$C_nH_{2n+1}O{-}\underset{\displaystyle S}{\underset{\|}{C}}{-}S^{\ominus}$$

where n = 6 to 12. A typical example is the dicyclohexyl-18-crown-6 potassium salt (Compound #14) of the anion which has the structure:

$$C_6H_{11}O{-}\underset{\displaystyle S}{\underset{\|}{C}}{-}S^{\ominus}$$

(7) A sulfonate having the general structural formula:

$$C_nH_{2n+1}\underset{\displaystyle R}{\underset{|}{CH}}{-}SO_3^{\ominus}$$

where n = 4 to 7, and R = H, CH_3, or C_2H_5. A typical example is the dicyclohexyl-18-crown 6$K^{\oplus}$ salt (Compound #15) of the anion which has the structure:

$$C_4H_9-\underset{\displaystyle C_2H_5}{\underset{|}{CH}}-CH_2-SO_3^{\ominus}$$

(8) A phenoxide having the general structural formula: $C_nH_{2n+1}O-\phi-CH=CH-\phi-O^{\ominus}$ where n = 1 to 5. A typical example is the dicyclohexyl-18-crown-6 potassium salt (Compound #16) of the anion which has the structure:

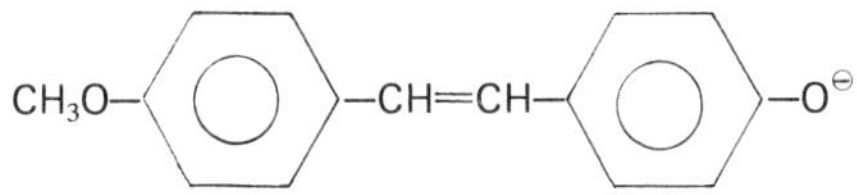

(9) A mercaptide having the structure:

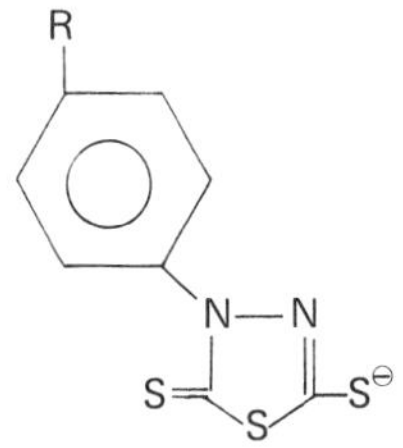

where R = H, CH_3 or C_2H_5. A typical example of a first subgroup compound containing this anion is dicyclohexyl-18-crown-6 potassium 3-phenyl-1,3,4-thiadiazole-2-thione-5-mercaptide (Compound #17). A further example of a compound of the first subgroup is dicyclohexyl-18-crown-6 potassium thiocyanate (Compound #18). A second subgroup of the abovementioned second class of compounds contain cations of potassium or sodium complexed by dibenzo-18-crown-6 ether, particular examples being:

Dibenzo-18-crown-6 potassium thiocyanate (Compound #19), and
Dibenzo-18-crown-6 sodium tetraphenylboride (Compound #20)

A third subgroup of the abovementioned second class of compounds are the following:

Tributylphosphine oxide lithium bromide (Compound #21), and
Triphenylphosphine oxide lithium bromide (Compound #22)

In carrying out the process, an appropriate amount of dopant in the quantity of 0.01 to 5% by weight of the total nematic liquid-dopant mixture is placed in a vial. To the dopant, there is added the proper amount of nematic liquid. The contents of the vial are heated with stirring to the point of isotropic transition. Thereafter, the heated mixture is permitted to cool to room temperature and the cooled mixture is then filtered, preferably under vacuum, the filter advantageously being a medium-coarseness sintered glass funnel.

The following table shows the effects of certain of the dopants. The electrical properties set forth in the table are the conductivity, cutoff frequency and turn-on or threshold voltage for dynamic scattering. The nematic liquid utilized in determining these effects is a 60:40 mixture of 4-butyl-4'-ethoxy-α-chlorostilbene and 4-octyl-4'-ethoxy-α-chlorostilbene.

	Concentration (percent by weight)	Conductivity (ω^{-1} cm.$^{-1}$)	Cut-off frequency (Hz.)	Threshold voltage for dynamic scattering (volts, RMS)
Compound #:				
1	Saturated	1×10^{-8}	2,000	9
7	do	1×10^{-9}	60	15
9	0.5	2×10^{-9}	350	7
9	5	1×10^{-7}	4,500	10
11	Saturated	5×10^{-9}	500	6
12	do	3×10^{-9}	500	15
18	do	6×10^{-8}	7,000	12
20	do	3×10^{-10}	300	12
21	do	1×10^{-9}	500	14
Blank	0	1×10^{-12}	<5	

Example 1: A quantity of 1% of compound 12 was added to an experimental mixture of nematic esters. The conductivity increased from 1 x $10^{-10}\Omega^{-1}$ cm^{-1} to 2 x $10^{-8}\Omega^{-1}$ cm^{-1}.

Example 2: A series of samples of varying concentrations of Compound #15 was dissolved in the 60:40 mixture of 4-butyl-4'-ethoxy-α-chlorostilbene and 4-octyl-4'-ethoxy-α-chlorostilbene. The AC conductivity transverse to the alignment axis was carefully measured at room temperature. The conductivity as a function of concentration is shown below:

Concentration (percent)	Conductivity (Ω^{-1} cm^{-1})
0	< 2 x 10^{-12}
0.05	4.8 x 10^{-11}
0.5	6.1 x 10^{-10}
1.3	2.3 x 10^{-9}
5.0	1.9 x 10^{-8}

Example 3: A quantity of 1% of Compound #12 was added under anhydrous conditions to a freshly purified sample of p-methoxybenzylidene-p-n-butylaniline. The resistivity and cut-off frequency were found to be 3 x $10^{-10}\Omega^{-1}$ cm^{-1} and 120 Hz, respectively, compared to 3 x $10^{-11}\Omega^{-1}$ cm^{-1} and 12 Hz for the undoped liquid crystal.

Example 4: A quantity of 1% of Compound #12 was dissolved in a distilled sample of Licristal Nematic Phase IV, an isomeric mixture of p-methoxy-p'-butylazoxybenzene. The resistivity and cut-off frequency were 7 x $10^{-8}\Omega^{-1}$ cm^{-1} and 40 kHz, compared to 3 x $10^{-10}\Omega^{-1}$ cm^{-1} and 270 Hz of the pure material.

Additives to Reduce the Turn-Off Time

A process described by *J.J. Wysocki; U.S. Patent 3,697,150; October 10, 1972; assigned to Xerox Corporation* relates to imaging systems where the imaging member comprises a liquid crystalline composition, and to compositions to facilitate the relaxation or switching time of such imaging systems.

It has been found that the inclusion of various additives, even in small amounts, drastically reduces the relaxation, retransformation, or turn-off response time in electrooptic cells and imaging systems. For example, it has been found that the inclusion of the additives in some examples reduces the relaxation or turn-off response time by orders of magnitude up to about 10.

The additives suitable for use in the process are believed to typically fall into two general groups: (1) electrophoretic-like (specifically including both electrophoretic and dielectrophoretic additives) centers which are capable of moving within the electrooptic liquid crystalline composition when electrical fields are placed across the thickness of the compositions; and (2) dipolar particles which realign themselves and may move within the electrooptic liquid-crystalline composition in response to the application of the electrical fields used in the process.

Materials suitable for use as the electrophoretic-type additives are typically particles or dispersed globules of substantially electrically insulating materials which become triboelectrically charged in the liquid crystalline composition-additive mixture. Insoluble or immiscible additives of this type include dispersed particulate carbon; polyvinyl chloride; polystyrene; Teflon, tetrafluoroethylene resins available from Dupont; silicon carbide; titanium dioxide; silica; sulfur, ammonium chloride; lead chromate; and various other insoluble salts and pigments; polymeric additives such as a copolymer of styrene and n-butyl methacrylate, and even a copolymer of styrene and n-butyl-methacrylate pigmented with carbon black.

Such materials are typically used in concentrations which are sufficiently low to allow good suspension or emulsification, without significant conglomeration of the additive material. Such materials are preferably used in concentrations of not greater than about 15%. Particles or globules of such materials are typically of average particle size not greater than about 5 microns, and particles of average size not greater than about 1 micron give preferred results in the system.

Where the additives are soluble or miscible, the additive is generally an organically soluble inorganic salt or organically soluble organic compound. Such soluble or miscible additives include alkyl ammonium halides such as tetraheptyl-ammonium iodide and hexadecyltrimethyl-ammonium bromide; alkyl phosphonium compounds including alkyl phosphonium sulfates, alkyl phosphonium bisulfides, alkyl phosphonium selenides, and alkyl phosphonium phosphates. Materials typically suitable for use as dipolar-type additives include various dispersed oils such as cottonseed oil, castor oil, silicone oils, linseed oil, mineral oil, polysulfone and others.

The following examples illustrate the process with respect to liquid crystalline electrooptic cells and imaging systems and compositions including electrophoretic-like, or dipolar additives in such compositions which are particularly suitable to control and minimize the turn-off response time in such systems. The parts and percentages are by weight unless otherwise indicated.

Example 1: A cholesteric liquid crystalline composition comprising about 35% cholesteryl chloride and about 65% cholesteryl nonanoate is prepared, and a 50:50% mixture of $PbCrO_3$ and a copolymer of styrene and n-butyl methacrylate pigmented with carbon black, is added in an amount comprising about 5% of the total resultant composition. This composition is mixed to suspend the additive in the liquid crystalline composition, and a layer of the composition is placed into a cell comprising a pair of substantially transparent electrodes, the electrode surface being a transparent chromium coating on a glass substrate, and the electrodes enclose the composition in a 1-mil-thick Mylar spacer gasket (Mylar is a polyethylene terephthalate resin).

A voltage of about 900 volts DC is provided across the thickness of the composition layer, thereby causing the cholesteric-nematic phase transition. Microscopic observation reveals that the included particulate additives bound back and forth between the electrode plates. Recovery or off-times are measured for the above cell and for a control cell which does not contain the additive. The cell exhibits off-times which are about one-half the duration of off-times in the control cell.

Example 2: A cholesteric liquid crystalline composition comprising about 59% cholesteryl chloride, about 39.4% cholesteryl nonanoate, and about 1.6% oleyl cholesteryl carbonate, is prepared, and polysulfone is added and dispersed therein in an amount comprising 15% of the total resultant composition. A layer of this composition is provided in a cell as in Example 1, except that the spacer gasket is of a thickness of about one-half micron. A voltage of about 1,500 volts DC is provided across the thickness of the composition layer, thereby causing the cholesteric-nematic phase transition to occur. Under a microscope, the dispersed additive is observed to move vigorously throughout the composition. Recovery or off-times are measured for the above cell and for a control cell which does not contain the additive. The cell exhibits off-times which are about one-third the duration of off-times in the control cell.

NEMATIC AND SMECTIC LIQUID CRYSTAL DISPLAY SYSTEMS AND DEVICES

NEMATIC

Rear Projection Screen

J.V. Cartmell and D. Churchill; U.S. Patent 3,674,338; July 4, 1972; assigned to The National Cash Register Company describe a rear projection screen where organic nematic mesomorphic compounds, i.e., liquid crystals, are employed. The liquid crystals have a threshold electric field which when exceeded causes the liquid crystals to scatter light. A thin film of the liquid crystal in the light scattering mode can be used for a screen in a rear projection system. Rear projection systems generally comprise a projector and screen, the screen being between the projector and viewer. The projector usually comprises a light source, image and lens, the image being between the light source and lens.

By using a rear projection or transmission screen, the image, i.e., an object or film, is projected onto the screen. Often, rear projection screens are made of ground glass. A more recent type is made up of a multitude of tiny lenses imbedded in a plastic. The primary optical characteristics of a rear projection screen are distribution, gain and resolution. Distribution is a measure of the light spreading character of a rear projection screen. Distribution is given as a plot which shows the intensity of a ray of light at the different angles to which the light is scattered. Screen gain is the ratio of the intensity of transmitted light normal to the screen to the intensity of light scattered from a perfect diffuser such as a magnesium carbonate block. Screen resolution is the ability of a screen surface to deliver fine image detail. Resolution usually is measured in lines per inch or lines per millimeter.

The quality of a rear projection system is determined by several parameters which are described in relation to the user reaction. Several of these parameters are dependent upon the properties of the screen itself either in conjunction with or independent from the rest of the system. Screen fall-off refers to a decrease in the screen brightness at positions to the sides of the screen center. This arises from a combination of nonuniformity of the projection system and poor distribution of the light scattered by the screen. Screen resolution relates to the detail visible on the screen independent of the resolution to the projection system. Screen breakthrough is manifested by a bright spot in the screen due to light which is not diffused.

Another effect which is not as easily defined but contributes greatly to eye fatigue is sparkle. This is the occurrence of multitudes of tiny bright spots in a light area of the image which seem to dance or scintillate. This is particularly troublesome for users who must observe a screen for long periods of time, e.g., parts catalogues recently have been put on film and are viewed on screens for periods of time ranging up to 8 hours a day. Also, instructional speed reading devices employing rear projection screens often are viewed by students for periods of time ranging up to 2 hours. In one commercially available rear projection screen sparkle is eliminated by employing two screens separated by a few mils. One of the screens is kept in constant motion relative to the other by a motor driven mechanism. This rather elaborate system is effective in reducing sparkle but adds considerable expense to the system.

It has been found that organic nematic mesomorphic compounds can be employed in rear projection screens. The screen comprises two substrates having adjacent parallel surfaces less than 500 microns apart, preferably less than 150 microns apart. The adjacent surfaces of each substrate are coated with a transparent electrically conductive film and the space between the adjacent surfaces of the two substrates is filled with liquid crystals. By passing focused light rays from a projector to the screen, an image is projected onto the screen. When the electric field between the substrates is zero, the light rays transmitted through the device are in the same direction as the incident light rays.

When the electric field between the two substrates is increased until a certain threshold value is reached, there is a sudden change in the optical properties between the two substrates. This change in optical properties is manifested as a scattering of the transmitted light in all directions. The observer now sees the image focused onto the screen. The distribution of the light scattered from the liquid crystal can be varied by adjustment of the voltage above the threshold value. With the use of rear projection screens employing liquid crystals, it has been found that sparkle has been essentially eliminated from rear projection systems. Furthermore, screen fall-off is much improved over standard screens.

The rear projection screen is described in greater detail in connection with Figure 4.1, in which the figure is a sectional view of a rear projection system. It has been found that the screens require low voltage and very little power. Generally, the voltage requirement ranges from 10 to 300 volts, preferably 10 to 100 volts. The rear projection system can employ either DC or AC current. AC current is preferred because it is more readily available, i.e., elaborate voltage supplies are not needed as the device merely can be plugged into any wall outlet. Also, the materials have a finite life expectancy which is extended when AC current is used.

FIGURE 4.1: REAR PROJECTION SCREEN

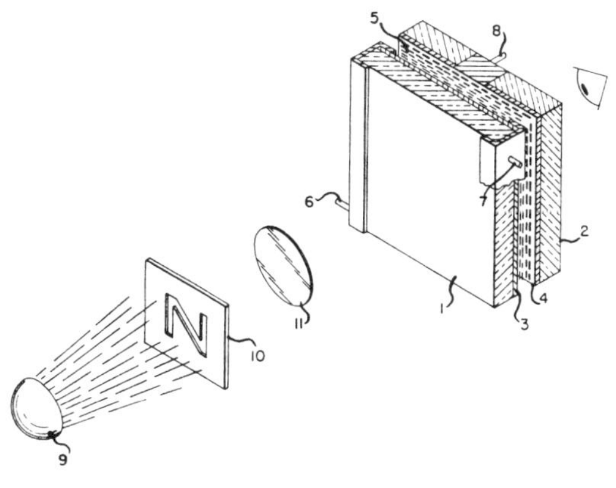

Source: J.V. Cartmell and D. Churchill; U.S. Patent 3,674,338; July 4, 1972

Example: In the system of this example, substrates **1** and **2** are transparent glass plates measuring 4" x 4". The substrates **1** and **2** are aligned so that their adjacent faces **3** and **4** are parallel and spaced about 50 microns apart. The faces **3** and **4** were coated with a thin layer of tin oxide. The substrates **1** and **2** can be positioned apart by means of shims or the spacing can be maintained by means of clamps or a frame-like holder. The space **5** between the two substrates is filled with a liquid crystal. In this example, a 50/50 blend of butyl p-(p-ethoxyphenoxycarbonyl)phenyl carbonate and butyl p-ethylphenyl terephthalate was employed. Note, the screen is sealed at both ends in order to retain the liquid crystal in space **5**. Electrical lead wires **6** and **7** are attached to substrate **1** as is electrical lead wire **8** attached to substrate **2**. This can be done by conventional techniques such as by silver paint.

Electric leads **6** and **7** run full length of substrate **1** and act as a buss bar heater. By applying voltage at leads **6** and **8**, the liquid crystal can be heated to its liquid crystal state if needed. In this example, the mixture was heated to 40°C by applying 20 volts AC to leads **6** and **7**. Leads **6** and **8** are connected to a voltage source ranging from 10 to 300 volts. By adjusting the voltage, e.g., 10, 50 or 100 volt increments, the ratio of transmitted light to scattered light can be controlled to suit the viewer. The light rays from the light source **9** pass through **10**, which then are focused by lens **11**. The projected image is focused on the liquid crystal material in space **5**.

Observance of the screen demonstrates the high degree of resolution obtained by employing liquid crystals. Fine image detail is delivered by the rear projection screen of this process as the resolution is 500 lines per inch. Observance of the screen demonstrates uniform brightness thereby indicating desirable light distribution or light spreading characteristics. Viewing the screen at wide angles demonstrates very little screen fall-off. Not only is this rear projection screen essentially free of sparkle, but it also provides the additional advantages of adjustable gain.

The ratio of transmitted light to scattered light can be varied by adjustment of the voltage to the conductive films. In this instance, the variable ratio of transmitted light to scattered light helps to adjust brightness at various angles. Through the use of a structure of a liquid crystal cell of parallel glass plates with a liquid crystal sandwiched between them, a thin conductive coating on the inside surfaces of the plates and electrode attached to each glass plate, a rear projection screen is provided which can be viewed at various angles with a high degree of clarity and brightness without fatiguing the viewer. Further, the gain can be adjusted to suit the viewer.

Pleochroic Dye for Color Display

J.A. Castellano; U.S. Patent 3,703,329; November 21, 1972; assigned to RCA Corporation describes a color display system capable of producing essentially all the colors of the visible spectrum. The system comprises three cells each including a solution consisting of a pleochroic dye in a nematic liquid crystal composition. Each of the solutions can change its transmission of polarized white light in response to an electric field so as to change the color appearance of the solution. One solution of the system can change in appearance from colorless to magenta, another from colorless to cyan, and a third from colorless to yellow. The system includes means for applying an electric field separately to each of the solutions and means for passing polarized white light successively through each solution.

Referring to Figure 4.2, there is shown a source of white light **12** which is preferably (but not necessarily) collimated by means of a collimating lens **14**. The collimated white light then passes through a polarizer **16** and thence in tandem through three liquid crystal electro-optical cells **18**, **20** and **22** respectively. Each liquid crystal cell comprises a solution **24** of a pleochroic dye dissolved in a nematic liquid crystal composition which aligns in an electric field. The solution **24** is contained between transparent support plates **26** and **28** having transparent conductive coatings **30** and **32** respectively on the inner surfaces thereof. In operation, each of the conductive coatings is connected to a voltage source so as to separately control and vary the electric field applied across the liquid crystal solutions of each liquid crystal cell. In this way the degree of alignment of each of the pleochroic dyes can be separately controlled. The specific pleochroic dye composition in each of the three solutions is different from that of the others. In order to obtain a system which is capable of transmitting substantially all the colors of the visible spectrum, the individual cells should comprise solutions which are capable of absorbing different portions of the spectrum. More particularly, the individual liquid crystal cells should be able to be varied in appearance from their unswitched or nonaligned state to their fully switched or fully aligned state as follows: one cell from cyan to colorless; one cell from magenta to colorless; and one cell from yellow to colorless.

FIGURE 4.2: THREE CELL COLOR DISPLAY DEVICE

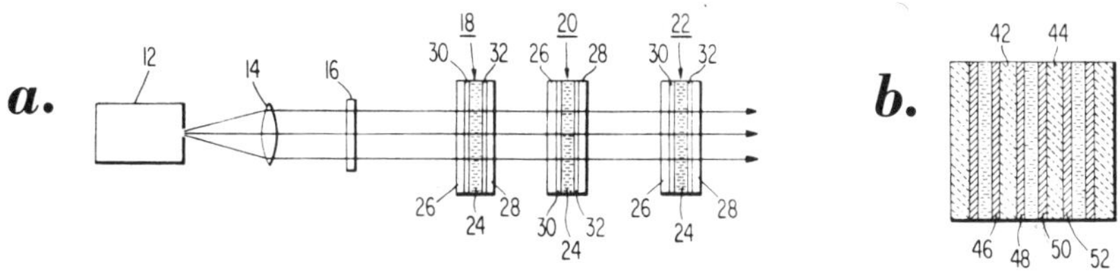

(a) Elevational View of a Multi-Cell System
(b) Cross-sectional View of Structure Useful in a Compact System

Source: J.A. Castellano; U.S. Patent 3,703,329; November 21, 1972

In operation, by varying the electric field of each cell from between zero to that which is necessary to produce maximum alignment of the dye molecules, one can obtain substantially every color in the visible spectrum, including black and white, when starting with polarized white light which passes in tandem through all three cells. Although only one color at a time will be transmitted by the system shown, a full color display system can be made by, for example, providing each device with conductors in the form of a cross-grid array. In such a system the color at any display element in the array can be separately controlled by controlling the applied field at each element of each of the cells thereby providing a full color display system.

In addition the system may be made more compact and the number of glass surfaces from which reflective losses may occur may be reduced by providing a structure as shown in Figure 4.2b where two inner support plates **42** and **44** each help support two solutions and electrodes therefor at the same time. In this instance, both sides of each of the inner support plates **42** and **44** are provided with transparent conductive coatings, **46** and **48** on plate **42** and **50** and **52** on plate **44**. Typically, the solutions consist of a nematic liquid crystal host material of the type which aligns in response to an electric field, into which is dissolved from about 0.5 to 5 weight percent of a pleochroic dye. A preferred dye concentration is from 1 to 2%. Examples of preferred nematic liquid crystal compounds useful as host materials in the solutions of the system include alkoxybenzylidene anils and acyloxybenzylidene anils such as those having the structural formula

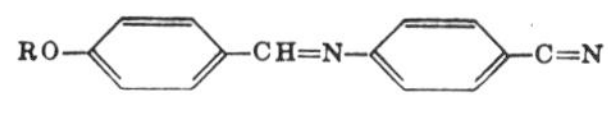

where OR is an alkoxy radical preferably having from one to seven carbon atoms or an acyloxy radical having from two to seven carbon atoms. Actually, mixtures of the nematic compounds are preferred. This is due to the fact that the temperature at which the mixture changes from a solid crystalline state to the nematic mesomorphic state, and hence, the temperature of operation of the solution, is lower in the case of mixtures than for the individual compounds within a given mixture. In order to obtain the color changes necessary for a broad color response, that is, to obtain solutions which switch from colorless to cyan, magenta and yellow, the following dyes are suitable.

Name and structure	Color change due to alignment
N,N'-dimethylindigo 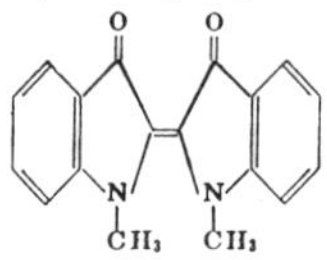	Colorless to cyan.
N,N'-dipalmitoylindigo 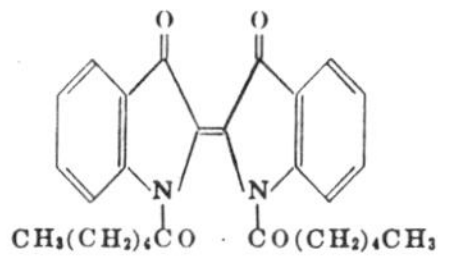	Colorless to magenta.
p-Nitrobenzylidene-phenylhydrazone 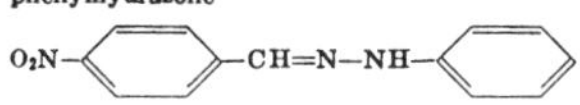	Colorless to yellow.
Derivative* of oil yellow (an azo dye) 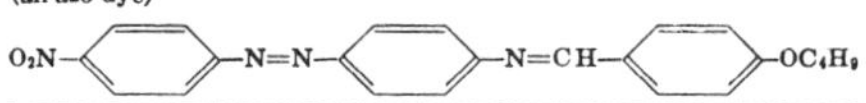	Colorless to yellow.

*The derivative of the azo dye is used so as to increase the solubility of the dye in the liquid crystal composition.

It should be pointed out that the preferred pleochroic dyes are nonionic in the nematic liquid crystal solution. The specific solutions employed in the preferred system consist of: (a) cyan to colorless; 0.65 weight percent indophenol blue

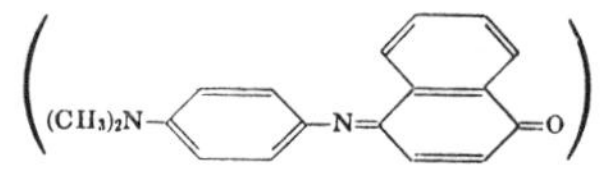

and 0.35 weight percent 4-butoxybenzylidene-4'-amino-4"-nitroazobenzene

dissolved in an equimolar nematic solution of p-hexoxybenzylidene-p'-aminobenzonitrile, p-hexoylbenzylidene-p'-aminobenzonitrile, and p-heptoylbenzylidene-p'-aminobenzonitrile.

(b) magenta to colorless; 1 weight percent of 2-amino-8-hydroxy-1-azonaphthyl-4'-benzonitrile

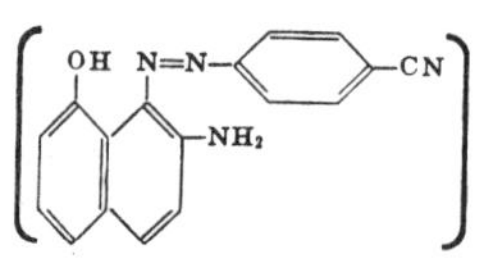

dissolved in the same equimolar nematic solution as recited in (a) above.

(c) yellow to colorless; 1.5 weight percent of 4-butoxybenzylidene-4'-amino-4"-nitroazobenzene dissolved in the same equimolar nematic solution as recited in (a) above.

T.B. Harsch; U.S. Patent 3,785,721; January 15, 1974; assigned to International Liquid Xtal Company describes a nematic liquid crystal device which is electrically controlled in order to produce colors which can be displayed for entertainment or advertising purposes. The liquid crystal material is sandwiched between a pair of parallel transparent plates coated with transparent conducting films. The coated plates are rubbed in the same direction such that the nematic molecules align parallel to the glass surface and point in the rubbed direction.

The liquid crystal sandwich structure is then placed between linear polarizers which are oriented either parallel or crossed with respect to each other and at 45° with respect to the rubbed direction. By applying an electric field across the liquid crystal material via the aforesaid transparent conducting films, the color observed through the polarizers can be made to vary, depending upon the magnitude of the electric field. Furthermore, by applying a continually varying voltage across the conducting films, the color can be made to continually vary also.

Electrooptical Elements

R. Williams; U.S. Patent 3,322,485; May 30, 1967; assigned to Radio Corporation of America describes an electrooptical element comprising two substrates having adjacent parallel surfaces less than ½-millimeter apart, such as planar plates positioned so that their adjacent opposing faces are substantially parallel. The space between the adjacent faces of the two substrates is filled with an organic nematic mesomorphic compound. Means such as conductive coatings on each of the adjacent faces are provided for applying an electric field between the faces and through the compound. A device may be constructed in which the field is applied selectively to change the light reflective or light transmissive properties of the element.

Example 1: An electrooptical device is described with reference to Figure 4.3. The device **10** comprises two substrates **11** and **12**, which are preferably plates each having plane opposed parallel faces. Substrates **11** and **12** are positioned opposite each other so that their adjacent faces **13** and **14** respectively are parallel. The distance between the inner faces **13** and **14** of the two substrates is critical, as will be explained below, and should be less than one-half millimeter. Preferably, the distance between faces **13** and **14** is about 5 to 300 microns.

The substrates are prepared with at least one electrically conductive region or strip **15** on inner face **13** of plate **11**, and at least one conductive region or strip **16** on the inner face **14** of plate **12**. The two plates **11**, **12** and their conductive regions are positioned so that the axis of the conductive path or paths on one plate are substantially transverse or perpendicular to the axis of the conductive paths on the other plate. An electrical lead wire **17** is connected to the conductive path **15** on wafer **11**, and another electrical lead wire **18** is connected to the conductive path **16** on wafer **12**, using known techniques. For example, the lead wires can be attached to the conductive strips by means of a droplet of silver paste. Lead wires **17** and **18** are connected to a voltage source **19**.

At least one of the two substrates or plates **11** and **12** is transparent. Furthermore, the conductive region or paths on the transparent substrate is also transparent. In this example, only one of the two substrates is transparent. A variety of transparent solids may be utilized for the transparent substrate, including the various types of glass, fused quartz, transparent varieties of corundum, and transparent plastics or resins. In this example, substrate **11** is transparent and consists of a glass microscope slide. The conductive portion **15** on the transparent substrate **11** is also transparent, as indicated above.

This may conveniently be accomplished by depositing thin layers of indium oxide or tin oxide on the desired regions of inner face **13** on transparent plate **11**. In this example, plate **12** is also a glass microscope slide, but is made reflecting by any convenient means, for example by depositing a metallic coating such as silver or aluminum on the outer face of plate **12**. The conductive region **16** on face **14** of plate **12** need not be transparent, and may for example be a film of copper or gold.

FIGURE 4.3: ELECTROOPTICAL ELEMENTS UTILIZING ORGANIC NEMATIC COMPOUNDS

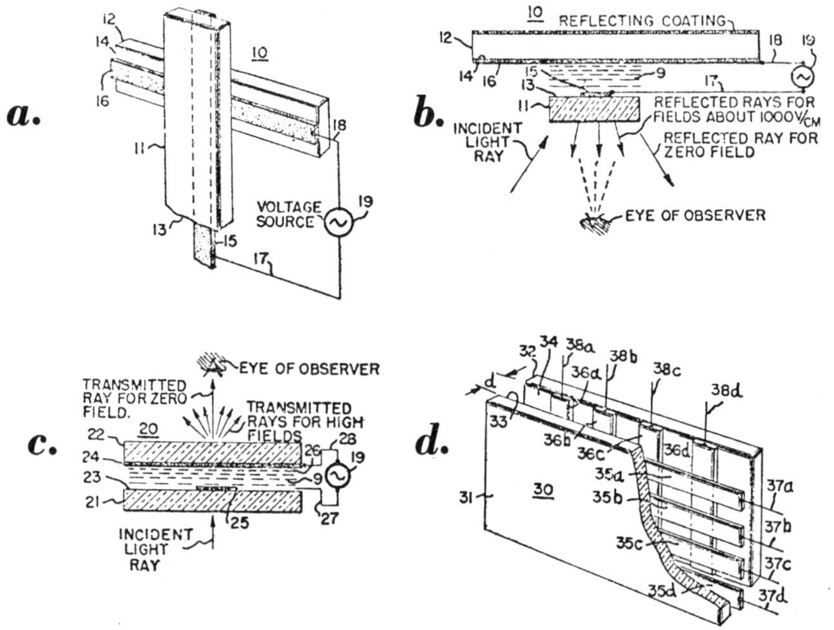

(a)-(d) Electrooptical Devices

Source: R. Williams; U.S. Patent 3,322,485; May 30, 1967

The operation of the device of this example will be described with reference to Figure 4.3b. The transparent plate **11** and the reflecting plate **12** of this example are positioned about 10 microns apart by means of shims. Alternatively, the spacing may be maintained by means of clamps, or by a suitable frame-like holder. The space between the two plates is filled with an organic thermotropic nematic compound **9** such as p-azoxyanisole. The device **10** is maintained in the temperature range in which the compound exhibits the nematic mesophase, that is, in the temperature range above the melting point of the solid compound but below the temperature at which the molten compound becomes isotropic.

The observer views the device through the transparent plate **11** and the transparent conductive region or electrode **15**. A light source is positioned so that a collimated beam of light rays is incident on the transparent substrate **11**. The angle between the incident light ray and the substrate **11** is less than 45°, and preferably less than 22°. When the electric field between the conductive strips or electrodes **15** and **16** is zero, most of the incident light is reflected by substrate **12** in a regular manner, so that the angle of incidence equals the angle of reflection. Under these circumstances, as shown in Figure 4.3b, very little light is reflected to the eye of the observer, whose line of sight is normal to the transparent plate **11** and transparent electrode **15** and the layer **9** of organic nematic thermotropic material. Accordingly, for zero applied field the observer sees the entire area of the transparent electrode over the organic nematic thermotropic compound as dark.

When a weak field is applied between electrodes **15** and **16** by the voltage source **19**, no change is seen by the observer. If the strength of the applied field is slowly increased, there is no visible change in the light intensity of the device viewing area, that is, the area of the transparent substrate over the nematic compound, until a certain threshold value is reached. This threshold value for the applied field varies with the particular organic nematic thermotropic compound and the particular distance between the two electrodes, but for most organic nematic thermotropic compounds is of the order of magnitude of about 1,000 volts per centimeter. When the voltage applied between the two electrodes reaches this value, there is a sudden change in the optical properties of layer **9** for that portion only of the organic nematic thermotropic compound **9** which is between the two electrodes and subject to the field.

As a result of this change, some of the incident light is scattered through the transparent electrode **15** and the transparent plate **11** in a direction normal to the plate **11**, and hence this scattered light reaches the eye of the observer. Thus, the observer, who has hitherto seen the entire area of the nematic layer **9** as uniformly dark, suddenly sees a particular portion of this area become brighter than the remainder of the area. The particular area which becomes bright is that area only of the nematic layer **9** which is directly between the intersection of the transparent conductive strip **15** and the conductive strip **16**.

While the exact nature of this light scattering effect is not definitely ascertained, it is theorized that groups of the parallel-oriented molecules of the nematic compound exist and in each group all the molecules tend to move in the same direction at the same time. Such groups or swarms of molecules may be regarded as domains. These domains tend to change their direction when the electrical field across the material reaches a certain threshold value. This change in the direction of groups or swarms of the molecules (all the molecules within each individual group or domain being similarly oriented) is thought to result in the scattering of light by the domain walls.

In the example just described, the effect is obtained by modulating light reflection, and the observer sees a predetermined area of the device become bright, while the remaining area remains dark. In the form described in Example 2, the effect is obtained by modulation of light transmission, and the observer sees a predetermined area of the device become dark, while the remaining area remains bright.

Example 2: In the device **20** of this example, both substrates or plates **21** and **22** (Figure 4.3c) are made of transparent materials. Fused quartz is one of the transparent materials suitable for this purpose. The quartz plates **21** and **22** are aligned so that their adjacent faces **23** and **24** respectively are parallel and spaced about 20 microns apart. The conductive region or electrode **25** on face **23** of plate **21** is transparent, as is the electrode **26** on face **24** of plate **22**. The transparent electrodes **25** and **26** may, for example, consist of films of indium oxide or tin oxide, and are substantially perpendicular to each other in a manner similar to those of electrodes **15** and **16** in Figure 4.3a.

An electrical lead wire **27** is attached to electrode **25** and another electrical lead wire **28** is attached to electrode **26**. Lead wires **27** and **28** are connected to a voltage source **19**. The space between the two substrates or plates **21** and **22** is filled with an organic nematic thermotropic compound, which in this example consists of anisaldazine. The observer views the device in a direction normal to the two transparent electrodes **25** and **26**. A light source is positioned to direct a collimated beam of light rays on the side of the device **20** opposite the observer so that the incident light rays pass through the device normal to the transparent plates **21** and **22** and the transparent electrodes **25** and **26** and the layer **9** of the organic nematic thermotropic compound.

When the electric field between the two conductive paths or electrodes **25** and **26** is zero, the light rays transmitted through the device are in the same direction as the incident light rays. The observer therefore sees the region of the device which contains the organic nematic thermotropic compound layer **9** as uniformly bright. When the electric field between the two conductive paths **25** and **26** is increased, no effect is observed until a certain threshold value is reached. As indicated above, this threshold value for the applied field is of the order of 1,000 volts per centimeter. When this threshold value for the applied field is reached, there is a sudden change in the optical properties of that portion only of layer **9** which is directly between the two electrodes **25** and **26**. This change in optical properties is manifested as a scattering of the transmitted light in all directions. The observer, has previously seen the entire area of the nematic layer **9** as uniformly bright, now suddenly sees a portion of this area become darker than the remaining area. The darker area is that area only which is in the intersection of the two conductive paths or electrodes **25** and **26**. In the devices of Examples 1 and 2, only one conductive path or electrode was shown. Devices may be constructed containing a plurality of conductive paths or electrodes on each substrate. One may thus fabricate an optical display device containing an x-y grid. Such a device may be used to display patterns or pictures, and is particularly suitable for the display of information in rectilinear coordinates. One such device will now be described.

Example 3: An optical display device **30** embodying an x-y grid comprises two planar substrates **31** and **32** (Figure 4.3d), which in this example consists of transparent plates of corundum (Al_2O_3 or transparent sapphire). The two plates **31** and **32** are parallel, and are separated by a distance **d** which is less than one-half millimeter, and preferably about 5 to 300 microns. In this example, the distance **d** is 40 microns. On the inner face **33** of plate **31** an array of transparent conductive strips are deposited, all running in one direction. In this example, only four such strips or electrodes (**35a, 35b, 35c, 35d**) are shown for greater clarity, but it will be understood that in practice, a much larger number of electrodes may be utilized.

The strips **35a** to **35d** have parallel longitudinal axes. On the inner face **34** of plate **32**, an array of transparent conductive electrodes are deposited, all running in the same direction substantially perpendicular to the direction of the conductive strips on plate **31**. Again only four such strips or electrodes (**36a, 36b, 36c, 36d**) are shown for greater clarity. The strips **36a** to **36d** have parallel longitudinal axes normal to the axes of strips **35a** to **35d**. The space between plates **31** and **32** is filled with an organic nematic thermotropic compound, which in this example, consists of 4,4'-dimethoxystilbene.

The optical display device **30** of this embodiment is utilized in a manner similar to that shown in Figure 4.3b, since both substrates and both sets of electrodes are transparent. A light source is positioned on one side of the device so that a collimated beam of light is directed through the device in a direction normal to the plates **31** and **32**. The observer is on the opposite side of the device. For zero applied field, the observer sees the entire area of the nematic thermotropic compound as uniformly bright. Now when a voltage pulse is applied between one electrode strip of the group **35a, 35b, 35c, 35d** and another electrode strip of the group **36a, 36b, 36c, 36d**, then provided the voltage applied is equal to or greater than the critical field, the observer will see a portion of his field of view become darker than the remaining area of the nematic compound.

The portion that becomes darker will be that portion corresponding to the intersection of the two electrodes which have been energized. For example, if the two electrode strips energized are **35a** and **36d**, the portion of the plate area which will suddenly darken is the area corresponding to the intersection of electrode strip **35a** and electrode strip **36d**. By energizing more than one electrode strip from each set, a plurality of predetermined areas are thus darkened. Scanning techniques known to the optical display art may be utilized to sequentially and cyclically energize the conductive strips of one array, for example the vertical columns **36a** to **36d**, at a relatively rapid rate, while the conductive strips of the other array, which are the horizontal rows **35a** to **35d** in this example, are sequentially and cyclically energized at a relatively slower rate.

Row selector circuits and column selector circuits for energizing selected rows and columns of an x-y grid are known to the mural television art, and need not be described in detail here. See, for example, U.S. Patent 2,928,894, issued March 15, 1960. In this manner various types of information may be optically displayed, for example, a pattern of an alphanumerical character. Furthermore, by using a large number of narrow electrode strips on each substrate, for example, electrode strips that are on the order of one mil wide and spaced one mil apart, a photograph or picture can be displayed on a device of this type. An important advantage of the optical display devices described in this example is that there is very little cross-talk or spurious display, since those portions of the organic nematic thermotropic compound which are not directly between the two energized electrodes do not receive the threshold field, and hence do not exhibit the optical effect of modulating either the transmission or the reflection of light. Still another advantage of the optical display devices described in this example is that they do not require polarized light, and may be utilized with any wavelength for which the substrate and the organic nematic thermotropic compound is transparent.

Fast Self-Quenching

G.H. Heilmeier; U.S. Patent 3,575,493; April 20, 1971; assigned to RCA Corporation describes a liquid crystal element which is operated within a critical frequency range such that upon removal of the exciting electric field, the dynamic scattering exhibited by the crystal decays within a matter of milliseconds, an interval which is substantially shorter than the usual natural decay time. The critical frequency range is related to the conductivity (which is temperature and voltage dependent) and dielectric constant of the liquid crystal.

Electronically Tunable Birefringence

F.J. Kahn; U.S. Patent 3,694,053; September 26, 1972; assigned to Bell Telephone Laboratories, Incorporated describes a low-cost, low-power, birefringence-tunable nematic liquid crystal device that is characterized by a long lifetime, zero retardation at zero voltage, a well-defined threshold and relatively short relaxation times. The process involves a nematic liquid crystal device exhibiting a birefringence that is electronically tunable over a range of applied voltages that lie below the threshold for light scattering due to hydrodynamic instabilities. In constructing the device, the molecules of a thin film of a member of a specified class of nematic liquid crystals are initially ordered in a homeotropic way. Moreover, the device is constructed to create a preferred anisotropy direction in the thin film. This in turn imposes a predetermined molecular orientation in the film in the presence of an applied electric field.

Hence when an electric field is applied to the device, the homeotropically ordered molecules are reoriented in the preferred direction. In this way a spatially uniform reorientation of a portion of the molecules in the thin film is achieved under the influence of an applied electric field. As a result, the birefringence of the device is thereby selectively altered in a controlled manner. A feature of the process is that the molecules of a thin film of a nematic liquid crystal are, in the absence of an electric field, homeotropically ordered. A preferred anisotropy direction is initially imposed on the molecules so that in the presence of an electric field any reoriented molecules are constrained to align themselves in the preferred direction.

The specific illustrative liquid crystal device shown in Figure 4.4a includes two parallel glass plates **12** and **14** having a thin film **16** of nematic liquid crystal sandwiched between them. Thin conductive coatings of, for example, $In_2O_3-SnO_2$ on the facing surfaces of the plates **12** and **14** provide electrodes by means of which a uniform electric field can be established across the depicted film. Polymeric spaces **18** and **20** made, for example, of Mylar maintain the thickness of the liquid crystal material at any desired value in a range of approximately 1 to 200 microns measured in the z direction. Thus, the liquid crystal is in effect a thin film lying in the x - y plane.

The active material **16** included in the device **10** of Figure 4.4a comprises a nematic liquid crystal of a specified type. In particular the material is a nematic liquid crystal of the type in which the electric susceptibility perpendicular to the long axis of a molecule thereof is greater than the electric susceptibility parallel to the long molecular axis. N-(p-methoxybenzylidene)-p-butylaniline (MBBA) is an example of one such nematic liquid crystal material suitable for inclusion in the device **10**. The rodlike molecules of the nematic liquid crystal **16** shown in Figure 4.4a are designated by short lines. These lines represent the long axes of the molecules. The long axes of the molecules of the thin film are initially oriented parallel to the z axis.

Such an orientation or homeotropic ordering is achieved either by mixing a surfactant additive with the nematic liquid crystal material or by directly coating the facing electrode surfaces of the planar members **12** and **14** with an appropriate surfactant.

FIGURE 4.4: NEMATIC LIQUID CRYSTAL DEVICE

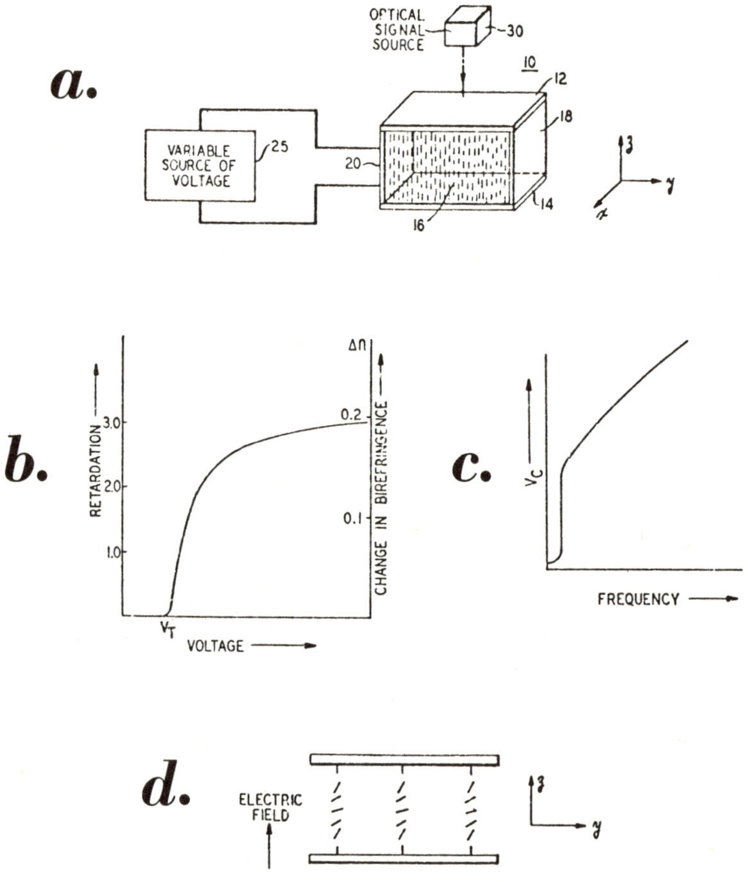

Source: F.J. Kahn; U.S. Patent 3,694,053; September 26, 1972

Surfactants which have been found to induce good homeotropic order in nematic liquid crystals include hexadecyltrimethyl ammonium bromide (HMAB), stearic acid, lecithin and polyamide resins such as, for example, Versamide 100. A tentative explanation of why such surfactants impose homeotropic order on nematic liquid crystals follows. The surfactant molecules generally each have one end which is polar in character and one end which is nonpolar, for example a hydrocarbon chain. The polar end tends to adhere to the electrode surface while the nonpolar end interacts with the liquid crystal molecules in such a way as to orient them homeotropically.

Thus, for example, homeotropic ordering of MBBA may be achieved by adding a small amount of HMAB. The resulting mixture is flowed between the planar members (by capillary action, for example). It is believed that the HMAB molecules then plate out on the planar members. This plating action can be accelerated by momentarily heating the liquid crystal material (to about 50°C, for example) and then allowing it to cool to room temperature. For 12-micron-thick samples the optimal concentration of HMAB in MBBA is about 0.005% by weight. If all the HMAB is assumed to plate out, this would correspond to a uniform coating on the order of a monolayer on each electrode surface. The effect of an additive to cause homeotropic ordering in the resulting nematic liquid crystal mixture is known in the art. In this connection, see, for example, "New Electro-Optic Effect in a Room Temperature Nematic Liquid Crystal" by W. Haas et al, *Physical Review Letters,* November 9, 1970, pp 1326-1327.

To create a preferred anisotropic direction in the nematic liquid crystal thin film, which in turn will give rise to a predetermined molecular orientation in the presence of an electric field applied thereto, a further preparation step is required in the course of fabricating the device represented in Figure 4.4a. This step involved selectively treating the electrode layers that are respectively deposited on the planar members **12** and **14**. Illustratively, the treatment comprises rubbing the electrodes, for example with lens paper, in a direction parallel to the direction in which the molecular axes of the thin film are to lie in the presence of an applied electric field.

Rubbing each of the electrode surfaces in the y direction about six times with moderate pressure has resulted in imposing the desired anisotropic direction in the thin film **16**. In carrying out this step neither the rubbing material, the number of strokes, nor the amount of pressure appears to be critical. The rubbing as described above is carried out prior to deposition of the surfactant.

The birefringence characteristic of the device **10** shown in Figure 4.4a can be controllably altered by establishing an electric field in the thin film **16**. A variable DC field whose value is below the threshold for dynamic scattering can be employed for this purpose. However, the range of tunable birefringence that is thereby obtained is relatively limited. The range of tunable birefringence is significantly expanded by using AC electric fields. Figure 4.4c illustrates for a particular sample the manner in which V_C, the voltage threshold for hydrodynamic instability, increases as the frequency of the voltage applied to a nematic liquid crystal is increased. Thus, by applying AC electric fields to the thin film **16** shown in Figure 4.4a, it is possible to make devices which are tunable over virtually the full available birefringence characteristics of the nematic liquid crystal material, but which operate at all times below the threshold for hydrodynamic instability.

Another advantage of AC operation of the device shown in Figure 4.4a is that the larger fields thereby made possible lead to relatively short turn-on times in the nematic liquid crystal film device **10**. A source **25** is shown in Figure 4.4a for applying variable voltages between the electrode surfaces respectively disposed on the members **12** and **14**. By that means, variable electric fields can be established in the nematic liquid crystal material **16**. In the presence of an applied field in the z direction, some of the homeotropically oriented molecules of the material are rotated in a spatially uniform manner (spatially uniform in the x-y plane) toward the y axis. This reorientation, which is represented schematically in Figure 4.4d, gives rise in turn to a change in the birefringence characteristic of the device **10**. Thus, optical signals directed at the device **10** by a source **30** (Figure 4.4a) see a bifringence characteristic whose value is a function of the voltage applied across the thin film **16**.

A typical curve of Δn (birefringence) versus applied voltage for a specific 16-micron thick film of nematic liquid crystal excited at 1 kHz with an electric field in the z direction is shown in Figure 4.4b. In such a specific device (of the type shown in Figure 4.4a) the voltage threshold V_r at which a change in birefringence is first observed approximates 3.85 volts rms and is well defined. This threshold corresponds to an electric Freedericksz transition. Theoretical analysis shows that this threshold is determined by the elastic constants and the dielectric anisotropy of the nematic liquid crystal. Observed values are typically in the range 3 to 4 volts for a variety of different samples. Figure 4.4b also indicates the variation of optical retardation with applied voltage for the above-specified example. The retardation is zero at zero applied voltage and approaches 3 microns at an optical wavelength of 546 nanometers when the applied voltage reaches about 24 volts rms. Just above V_r the differential change in retardation is about 1.25 microns per volt rms. The advantageous low-voltage nature of the described device is apparent. For example, retardations greater than 1 micron can be obtained with applied voltages less than 5 volts rms. A tunable birefringence device of the particular type described has a number of useful and practical applications. For example, if such a device is positioned between either crossed or parallel polarizers whose axes are disposed at 45° with respect to the axes of the tunable birefringence device, a color and/or intensity modulator results.

If the nematic liquid crystal molecules included in the device **10** of Figure 4.4a are in turn used to control the orientation of pleochroic dye molecules or other optically anisotropic particles (in this connection see G.H. Heilmeier, U.S. Patent 3,551,026, issued December 29, 1970), a variable density filter is obtained. Furthermore, by operating the device **10** in either conventional transmission or reflection modes, other arrangements of practical value may be formulated.

Thermally Addressed Display Cell for Optics Equipment

H. Melchior; U.S. Patent 3,836,243; September 17, 1974; assigned to Bell Telephone Laboratories describes a system which utilizes a thermally addressed liquid crystal display cell as an imaging device in, for example, a schlieren optics projection arrangement. The display cell includes a layer of liquid crystal substance, sandwiched between a first optically transparent heat-producing electrode, a second electrode, and transparent support substrates. The electrodes and the liquid crystal layer are supported by and contained between the support substrates. A controlled laser beam selectively generates hot spots in the heat-producing electrode which, in turn, locally heat the liquid crystal substance above its transition temperature. When the liquid crystal substance is subsequently cooled to its mesophase, long-lasting light-scattering centers remain. These long-lasting light-scattering centers are erased by applying a relatively high-frequency AC voltage to the electrodes.

Figure 4.5a shows a projection system in accordance with the process. In particular, a source of viewing light **100**, illustratively, a 150-watt xenon arc lamp, emits light which is focused by lens **102** such that the light converges at aperture plate **103**. Lens **110** is arranged to project the nonscattered light passed by aperture plate **103** onto screen **112**. Liquid crystal cell **104** is positioned between lens **102** and aperture plate **103** so that an image defined by the light-scattering regions on cell **104** will be projected onto screen **112**. Cell **104**, as mentioned, is a liquid crystal cell constructed to operate in the transmissive mode. Reflecting surface **106** is transparent to light from source **100**.

Figure 4.5b illustrates a liquid crystal cell for use in the projection arrangement as shown in Figure 4.5a. The cell of Figure 4.5b includes transparent support substrates **201** and **202** typically made of quartz, glass or any other suitable material. Adjacent these support substrates are electrodes **203** and **204**, and sandwiched between electrodes **203** and **204** is a liquid crystal substance **205**. Figure 4.5b depicts a typical cell in cross section. Spacers **206** and **208**, which extend around the periphery of the cell form a cavity with the electrode-coated substrates **201** and **202** for containing the liquid crystal substance. The spacers **206** and **208** typically comprise Mylar, or similar material exhibiting low electrical conduction and high dielectric strength. Electrodes **203** and **204** are in close electrical and thermal contact with liquid crystal substance **205**.

In preferred examples, such a cell includes two quartz support substrates (**201, 202**) with a layer of indium tin oxide, $In_{2-x}Sn_xO_{3-y}$, deposited on each substrate to form the heat-producing electrodes **203, 204**. These indium tin oxide electrodes are sputter-deposited on the substrates **201, 202** which are maintained above 500°C in an oxygen-deficient environment during sputtering. Electrodes produced in this manner are of the order of 2,000 to 10,000 angstroms thick and exhibit good light transmission properties at visible wavelengths with little scattering. Further, the electrodes have a sufficiently high carrier density to absorb light by free-carrier absorption in the 0.8 to 1.6 μm wavelength range. The liquid crystal substance sandwiched between the electrode-deposited surfaces of the quartz substrates is, for example, a cholesteric liquid crystal with a negative dielectric anisotropy, such as a mixture of N-(p-methoxy benzylidene)-p butylaniline (MBBA) with approximately 10% by weight of cholesteryl nonanoate.

FIGURE 4.5: THERMALLY ADDRESSED DISPLAY CELL

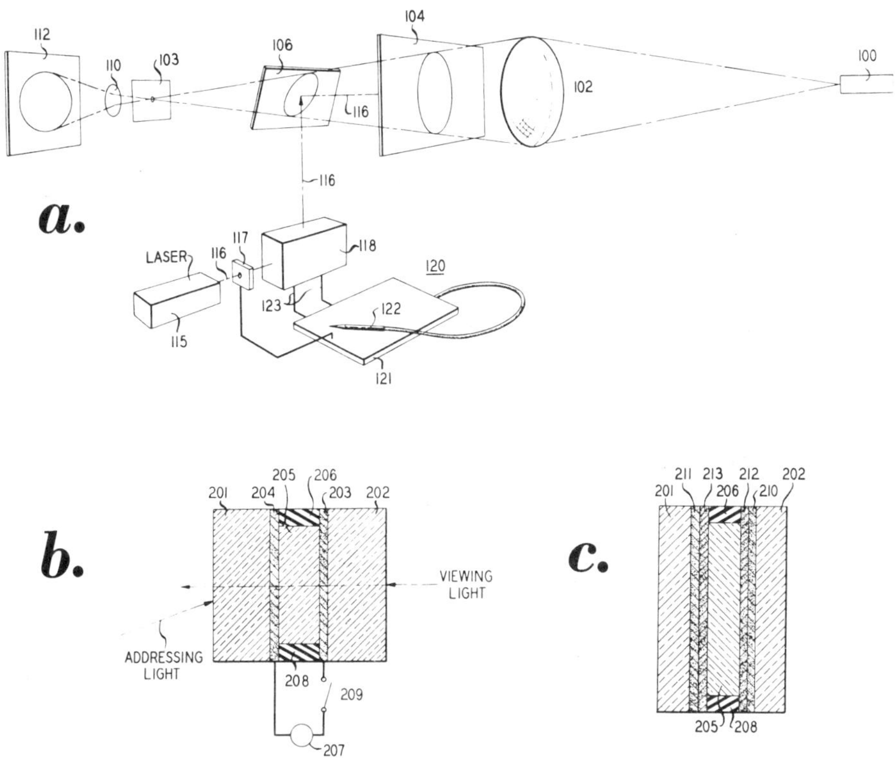

Source: H. Melchior; U.S. Patent 3,836,243; September 17, 1974

In accordance with this preferred cell arrangement, coherent addressing light shone on the liquid crystal cell of Figure 4.5b, and illustrated by the arrow designated addressing light in Figure 4.5b, passes through the cell from left to right. It is understood that the terms coherent light and laser light, used interchangeably, mean a narrow beam of laser light. In particular, substrate **201** transmits essentially all of the incident light through it, as does substrate **202**. A portion of the addressing beam is absorbed, however, by the heat-producing electrode layers **203** and **204**. This absorbed radiation stimulates the electrode layers to produce heat. The heat thus generated is advantageously arranged to heat the liquid crystal material between the stimulated areas of the electrodes to a temperature sufficient to transform the liquid crystal substance from the mesophase to the isotropic liquid state.

This transition temperature for the illustrative liquid crystal materials specified above is approximately 38° to 45°C. The addressing beam is then removed and the liquid crystal cooled to its mesophase, thus inducing a long-lasting light-scattering area in the liquid crystal. In addition, a source of AC voltage **207** is advantageously connected to heat-generating (erasing) electrodes **203** and **204** by means of switch **209**. When switch **209** in Figure 4.5b is closed, the voltage from source **207** is applied to those erasing layers. Specifically, an AC voltage with a frequency higher than the relaxation frequency of the liquid crystal material applied across the liquid crystal substance removes any scattering areas between those electrodes. As is known, the dielectric relaxation frequency is advantageously equal to $\sigma/2\pi D$, where σ is the conductivity in (ohm-centimeters)$^{-1}$ of the liquid crystal substance, $D = \epsilon\epsilon_0$, ϵ is the dielectric constant of the liquid crystal substance and $\epsilon_0 = 8.86 \times 10^{14}$ farad/centimeter, the dielectric constant of free space. Illustratively, an erasing voltage having a magnitude of 50 to 150 v peak-to-peak and a frequency of the order of 0.5 to 8 kHz is convenient to erase the scattering centers of a layer of the above-mentioned MBBA substance of the order of 6 to 30 μm thick.

Uniformly ordered crystals in the mesophase scatter little or no light in the absence of addressing stimuli. A number of liquid crystal substances having well-ordered mesophases occur naturally. Alternatively, an ordered condition of the molecules can be imposed on a substance having a disordered mesophase by appropriate treatment of the substrate surfaces or electrodes. Once a well-ordered condition has been established in the liquid crystal substance, light-scattering regions are introduced by disordering molecules or groups of molecules with respect to each other in those regions. By application of an appropriate erase voltage, these scattering areas can be reordered again to transmit light without scattering.

For example, if an erase voltage between the electrodes **203** and **204** is to erase light-scattering regions in a liquid crystal mesophase with a negative dielectric anisotropy, in the ordered state of this material the long axes of the molecules are preferably aligned in directions parallel to the surfaces of the electrodes. As a further example, in a liquid crystal material with positive dielectric anisotropy, an erase voltage between electrodes **203** and **204** erases light-scattering regions if the molecules in the ordered mesophase are aligned with their long axes of the molecules directed substantially perpendicular to the large surfaces of the electrodes **203** and **204**. Referring to Figure 4.5a, it should be noted that a laser addressing arrangement is included. In accordance with this arrangement, a source of laser radiation **115** produces a beam of coherent laser light. In particular, laser source **115** is arranged to generate a beam of laser radiation designated **116** in Figure 4.5a.

Before continuing with the discussion of the overall arrangement of Figure 4.5a, it is well to consider typical laser sources suitable for use in the arrangement of Figure 4.5a. Thus, consider a cell such as that shown in Figure 4.5b, comprising quartz substrates, indium tin oxide electrodes and a layer of MBBA with 10% by weight of cholesteryl nonanoate (CN) liquid crystal. The indium tin oxide is transparent to visible (projection) light and absorbs laser light in the infrared region, specifically, light of a wavelength in the range from 0.8 to 1.3 μm. Hence, a gallium arsenide (GaAs) laser source, emitting light of a wavelength 0.9 μm, an yttrium aluminum garnet (YAG) laser source (1.06 μm) or the 1.15 μm emission line of a helium neon (HeNe) laser are all suitable for addressing such a cell.

In addition, since indium tin oxide absorbs light in the ultraviolet range, ultraviolet laser sources such as a helium cadmium source emitting light having a wavelength of 0.325 μm or a krypton laser source emitting light having a wavelength of 0.35 μm are also suitable for use in the arrangement of Figure 4.5a. The laser beam **116** is controlled in standard fashion by shutter **117** which is arranged to either essentially completely stop or transmit the laser beam. If shutter **117** is open such that it transmits laser beam **116**, the beam is further controlled by control apparatus **118**. Control apparatus **118** includes apparatus, e.g., movable mirrors for redirecting the laser beam to intersect reflecting surface **106**. (Reflecting surface **106**, in this arrangement, is a dichroic mirror which reflects laser light but transmits viewing light.)

Clearly, from Figure 4.5a, it is apparent that the point of incidence of laser beam **116** on inclined reflecting surface **106** determines the point on liquid crystal cell **104** to be intersected by and, hence, addressed by laser beam **116**. Thus, by changing the angle of deflection of the mirrors of control apparatus **118**, for example, any point on the image area of liquid crystal cell **104** can be addressed by laser beam **116**. Further, by continuously changing the angle of deflection of the control mirrors, the laser beam can be continuously moved about the image area of cell **104**. In a typical arrangement, the laser beam **116** is arranged to reproduce rapidly applied handwriting on the display cell image area. Toward this end, Figure 4.5a also includes an electronic writing graphics input arrangement **120**. Such graphics input writing arrangements exist in a number of forms and in varying degrees of sophistication in the prior art. For example, for purposes of discussion, consider a typical arrangement comprising a writing pad such as pad **121** of Figure 4.5a including electrical position-specifying sensors. Pen **122**, in turn, includes means for stimulating the position-specifying sensors of pad **121**.

As pen **122** is, in a typical arrangement, touched to the surface of pad **121**, a signal unique to the position on the pad of pen **122** is generated. In the arrangement shown in Figure 4.5a, the position signals generated by arrangement **120** are conveyed to control apparatus **118** via leads **123**. Control apparatus **118** includes circuitry and other well-known apparatus for controlling the laser-beam redirecting apparatus in response to the position-specifying signals received from graphics apparatus **120**. In addition, indication signals are communicated from graphics unit **120** to shutter **117** which are arranged to, in one instance, close and hence block laser beam **116** when pen **122** is not in contact with the surface of pad **121** and to open and transmit the addressing beam when pen **122** is in contact with pad **121**.

From the foregoing discussion, it is evident that, in accordance with the arrangement of Figure 4.5a, information such as a signature, can be written on pad **121** (which, incidentally, typically includes paper placed on the surface of pad **121** and ink depositing means within pen **122** to provide a visual aid to the operator). Laser beam **116**, controlled by graphics unit **120**, shutter **117**, control means **118** and reflecting surface **106**, accurately transcribe the signature written on pad **121** onto liquid crystal cell **104** by means of the heat-stimulated scattering area generation.

At the same time, if viewing light **100** is on, the information transcribed onto cell **104** is simultaneously projected on screen **112**. As mentioned, reflecting surface **106** is transparent to viewing light. In addition, aperture plate **103** blocks scattered light (that is, light scattered by the light-scattering regions of the liquid crystal cell), rendering the addressing information black on a white background. Clearly, a stop could be used to block the nonscattered light to produce white writing on a black background.

The cell of Figure 4.5b, suggested as a typical cell useful in the display arrangement of Figure 4.5a, illustrates a transmission mode cell in which the electrodes **203, 204** which are transparent to viewing light not only perform the erasing function but absorb addressing light to produce heated areas. These heated areas, in turn, are arranged to induce scattering centers of light in a layer of liquid crystal substance. Figure 4.5c illustrates a cell including electrodes **212, 213** which perform the erasing function only. (The voltage source and switch have been omitted from Figure 4.5c for convenience.) In addition, the cell of Figure 4.5c includes heat-producing layers **210, 211** which absorb addressing light and produce heat in response thereto.

The electrodes **212** and **213** are arranged to efficiently conduct heat from the heat-producing layers **210** and **211** to the liquid crystal layer **205**. In a typical embodiment, the layers **210** and **211** are composed of pale yellow zinc cadmium sulfide. The source of laser radiation, in this embodiment, is a helium cadmium laser which emits light at a wavelength of 0.4416 μm which is advantageously absorbed by the zinc cadmium sulfide, $Zn_xCd_{1-x}S$. Although the above detailed description has proceeded in terms of a transmission cell, it is often desirable to provide liquid crystal projection systems which include a reflection mode cell. A liquid crystal reflection display cell is described in detail in the patent.

Kerr Effect Cells

A. Boller and H. Scherrer; U.S. Patent 3,795,436; March 5, 1974; assigned to Hoffmann-La Roche, Inc. describe an electrooptical light-modulation cell having a dielectric disposed between electrodes and being comprised of a nematogenic fluid in the isotropic state. The dielectric, upon application of a voltage to the electrodes, becomes double refracting. More particularly the nematogenic fluid comprises compounds of the general formula

(1) 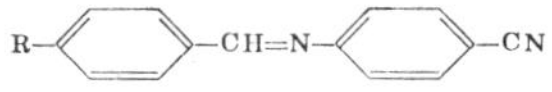

where R is ethyl, n-propyl, n-butyl, n-pentyl, n-hexyl, isohexyl, n-heptyl or n-octyl or compounds of the general formula

(2) 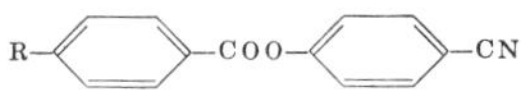

where R is n-butyl, n-pentyl, n-hexyl, n-heptyl or n-octyl. Light modulation cells or Kerr cells are known and until now nitrobenzene, because of its high Kerr constant, has been the only substance used as the dielectric. However, Kerr cells having a nitrobenzene dielectric have the disadvantage of requiring operating voltages of several kilovolts. Operation of such a cell places exacting voltage breakdown strength of an associated electronic control system. Consequently, when the cells are intensively modulated, it is impossible to make use of the theoretically very short response time of nitrobenzene.

A further disadvantage of Kerr cells having a nitrobenzene dielectric comprises the fact that they need to be relatively thick in the direction of the light beam. For example, the thickness of the dielectric parallel to the light beam must be 5 cm in order to produce a phase difference of $\delta = 0.835\lambda$ ($\lambda = 583\ \mu m$, T = 24°C) at a field strength of 24 kv cm^{-1}. Cells of this thickness are unsuitable for a number of practical applications. This process provides an electrooptical light-modulation cell which is very quickly responsive and one requiring a control voltage in a range which only necessitates the utilization of relatively inexpensive circuitry, using transistors where possible.

In particular, the process relates to an electrooptical light-modulation cell having a dielectric disposed between electrodes which becomes doubly refracting upon an application of a voltage to the electrodes, in which the general formula (1) and/or (2) and/or mixtures with other nematogenic substances.

As shown in Figure 4.6, a light-modulation cell **10** comprises two parallel rectangular electrodes **16** and **17** between which is disposed a dielectric **13** composed of a nematogenic substance. The electrodes are made of stainless steel, but can be made of any other metal, which does not react with nematogenic substances, for example, gold or the like. The distance between the electrodes depends on the diameter of a light beam which is to be modulated. In the example, this distance is so small that the nematogenic substance is held between the electrodes by capillary forces. In such a case, no additional walls are needed for laterally binding the space between the electrodes.

If the distance between the electrodes is greater, the space for the dielectric must be sealed on all sides, using plates of suitable size and made of electrically insulating material. At least two opposite insulating plates must be transparent to the light beam, i.e., they should advantageously be made of glass and acrylic glass. The dielectric is made of a nematogenic substance, i.e., a substance having a nematic liquid-crystal phase over a certain temperature range. Specially good results are obtained with nematogenic substances which are dielectrically and optically strongly anisotropic and which have a pronounced electric dipole moment extending parallel to the direction of maximum polarizability of the molecules.

The substances must also have a marked tendency for parallel orientation over a number of molecular spacings, i.e., the tendency to form clusters. Since the tendency to form clusters is particularly marked in the immediate neighborhood of the temperature T_c at which there is a change from the isotropic to the nematic state, the cell is desirably operated in the temperature range $T_c < t \leqslant T_c + 1°C$, where t is the temperature of the nematogenic fluid. Particularly preferred are mixtures of p-[(p-n-hexylbenzylidene)amino] benzonitrile and p-[(p-n-butylbenzylidene)amino] benzonitrile in a molar ratio of 2:1.

FIGURE 4.6: ELECTROOPTICAL LIGHT MODULATION SYSTEM

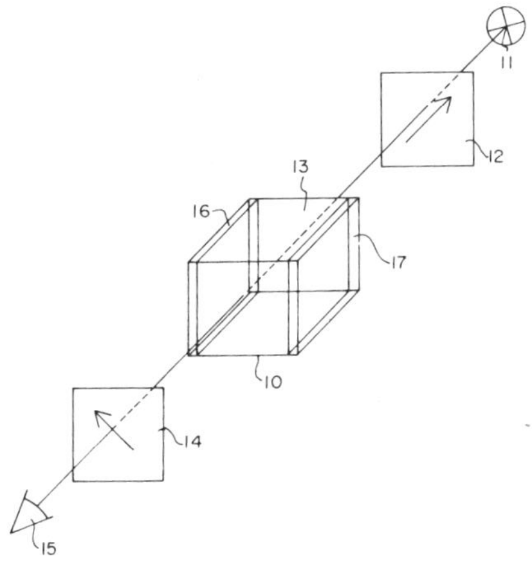

Source: A. Boller and H. Scherrer; U.S. Patent 3,795,436; March 5, 1974

The light modulating arrangement shown in Figure 4.6 comprises, in addition to cell **10**, a polarizer **12** disposed upstream of the cell and an analyzer **14** disposed downstream of the cell. A beam of light from a source **11** travels through polarizer **12** and cell **10** parallel to the electrodes and strikes analyzer **14**. If the polarizer **12** and analyzer **14** are rotated 90° relative to one another as shown by the arrow in Figure 4.6 and if the dielectric of cell **10** is isotropic, the light beam does not reach the eye of an observer downstream of the analyzer **14**.

If a DC voltage is applied to the electrodes of cell **10**, the dielectric **12** becomes doubly refracting, so that the linearly polarized light from polarizer **12** is converted to elliptically polarized light, some of which can travel through the analyzer **14**. The observer therefore notices a brightening of his field of view through the analyzer. Maximum brightening occurs at an optical phase difference of $\delta = (2i + 1)\ \lambda/2$, with i = 0, 1, 2.... The optical phase difference is directly proportional to the wavelength, the thickness of the cell or dielectric parallel to the light beam, the Kerr constant and the square of the electric field strength.

An important criterion of the capacity of electrooptical light-modulation cells is the attainable contrast between the light intensity when no voltage is applied to the electrodes and the light intensity at maximum drive. The contrast obtainable in a cell embodying the process is of the order of 1:1,000, and varies with the fourth power of the field strength. Electrooptical light-modulation cells of this type are suitable for a number of applications. In general, they can be used for controlling and modulating light far beyond the acoustic frequency range. An important application is to electronic shutters, particularly in cases in which large apertures are required. In very specific applications the double refraction may be optically induced, e.g., by the field of high-powered light sources.

In related work *W. Helfrich, M. Schadt and H. Scherrer; U.S. Patent 3,796,479; March 12, 1974; assigned to Hoffmann-La Roche Inc.* describe an electrooptical light-modulation cell having a dielectric disposed between electrodes and being comprised of a nematogenic fluid in the isotropic state. The dielectric, upon application of a voltage to the electrodes, becomes doubly refracting. The dielectric is made of a nematogenic substance, i.e., a substance having a nematic liquid-crystal phase over a certain temperature range. Specially good results are obtained with nematogenic substances which are dielectrically and optically strongly anisotropic and which have a pronounced electric dipole moment extending parallel to the direction of maximum polarizability of the molecules.

The substances must also have a marked tendency for parallel orientation over a number of molecular spacings, i.e., the tendency to form clusters. Since the tendency to form clusters is particularly marked in the immediate neighborhood of the temperature T_c at which there is a change from the isotropic to the nematic state, the cell is desirably operated in the temperature range $T_c < t \leqslant T_c + 1°C$. A particularly suitable mixture of nematogenic substances might be a 1:1:1 mixture of N-(4'-butoxybenzylidene)-4-aminobenzonitrile, N-(4'-hexoxybenzylidene)-4-aminobenzonitrile and N-(4'-octoyloxybenzylidene)-4-aminobenzonitrile.

Scintillation-Free Image from Coherent Light Source

A process described by *C.E. Baker; U.S. Patent 3,650,608; March 21, 1972; assigned to Texas Instruments Incorporated* provides an apparatus comprising a source of a coherent light beam, a screen means for scattering a coherent light beam, means for directing the light beam toward the screen means, the screen means including an optically transmissive layer of an organic nematic mesomorphic compound less than **20** mils thick and means for continuously impressing a voltage gradient across a functional portion of the layer having a value greater than the threshold value which causes the compound to scatter light.

The procedure for carrying out the process includes displaying a coherent light beam on a screen comprising directing a coherent light beam toward an optically transmissive, relatively thin layer of an organic nematic mesomorphic compound having a threshold electrical field which when exceeded causes the compound to scatter light, and impressing a voltage gradient across the layer of the organic nematic mesomorphic compound.

Referring to Figure 4.7, a coherent light source **10**, illustrated schematically as a laser, directs a coherent light beam **12** through light modulator **14** in which the intensity or brightness of the light beam is varied in accordance with a predetermined input signal. From the light modulator **14** the intensity modulated laser beam **16** is directed into a scanning mechanism or light beam deflector which causes the light beam to be deflected or scanned in a predetermined pattern. This pattern is projected or displayed upon a screen, generally designated **18**. The rear projection display screen **18** includes a first transparent substrate **20**, for example a planar sheet of glass.

The substrate is coated on one side with a layer or coating **22** of material which is optically transmissive and electrically conductive. The display screen also includes a second substrate **24** similar to substrate **20** which is also coated with a layer or coating **26** of material which is electrically conductive, and in this embodiment also optically transmissive. Sandwiched between the two layers **22** and **26** and in electrical contact therewith, is a relatively thin layer **30** of an organic nematic mesomorphic compound. The layer has an optimum thickness of less than 20 mils, preferably less than 10 mils, and most preferably has a thickness in the range of 1 to 4 mils. Leads **32** and **34** are connected respectively to the electrically conductive layers **22** and **26**.

The leads are also connected to a power source **36** which impresses a sufficient voltage when energized to cause the organic nematic mesomorphic compound to diffuse light which strikes it, as will be explained hereinafter. The power source is preferably a direct current source, although alternating current is usable. To the eye of the observer then, the beam of coherent light **38** striking the rear side of the projection screen **18** appears to be a bright stationary display point, not exhibiting the heretofore troublesome phenomenon of speckling or scintillation. The image created on the screen, of course, can be of a variety of configurations, depending upon the particular manner in which the coherent light beam is modulated and/or deflected. An exemplary image can be produced by a raster scan of the screen, displaying, for example, a dynamic image such as a conventional line television image. This type of display would be of great importance in a large screen dynamic display such as that utilized in closed circuit television, conferences, and public information displays. Previously, this type of display has been subject to the scattering or scintillation effect which increases with the distance of the observer from the display screen.

FIGURE 4.7: APPARATUS FOR DISPLAYING COHERENT LIGHT IMAGES

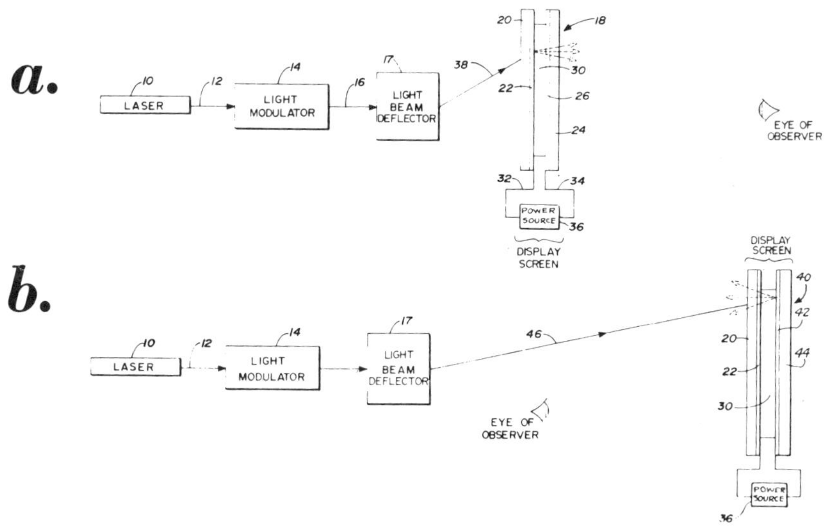

Source: C.E. Baker; U.S. Patent 3,650,608; March 21, 1972

In Figure 4.6b, a second example of the process is shown where a front projection screen **40** (rather than a rear projection screen) is shown. A coherent light beam **12** from a laser **10** is modulated and deflected according to a predetermined pattern. The display screen **40** is similar to that previously described except that the layer of coating **42** on substrate **44** is optically reflective rather than transmissive. Light beam **46** is directed towards display screen **40**, transmitted through substrate **20**, layers **22** and **30** where it strikes the coating **42** and is reflected back toward the eye of the observer. The coherent light beam is dynamically diffused resulting in a stationary, nonscintillating image.

The display screen must include two substrates with coatings of uniform thickness which are spaced equidistant along the entire surface which will be scanned or used for display. This aids in control of the thickness of the organic nematic mesomorphic compound layer and also assists in maintaining an equal voltage drop across the entire layer. The voltage is continuously impressed across the organic nematic mesomorphic layer to effect a continuous stationary image as seen by an observer. The substrates comprising the display screens **18** or **40** are positioned less than 20 mils apart by means of spacers such as a frame. Substrates **20** and **24** can be composed of any variety of optically transmissive solids which include transparent and translucent materials.

The transmissive coating on the substrates is also transparent in the case of coatings **22** and **26**. Such a coating can be composed of layers of indium oxide or tin oxide across the entire surface of substrate **20** which is contacting the nematic mesomorphic compound. If a reflective surface is desired, a thin layer of a metallic coating such as silver or aluminum, can be deposited on substrate **44.**

Storage Device Employing CdS Photoconductor Layer

M. Braunstein and W.P. Bleha, Jr.; U.S. Patents 3,732,429; May 8, 1973; and 3,811,180; May 21, 1974; both assigned to Hughes Aircraft Company describe a liquid crystal device which comprises a liquid crystal layer, a thin photoconductor layer, and a thin insulator layer sandwiched between the transparent supporting covers having a transparent electrically conductive layer thereon. The photoconductor layer is cadmium sulfide (CdS) and the insulator layer is zinc sulfide (ZnS).

Figure 4.8a is an isometric view of a portion of the basic cell structure showing a vertical section together with schematic electrical connections. Figure 4.8b shows a cell, similar to the cell of Figure 4.8a, and its schematic electrical connection. Figure 4.8c shows the manner in which either of the cells of Figures 4.8a and 4.8b can be employed in conjunction with a projection light source where the image information on the liquid crystal cell is projected onto screens.

Figure 4.8d is similar to Figure 4.8c but shows a different manner of application of the image information containing exposure light source. Figure 4.8e is a schematic showing of the manner in which either of the cells of Figures 4.8a and 4.8b can be employed with a cathode ray tube used for image information together with reflective display. Figure 4.8f is another schematic arrangement similar to Figure 4.8e. Figure 4.8g is a schematic drawing of the CdS thin film vacuum deposition chamber. Figure 4.8h is a schematic drawing of a postdeposition thermal processing furnace and processing gas controls.

Figure 4.8a illustrates a basic cell **10** arrangement. The cell is a laminar structure. Glass substance **12** is ordinary clear glass, and since it can be used for transmitting light therethrough, it is preferably clear and flat, with its outer surfaces substantially parallel. Transparent, electrically conductive coating **14** is applied to one side of the glass. This coating is specifically antimony doped tin oxide, known as NESA glass, or tin doped indium oxide. The glass substance is conveniently one-eighth inch thick while the electrically conductive coating thereon is of such thickness as to give a resistance in the range from 1 to 10^3 ohms per square. Instead of employing NESA glass, another convenient, substantially transparent and electrically conductive coating, as described in U.S. Patent 3,698,946, can be employed.

FIGURE 4.8: LIQUID CRYSTAL DEVICE EMPLOYING CdS PHOTOCONDUCTOR

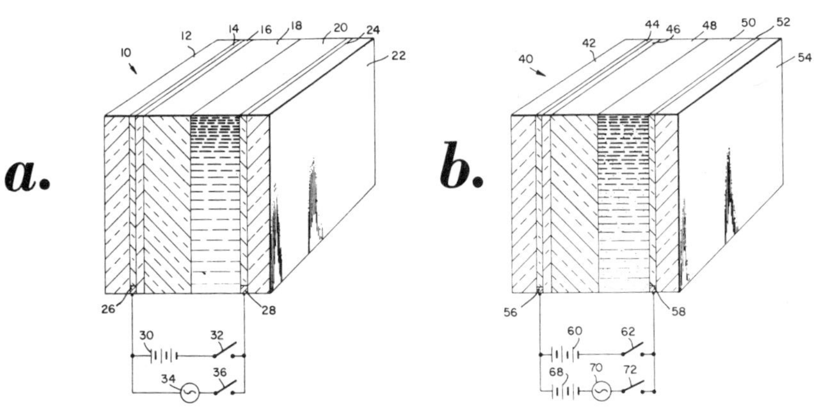

(continued)

FIGURE 4.8: (continued)

c.

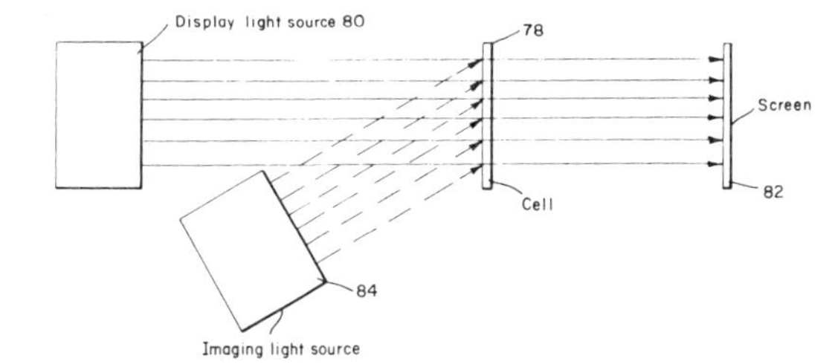

d.

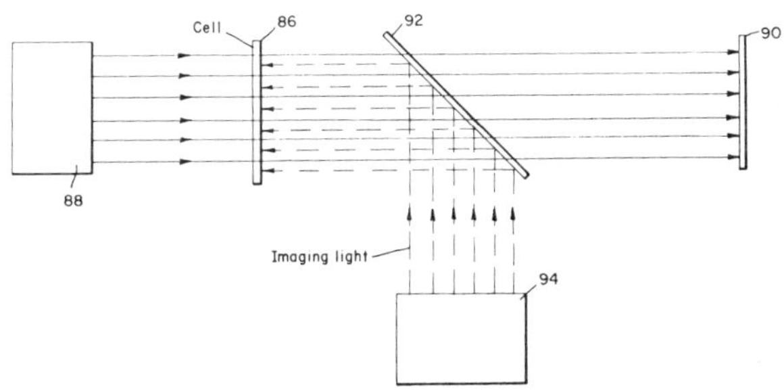

e.

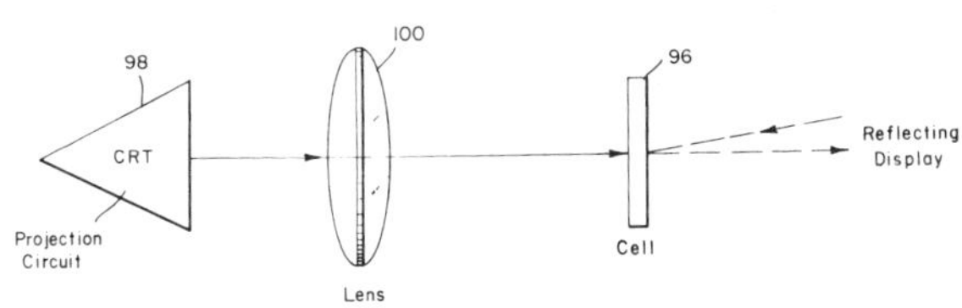

f.

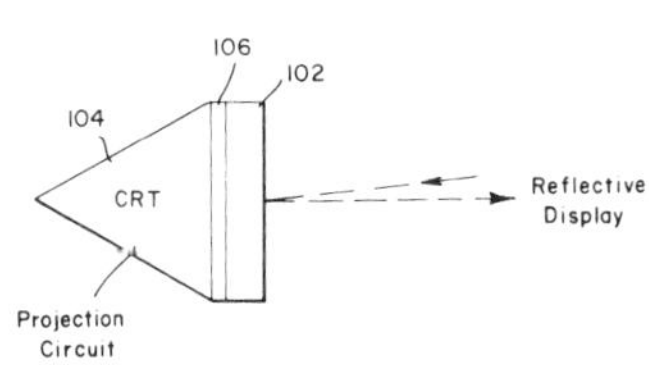

(continued)

FIGURE 4.8: (continued)

g.

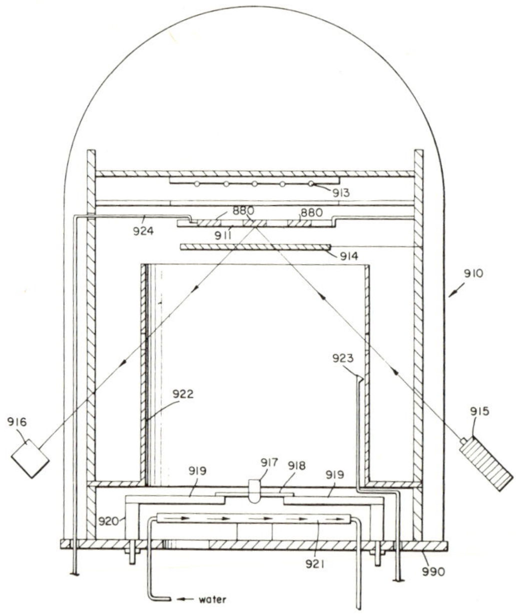

h.

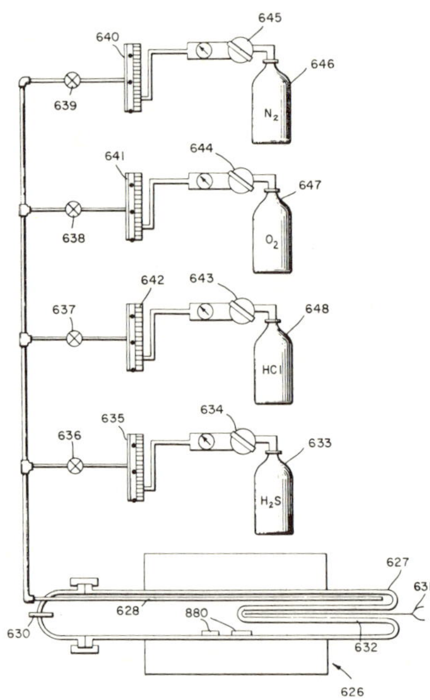

Source: M. Braunstein and W.P. Bleha, Jr.; U.S. Patent 3,732,429; May 8, 1973

Vacuum deposited onto the coating **14** is a thin layer of the insulator zinc sulfide, illustrated as the layer **16.** Thin insulator films of materials such as silicon dioxide or silicon monoxide could also be used. The insulator **16** is in the range from 10 to 10^2 nm in thickness, and the optimum thickness is 50 nm. Over the zinc sulfide layer **16** there is vacuum deposited layer **18** of the photoconductor cadmium sulfide. Photoconductor films of cadmium selenide could also be used. The photoconductor layer is vacuum deposited to a thickness of from 2.0 to 12.5 microns.

On top of the photoconductor layer is deposited a liquid crystal layer **20.** The liquid crystal layer is a nematic cholesteric liquid crystal of the type defined in "A New Electric Field Controlled Reflective Optical Storage Effect in Mixed Liquid Crystals Systems," by G.H. Heilmeier and J.E. Goldmacher, *Applied Physics Letters,* Volume 13 No. 4, p. 132 and 133 (August 15, 1968). The thickness of the liquid crystal layer is typically 12.5 microns, and as in conventional liquid crystal practice can range from 4 to 25 microns. The thickness is contained by spacers positioned around the edges of the cell to maintain the volume in which the liquid crystal is positioned.

The cell is completed by glass **22** on which is coated transparent conductive layer **24.** This structurally corresponds to the glass **12** in transparent conductive layer **14** described above. The two glass layers **12** and **22** thus define the outer surfaces of the cell. Electric contacts **26** and **28** are respectively connected to the electrically conductive substantially transparent layers **14** and **24.** Source **30** of direct current, together with its series control switch **32,** are connected across the contacts **26** and **28.** The closure of the switch applies a DC field across the layers between the electrically conductive layers **14** and **24.** AC source **34** and its serially connected switch **36** are also connected across contacts **26** and **28,** in parallel to the DC source and switch. The AC source provides AC current from 1,000 to 100,000 hertz.

Figure 4.8b illustrates a cell **40** which is a second example of the liquid crystal device in accordance with this process. This cell is built of a plurality of layers comprising glass layer **42,** transparent conductive layer **44,** zinc sulfide insulating layer **46,** cadmium sulfide photoconductive layer **48,** liquid crystal layer **50,** transparent conductive layer **52,** and glass cover **54.** The layers **42, 44, 46, 48, 52** and **54** are respectively of an equal thickness to layers **12, 14, 16, 18, 24** and **22** of the cell **10** shown in Figure 4.8a. The method of formation of these various layers is the same in both cells. The principal difference between the cells is that the layer **50** is a nematic liquid crystal.

Electric contacts **56** and **58** are respectively connected to substantially transparent conductive layers **44** and **52.** Electrically connected to these contacts are the series combination of DC voltage source **60** and its switch **62.** This is identical to the DC source **30.** In addition, in Figure 4.8b a combined DC and AC source is connected across the contacts **56** and **58,** in parallel to the DC sources **60.** Thus, DC source **68** is connected to one of the contacts, and through AC source **70** and switch **72** to the other of the contacts. By closure of switch **72,** an AC field with a superimposed DC field is applied between the substantially transparent layers **44** and **52.**

Thus, the cells **10** and **40** are the same, except for the differences between the liquid crystal layers, and the fact that the nematic liquid crystal layer **50** also has the capability of having a combined field applied thereacross. The method of manufacture of the insulating layer **16** and the photoconductive layer **18,** and their corresponding layers **46** and **48**, is as follows. In the first step of manufacture, the transparent conductive coating-glass combination **12** and **14** is cleaned for use as a substrate in a high vacuum deposition process.

There are many methods, well-known in the art, to accomplish this. In the second step the zinc sulfide film is vacuum deposited on the conductive coated glass substrates in a high vacuum coating station. The deposition of zinc sulfide is also well-known in the art, and the procedure described here is not unique to the performance of the device. The method used is to evaporate electronic grade powder or high purity crystalline pieces of zinc sulfide from a resistance heated molybdenum boat or by electron bombardment from an electron beam evaporation unit. The pressure in the bell jar during deposition is between 1 x 10^{-5} and 1 x 10^{-6} torr and is not critical. The substrates are placed 10" above the evaporation source along with a thickness monitoring instrument, such as quartz crystal deposition thickness transducer. After glow discharge cleaning of the substrates, the ZnS evaporation commences. The films, typically 50 nm thick, are deposited in 1 hr, although times of 10 min to 2 hr have been used. The films have also been varied in thickness from 10 to 100 nm with proportional changes in deposition time. After deposition the films are transparent (absorbing no light in the visible spectrum) and thus are characteristic of near stoichiometric zinc sulfide.

The third step in the fabrication is the vacuum deposition of the CdS film **18** on top of the zinc sulfide layer **16.** The deposition is done in the bell jar of a conventional high vacuum system in the pressure range of 1 x 10^{-7} to 1 x 10^{-5} torr. The pressure, however, does not appear to be critical. A cross section drawing of the instrumentation is shown in Figure 4.8g. The vacuum is enclosed by a baseplate **990** and a glass bell jar **910.** The structures **880** with the zinc sulfide film are held by a stainless steel substrate holder **911** and heated by quartz lamps **913.** A removable shutter **914** shields the substrates until deposition on them is to be commenced. The thickness of the CdS films is directly and continuously monitored on a substrate by the use of optical interference. This is accomplished with the use of a laser **915** and detector **916** positioned outside the bell jar **910.**

The electronic grade CdS powder, in the form of a pressed cylindrical pellet **917,** is evaporated from a formed tantalum boat **918** which is resistively heated by current passing through bus bars **919** and current feedthroughs **920.** The boat is designed such that as the CdS evaporates, the pressed pellet **917** settles down into the boat. This gives an efficient thermal evaporation over the long period of time required for the deposition of the CdS films.

While this particular configuration gives the characteristics desired, other forms of CdS, such as crystals, and other thermally efficient evaporation sources, such as baffle boxes, could be used to get the same results. The evaporation rate is controlled by controlling the current to the tantalum boat. The current is set so that 2.5 microns of CdS as monitored by optical interference is deposited on the structures **880** in 1 hr. Typical thicknesses of CdS films are 5 to 12.5 microns so that deposition times of 2 to 5 hr are required. Successful results have been obtained with thicknesses from 2 to 15 microns and evaporation rates from 0.5 to 10 microns/hr.

To avoid the heating of the various elements in the deposition chamber by radiation from the tantalum boat a water cooled plate **921** is positioned beneath the boat and extending to the diameter of a cylindrical stainless steel deposition chamber **922** positioned around the boat. The water-cooling is also used to maintain the temperature of the wall of chamber **922** as measured by a thermocouple **923** below 60°C. This low temperature, as compared with the structure **880** temperature of 130°C, as measured by a thermocouple **924,** is necessary to obtain the desired characteristics in the films.

It should be noted, however, that the important fact is that the chamber **922** and baseplate **921** are maintained at a lower temperature than the structure **880** and methods of achieving this can be determined by one skilled in the art. Also the temperatures given can be changed to vary the conductivity and current-voltage characteristics of the CdS films. A range of substrate temperatures from 100° to 200°C and chamber wall temperatures from 40° to 90°C have been used to make CdS films of the given characteristics.

The fourth step in device fabrication is the postdeposition thermal processing of the structure **880** as it emerges from step three. The preferred process can be understood with reference to Figure 4.8h. A controllable furnace **626** has a quartz processing tube **627** of suitable diameter. A gas inlet tube **628** introduces gas which is preheated by passing through the core of the furnace. The gas exits through short exit tube **630.** The temperature for monitoring and control near the center of the tube and also near the center of the hot zone is sensed by a thermocouple **631** sheathed in a quartz tube **632.**

The structures **880** are placed in the tube near the center of the hot zone. A controllable flow of a single gas or a mixture of gases is provided by valve **636**, flowmeter **635**, regulator **634** for cylinder **633** of H_2S; by valve **637**, flowmeter **642**, regulator **643** for cylinder **648** of HCl; by valve **638**, flowmeter **641**, regulator **644** for cylinder **647** of O_2; and by valve **639**, flowmeter **640**, regulator **645** for cylinder **646** of N_2.

In operation the following procedure is followed. First the structures **880** are inserted in the processing tube **627** and the tube is flushed out using only the N_2 gas from cylinder **646**. The furnace is then brought to 300°C, as indicated by thermocouple **631**, with N_2 flowing at a rate 0.1 to 20 cfh (STP). Nitrogen is used only as a purge medium. At this point the N_2 is stopped and the processing gas or gas mixture is introduced into the tube. The percentage by volume of the gases that have been used for processing are summarized as follows: H_2S, 95 to 100%; HCl, 0.1 to 2%; N_2, 0.1 to 5%; and O_2, 0.1 to 0.5%.

The total flow rate is from 0.1 to 20 cfh (STP). It should be noted that the main constituent gas is H_2S and that pure H_2S can be used as the processing gas. The furnace is brought to the desired temperature, typically 500°C, and kept at that temperature for the desired time, about 30 min. Temperatures from 385° to 550°C and times from 1 to 60 min have been successfully used. The particular time and temperature used depends on the thickness of the CdS film **18** or **48**, the substrate **12** or **42** material, and the mixture of gases used. Also the device characteristics, for a given thickness of CdS film, substrate material, and gas, can be altered by changing the temperature and time.

After the desired time has elapsed the outside of the quartz tube **627** is exposed to the ambient and allowed to cool in approximately 10 min to 300°C, as measured by thermocouple **632**, at which point the processing gas or gas mixture is turned off and N_2 only is allowed to flow through the tube. When the thermocouple indicates 70°C, the structures are removed from the tube and exposed to the ambient.

While this rapid cooling, caused by exposing the outside of the quartz tube **627** to the ambient, produces superior results, devices exhibiting the desired characteristics can also be obtained by allowing the quartz tube to remain completely in the furnace **626** and turning off the power to the furnace. Under these circumstances, the structures cool down at a rate which is slower by a factor of about 10.

The fifth and final step in the device fabrication is the placing of the liquid crystal layer **20** or **50** and counter-electrode-glass substrate combination **22** and **24** or **52** and **54** on the subassembly completed in step four. To define the area and thickness of the liquid crystal layer a spacer with an open area in the center, typically made of the insulator Mylar, is placed on the subassembly. The thickness of the spacer is typically 12.5 microns but can be from 6 to 25 microns.

The liquid crystal is then spread in the open area in the center of the spacer. An excess of the viscous liquid crystal remains until the counter-electrode glass assembly is placed down on the liquid crystal and spacer thereby squeezing the excess liquid crystal out of the center, and maintaining a layer of liquid with no voids. A supporting clamp is then placed around the device and wires are bonded to the electrodes for electrical contact. The applied DC voltage for operation with either liquid crystal layer **20** or **50** is with the liquid crystal biased negatively with respect to the photoconductor.

The double layer structure consisting of the CdS photoconductor and ZnS insulator allows a higher dark impedance to be obtained in conjunction with the liquid crystal layer than the use of just the CdS film. The higher dark impedance is obtained without significantly altering the light-to-dark impedance ratio. As a result higher contrast between scattering and nonscattering areas in the liquid crystal layer is obtained because of the better impedance match to liquid crystal impedance levels which give efficient scattering. It can be hypothesized that the reason for the improvement of the performance of the photoconductor with the insulating film is due to the improved nucleation of the CdS film on the ZnS film which results in fewer defects through the CdS film.

It is also possible that a heterojunction or blocking contact is formed. It is expected that the photoconductor-insulator structure could be used to improve the performance of photoconductor films of cadmium selenide which has a lower intrinsic resistivity than cadmium sulfide. In the use, a light having an intensity pattern in accordance with the desired image is projected onto the cell. This will be called the imaging light, in particular sources and arrangements of imaging will be described hereinafter. After imaging is completed, the liquid crystal stored image can be viewed. This viewing can be accomplished either by a projected display light, or by reflective observation of the image by frontal illumination. In either case, a display light is also utilized.

Considering the structure of Figure 4.8a, both switches **32** and **36** are open. Thereupon, the imaging light is turned on. The imaging light projects the desired image into the cell, both through the nematic-cholesteric mixture liquid crystal and the photoconductive layers. With the imaging light on, the switch **32** is closed for a short time, for example, from 0.1 to 1 sec. When the voltage, typically from 20 to 150 v, is applied, the direct current potential field is divided across the liquid crystal layer **20,** and the insulator layer **16** and photoconductive layer **18.** The imaging light is of such wavelength where the photoconductor is sensitive. For cadmium sulfide, the imaging light is thus between 400 and 520 nm. Where there is no imaging light the photoconductor and insulator layers totally comprise a higher impedance than the liquid crystal.

Where the imaging light is projected onto the cell the resistance of the photoconductor decreases so that it is less than the liquid crystal. Thus, in the exposed area of the photoconductor, the current is caused to flow to induce change in the scattering properties of the liquid crystal. In the regions where there was no imaging light, the voltage pulse caused by closure of switch **32**, the voltage division is such that the principle voltage drop was across the resistor and the photoconductor and the liquid crystal adjacent thereto is uneffected. Thus, this area remains clear. After switch **32** is open, and the imaging light removed, the image is retained in the liquid crystal because of the scattering storage properties of the nematic-cholesteric mixture.

At a later time, when it is desired to remove the image stored in the liquid crystal, the liquid crystal layer can be brought to its nonscattering state in all regions by closing the switch **36** and applying the AC field across the liquid crystal from source **34.** A frequency greater than about 1,000 cycles/second is required to effect erasure and return the liquid crystal to the clear state. Closure of the switch **36** for 0.5 to 2 sec is adequate to obtain this erasure.

In the structure of Figure 4.8b, the liquid crystal **50** is a nematic material. There is no optical memory, in the sense of scattering condition of such a material, in the absence of a continuously applied field. The employment of the DC field provided by DC source **60** and closure of its switch **62** is alternative to the application of the DC field superimposed upon the AC field, by closure of switch **72.** The DC field provided by source **60** is from 10 to 100 v.

On the other hand, the combined field comprises a DC potential from source **68** of from 10 to 100 v, together with an AC potential from source **70** of from 10 to 100 v rms at a frequency in the range of from 1,000 to 20,000 cycles per second. With either of the switches **62** or **72** closed, the imaging light is turned on. The imaging light is of such wavelength that it affects the photoconductor. In the case of the cadmium sulfide, the imaging light has a wavelength from 400 to 520 nm.

The illumination light, either for frontal or projective viewing of the images of the liquid crystal, is of such wavelength that the photoconductor is not photosensitive thereto. In the case of cadmium sulfide, the projection light has a wavelength above 520 nm. In this mode, the information supplied by the imaging light is displayed in real time by the viewing light. This is accomplished because the regions where the photoconductor is of lower impedance due to the incidence of light, the voltage drop is substantially taken across the liquid crystal, and this causes dynamic scattering in that region.

In the regions where the imaging light does not occur, the principle part of the voltage drop is across the photoconductor, so that the voltage drop across the nematic liquid crystal is not sufficient to cause scattering. Therefore, in those areas the liquid crystal remains transparent. In order to achieve real time imaging, the use of the combined DC and AC field is preferred to the ordinary DC field provided by the DC source **60.** The normal decay time of the nematic liquid crystal under the influence of the DC field, and without illumination, may not be sufficient to provide the real time display. Therefore, the superimposed AC wave quickly erases the image in areas where the imaging light has been removed.

In general, a cadmium sulfide photoconductor film of the type specified, when used alone and when used in the mode necessary for the operation described, would not have the sufficiently high impedance or sensitivity necessary for the described operation of either the cells **10** or **40.** By the use of the double layered structure described, the necessary properties can be obtained. A summary of some of the characteristics measured on laboratory models of the nematic-cholesteric device are summarized below.

Resolution	>25 line pairs/mm
Sensitivity	<0.1 mj/cm^2 (466.0 to 520.0 nm)
Turn on	0.1 sec
Erase time	~1 sec
Storage time	Many days
Contrast	>5:1 (no half-tone reproduction)

The sheet resistivity of the zinc sulfide insulator layer is greater than 10^{15} ohms/square. The sheet resistivity of the cadmium sulfide photoconductor layer in the dark state is greater than 10^{12} ohms/square. The dark resistivity of the liquid crystal layer is greater than 10^{12} ohms/square. The complete cell has a light to dark current ratio of about 20 to 1. The complete cell has a conductance of about 10 microamperes/square centimeter at 40 v DC in the dark state. The manner of use of the cells **10** and **40** is described with respect to Figures 4.8c through 4.8f. Referring to Figure 4.8c the cell **78** could be either of the cells **10** or **40**, equipped with the electric application means described with respective Figures 4.8a and 4.8b. The cell **78** is positioned between a source of visible light **80** which serves as a projection light and screen **82.**

A source of visible imaging light **84** provides visible radiation for photoactivating the photoconductor layer **18** inside the cell **78,** the photoconductor layer in this instance being on the left side of the cell facing the sources **80** and **84.** By this arrangement, images formed inside the liquid crystal layer of the cell may be portrayed on the screen **82.** As is typical, the imaging light from the source **84** is of a shorter wavelength than the display light from the source **80.** If the nematic liquid crystal layer is transparent to imaging light, the cell **78** can be turned around so that the photoconductor layer is on the right side of the cell **78** facing the screen **82** with the source **84** to the left of the cell thus being arranged to expose the photoconductor material layer by transmission of light from source **84** through the liquid crystal layer to the photoconductor material layer.

Figure 4.8d shows the cell **86** arranged between a projection light **88** and a screen **90.** This is similar to Figure 4.8c except that a partial mirror **92** is used in the optical path between the imaging source **98** and the cell **86,** and in the optical path between the projection of display light **88** and the screen **90.** In this instance, the photoconductor layer inside the cell is disposed on the side of the cell closest to the screen with the liquid crystal layer being disposed on the side of the cell closest to the source **88.**

The partial mirror **92,** as indicated, has the property of transmitting light from the source **88** to the screen **90** and reflecting light from the imaging source **94** to the cell **86.** A half silvered mirror may be used for this purpose. In the alternative, which is preferred, a dichroic mirror is used for the partial mirror **92** in which case most of the light from the source **88** reaching the mirror is transmitted therethrough toward the screen **90** and most of the light from the source **94** is reflected by the mirror toward the cell **86.** The cell **86** can be either cell **10** or **40.**

In Figure 4.8e, the cell **96** is disposed in front of a cathode ray tube **90** and a lens **100**. Image light from the fluorescent screen of the cathode ray tube is focused onto the photoconductor material layers, of the cell **100** to write an image on the photoconductor material layer. Display light, provided by ambient light or by a display light source, is reflected from the cell **96** to display the resultant image written into the nematic liquid crystal layer. In Figure 4.8f, the cell **102** is attached to the front panel of the cathode ray tube **104** by a panel **106** of optical fibers which convey imaging light from the fluorescent display screen of the cathode ray tube to the cell's photoconductor layer.

The cell **96** and the cell **102**, with their electric fields, can be either the type described with respect to Figure 4.8a, or the type described with respect to Figure 4.8b. With either of these cells the imaging light is provided by the cathode ray tube, which provides an image in accordance with the cathode ray tube display driven by the cathode ray tube drive information, identified in Figures 4.8e and 4.8f as the projection circuit. The cells of Figures 4.8e and 4.8f have a reflected display, so that display light is provided from the face of the cell, as contrasted to the display light of Figures 4.8c and 4.8d which was projected through the cells.

Good image quality is a high contrast ratio in the projected image. This implies high light scattering in the liquid crystal film next to the activated regions of the photoconductor, and zero or low light scattering in the unactivated regions. With a single layer of CdS it is not possible to get as good an image quality and high sensitivity in conjunction with a liquid crystal layer. The reason for this is that the CdS film by itself is not of high enough dark impedance to effectively match the impedance of efficient scattering liquid crystal material.

With the double layer structure consisting of the ZnS insulator film and the CdS photoconductor film constructed in the manner disclosed, the dark impedance is raised without reducing the light-to-dark current ratio which determines the sensitivity of the device. Thus the benefits to be accrued are improved device performance consisting of improved contrast ratio in the projected image, and improved sensitivity to the input (write) light signal.

Electrooptic Shutter Having Thin Glass Layer

J.V. Cartmell, D. Churchill and D.E. Koopman; U.S. Patent 3,700,306; October 24, 1972; assigned to The National Cash Register Company describe an electrooptic shutter employing nematic liquid crystal material. The electrodes of the shutter are protected with a thin overcoat of an insulating material such as glass. This thin overcoat essentially prevents rapid electrode failure due to conduction of current through the liquid crystal material.

Electrooptic devices employing nematic liquid crystals are known in the art. These devices generally comprise glass plates with conductive planar electrodes on the interior surfaces, wherein at least one of the electrode planes is transparent. A thin layer of liquid crystal material is sandwiched between the glass plates. When the liquid crystal has a net positive dielectric anisotropy, the application of an electric potential across the cell electrodes causes the liquid crystal to be uniformly aligned with the molecular axes perpendicular to the cell surfaces.

Often, a desired orientation of the liquid crystal in the quiescent or field-off state is obtained by rubbing the electrodes with a material such as a cotton swab prior to construction of the cell. In some applications, the optical contrast is enhanced by placing the cell between two linear polarizers oriented either parallel or perpendicular to the direction of rubbing. Such devices are described in U.S. Patent 3,499,702 and in an article by M. Schadt and W. Helfrich in *Applied Physics Letters,* Volume 18, No. 4, p. 127 and 128 (1971).

A major problem with such electrooptical devices is rapid electrode failure due to electrolysis of the electrode material as a result of current conduction through the cell. One solution to this problem has been the employment of high purity liquid crystal material. It has been found that the electrodes can be protected with a thin overcoat of an insulating material such as sintered or fused glass, silicon oxide or the like. This thin overcoat removes the criticality of the purity of the liquid crystal material as current conduction is prevented even if the liquid crystal material is not pure. The liquid crystal materials employed are nematic liquid crystals. In particular, these liquid crystals are characterized by a positive dielectric anisotropy, that is, in an electric field they tend to align with the molecular axis parallel to the field direction. Compounds which exhibit this property are p-n-alkoxybenzylidene-p-aminobenzonitriles (see U.S. Patent 3,499,702).

The property of positive dielectric anisotropy often is realized by mixtures of one or more of these or similar compounds with materials which do not normally exhibit positive dielectric anisotropy. Often, to obtain liquid crystal materials which are liquid through a desired temperature range, it is necessary to form mixtures of liquid crystals.

A thin sandwich cell is employed. The cell comprises two glass plates coated on the interior surfaces with a transparent conductive material. Indium oxide electrodes are most often used. Other electrodes include tin oxide and gold. The electrode patterns are etched by conventional photoetch techniques using hydrochloric acid for dissolution of the indium oxide. Other suitable etchants are employed for different electrode material. The electrode pattern can be any desired configuration such as a 7-bar matrix, a 9-bar matrix, X-Y line pattern and the like.

After etching of the electrodes, the protective insulating overlay is applied. One means for applying such a film is to vacuum evaporate silicon oxide on the surface containing the etched electrodes. Insulating films of 0.1 to 1.0 micron can be deposited in this manner. A thin layer enhances the electrooptic characteristics of the cell but the layer must not be so thick that complete isolation of the electrodes is obtained. Another technique for applying an insulating layer is to silk screen a thin film of overlay on the surface and heat the plate in a kiln to form a glaze. This is preferably done with tin oxide electrodes because of their superior thermal characteristics.

The interior surfaces of the cell comprising the electrodes and the protective overlay are rubbed uniformly along one direction. When the plates are brought together to form the cell, they are positioned so that the rubbing directions are at right angles to each other. The finished cell contains a thin layer of nematic liquid crystal material between the plates. A typical example of this material is a mixture containing, by weight, 32% methoxybenzylidene-n-butylaniline, 58% ethoxybenzylidene-n-butylaniline and 10% p-n-butoxybenzylidene-p'-aminobenzonitrile. Stroking the electrode's interior surface causes the molecular axis of the liquid crystal material to align in that direction. This can be done by rubbing the clean surface with a dry cloth such as a dense polishing cloth.

The liquid crystals align with their molecular axes parallel to the surfaces of the plates and along the rubbing direction with the field off and change their orientation to perpendicular to the plates (along the field direction) when the field is on. These effects are obtained with liquid crystals having a strong permanent dipole at one end of the molecule. A typical example is a mixture containing, by weight, 40% methoxybenzylidene-n-butylaniline (MBBA), 40% ethoxybenzylidene-n-butylaniline (EBBA), and 20% p-n-butoxybenzylidene-p'-aminobenzonitrile. Normally, these liquid crystals are liquid at room temperature.

Figure 4.9a is a sectional view of the electrooptic shutter of this process and Figure 4.9b details the electrodes of the shutter. In Figure 4.9a, light rays pass from light source **1** through polarizer **2**. Substrates **3** and **6** are aligned so that their adjacent faces **4** and **7** are parallel and spaced about 0.5 to 1.0 mil apart. Shims can be used to maintain this spacing.

Substrates **3** and **6** then are sealed. Prior to sealing, space **5** is filled with the desired liquid crystal materials. Light transmitted through substrates **3** and **6** then is transmitted through polarizer **8** and is seen on light scattering screen **9** by viewer **10**. In Figure 4.9b, electrodes **11** and **12** are shown. It should be noted that electrode **11** is a 7-bar matrix and electrode **12** is a coating.

FIGURE 4.9: ELECTROOPTIC SHUTTER

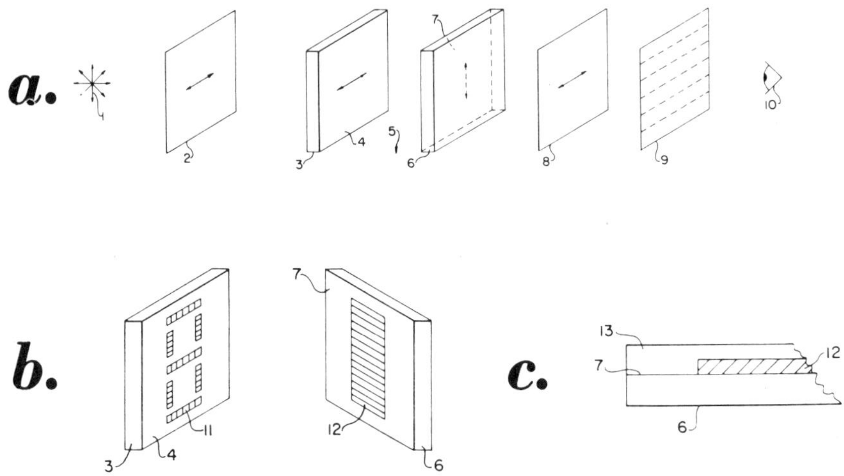

Source: J.V. Cartmell, D. Churchill and D.E. Koopman; U.S. Patent 3,700,306; Oct. 24, 1972

Electric leads are attached to the electrodes. It is these electrodes which are coated with the protective overlay of this process. In Figure 4.9c a side view of substrate **6** is shown with electrode **12** and face **7** coated with thin overcoat **13**. A feature of the process is the cell construction. After the plates have been etched and rubbed, a thin ribbon of ethylene-acrylic acid copolymer is laid around the edge of the plates on the rubbed surface. The liquid crystal is placed between the plates and they are brought together with pressure and heat (to 100°C). The copolymer wets the cell surfaces near the edges and effectively seals the cell.

Typically, light passing through the first polarizer is polarized in the horizontal plane (arrow). As it passes through the cell it is rotated 90° so that it emerges polarized in a vertical direction. If polarizer **2** is aligned to transmit light only in the horizontal plane, the light is extinguished and the observer sees no light. When the field is turned on, the cell does not affect the plane of polarization and the light emerges from the cell polarized in the horizontal direction, is transmitted by the second polarizer and the viewer sees light on the screen. A light image of the character in a dark background is observed.

Alternately, the second polarizer may be rotated 90° and the field off mode will transmit light so that when the field is applied a dark image on a lighted background is observed. Contrast values of 30:1 are typically obtained. In one case the electrodes have been etched in a 7-bar configuration. An example is where these bars necessary to generate a figure four are activated. The figure then is viewed on the screen.

The scattering screen is used to improve the quality of the image. Also, since the effect is that of a shutter, the angle of view is very restricted and the screen acts as a diffuser. Without the screen, and with proper lenses, the system could be used as a dynamic projection image. The device can be operated by application of a direct current potential across the cell but the insulating character of the leach requires a high potential. The preferred mode of operation is with a pulsed, unipolar DC, bipolar DC or AC potential. The exact potential required is a function of the thickness and dielectric constant of the insulating layer.

Typical threshold voltages are 4 v for 250 pulses per second unipolar DC and 2 v for 60 hertz AC. Since this is a field orientation the current is very low. The AC frequency can be varied from 20 to several thousand Hz. The switching time is about 10 milliseconds on and 40 milliseconds off. The following examples illustrate the process.

Example 1: Two identical electrooptic shutters were operated according to this process on a 10 v pulsed 250 cycles per second, unipolar DC potential. A shutter such as described in Figures 2.9a and 2.9b was employed. Substrates **3** and **6** were 4" x 4" glass plates. Electrode **11** was etched as a 7-bar matrix and electrode **12** was etched panel. The electrodes were indium oxide. In one shutter, a thin, approximately 1 micron, overlay of vacuum deposited SiO was placed over the electrodes, while no overlay was employed in the second shutter. Employing a polishing cloth, the interior surface of substrate **3** was rubbed in a horizontal direction and the interior surface of substrate **6** was rubbed in a vertical direction.

Space **5** of both shutters was filled with a mixture containing by weight 40% methoxybenzylidene-n-butylaniline (MBBA), 40% ethoxybenzylidene-n-butylaniline (EBBA) and 20% p-n-butoxybenzylidene-p'-aminobenzonitrile and then sealed. Cross polarizers **2** and **8** were oriented in a horizontal direction. In the shutter with no protective overlay, the electrodes failed after 20 hr of continuous operation. In the shutter with the SiO protective overlay, the electrodes showed no sign of failure after 2 mo of continuous operation. Note, the only difference between the two shutters was the SiO overlay.

Example 2: A second cell was constructed using transparent tin oxide electrodes. A layer of Vita 1001 glass fritt was deposited over the electrode areas by a silk screen technique. The plates were then heated to 615°C to fuse the fritt to a clear layer of glass 1 mil thick. The surfaces were then rubbed and the nematic mixture of Example 1 was sealed in the cell. The lifetime and operating characteristics were similar to the cell of Example 1 having the protected electrodes.

Electroluminescent Element

M. Morikawa; U.S. Patent 3,775,631; November 27, 1973; assigned to Ise Electronics Corp., Japan describes a dispersion type electroluminescent element comprising a pair of electrodes and an electroluminescent layer sandwiched between the electrodes and including an electroluminescent phosphor dispersed in an organic or inorganic binder.

The luminescent layer comprises a mixture of 100 parts, by weight, of an electroluminescent phosphor; 30 to 50 parts, by weight, of a liquid crystal; and 10 to 40 parts, by weight, based on the sum of the parts of the electroluminescent phosphor and the liquid crystal, of an inorganic binder. The liquid crystal used in the process may be MBBA and/or EBBA.

Light Valve Using Nematic and Cholesteric Mixtures

G.H. Heilmeier and J.E. Goldmacher; U.S. Patent 3,650,603; March 21, 1972; assigned to RCA Corporation describe an electrooptic light valve which comprises a mixture of a nematic liquid crystal composition of the type whose molecules align in an electric field with at least one member of the group consisting of cholesterol, a cholesterol derivative and a cholesteric liquid crystal. The light valve includes means for applying an electric field to this mixture.

It has been found that by mixing cholesterol, and/or a cholesterol derivative, and/or a cholesteric liquid crystal with a nematic liquid crystal composition of the type whose molecules align in an electric field, a mixture which is highly light scattering is achieved. The mixture also results in depression of the crystal-nematic transition temperature of the nematic liquid crystals. When an electric field is applied to such a mixture, so as to align the nematic liquid crystal molecules of the mixture with their axes substantially parallel to the direction of the incident light, the mixture essentially no longer scatters the light as it would in its unexcited state and becomes transparent to the incident light.

This discovery makes possible light valves and display devices having improved contrast ratios as compared to prior art liquid crystal devices of the type which depend upon alignment of nematic liquid crystal molecules in an electric field. A preferred form of the liquid crystal element is shown in Figure 4.10 in the form of a crossed grid reflective electrooptic display device **30.** This device consists of a back and a front transparent support plate **31** and **32,** respectively. These support plates are parallel and are separated by a distance of about ½ mil. On the inner surface **33** of the back plate **31** is an array of parallel spaced conductive back electrode strips **35.**

FIGURE 4.10: ELECTROOPTIC DEVICE

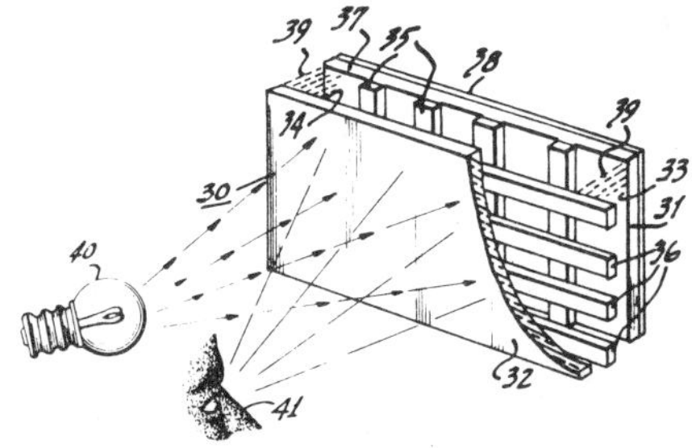

Source: G.H. Heilmeier and J.E. Goldmacher; U.S. Patent 3,650,603; March 21, 1972

On the inner surface **34** of the front plate **32** is an array of parallel spaced transparent conductive front electrode strips **36.** The front conductive strips **36** are perpendicular to the back conductive strips **35.** The outer surface **37** of the back plate **31** is coated with a reflective material **38,** such as aluminum. In the device the space between the front and back plates is filled with a mixture **39** comprising 80 wt % of a nematic liquid crystal composition and 20 wt % of cholesterol derivatives. The nematic composition consists essentially of equal weights of p-n-ethoxybenzylidene-p'-aminobenzonitrile and p-n-butoxybenzylidene-p'-aminobenzonitrile.

The cholesterol derivatives consist essentially of 23 wt % of cholesteryl chloride and 77 weight percent of cholesteryl oleate. The mixture may be sealed in the device by using epoxy cement around the edges of the device. The nematic composition by itself has a crystal-nematic transition temperature of about 41°C. However, this transition temperature is reduced to about room temperature when mixed with the cholesteryl compounds.

In operation, in its reflective mode, a light source **40** and an observer **41** are both on the same side of the device as the front support plate **32.** With no field applied to the device, the device appears uniformly bright to the observer. When a DC or AC field of, for example, about 10 v is applied to any portion of the liquid crystal composition, that portion appears darker than the remainder of the device. Desired or specific portions of the device can be darkened by applying proper voltages to particular pairs of front and back electrode strips.

The device can operate in a light transmitting mode when there is no reflective coating **39** on the back plate and when the back electrode strips **35** are transparent. In this mode, an observer is on the side of the device opposite a light source and will observe transmitted light rather than reflected light. In this mode of operation the device appears brighter to the observer in the region of an applied electric field when the incident light is essentially parallel to the direction of the field. This effect is due to the reduction of scattered light upon the application of the electric field and the direct passage of the light to the observer.

SMECTIC

Smectic-Nematic Liquid Crystals

L.T. Creagh, D. Jones and S. Lu; U.S. Patent 3,716,289; February 13, 1973; assigned to Texas Instruments Incorporated describe a display device which includes a layer of a smectic-nematic liquid crystal composition that is normally transparent to light. The composition is effective to scatter light in response to an applied voltage. The smectic-nematic compositions exhibit the mesomorphic state through a broad temperature range that preferably includes room temperature. The nematic-smectic mixture used in the electrooptical display device exhibits satisfactory switching characteristics and contrast ratios.

The electrooptical display device includes a first optically transmissive substrate and a first optically transmissive and electrically conductive coating on a face of the first substrate. A second substrate with an electrically conductive coating on a surface is positioned such that the coatings of the two substrates are adjacent and substantially equidistantly spaced apart.

A light source means is positioned to direct light toward the substrates and means are provided for impressing a voltage between the coatings. A nematic-smectic liquid crystal composition is placed in the space between the coatings, the composition having a threshold electrical field which when exceeded will cause the composition to scatter light. The composition has a broad temperature range in which it exhibits the mesomorphic state, the temperature range preferably encompassing room temperature.

Figure 4.11a is an exploded schematic view of a display device and Figure 4.11b is a side view of a device similar to that shown in Figure 4.11a. In Figure 4.11a, two substrates or plates **10** and **12** are positioned substantially parallel to each other. Substrate **10**, shown partially broken away, is optically transmissive and has deposited on a face thereof a coating **14** of a material which is optically transmissive and electrically conductive. Substrate **12** is also optically transmissive and has coated thereon a plurality of electrically insulated electrodes **16, 18, 20, 22, 24, 26** and **28.** These electrodes are also optically transmissive and electrically conductive.

Each of the electrodes **16** through **28** has attached thereto a lead **30** which is optically transmissive and which has been electrically insulated by an optically transmissive insulating composition. The leads **30** are connected to switches **34, 36, 38, 40, 42, 44** and **46** which correspond respectively to specific ones of electrodes **16, 18, 20, 22, 24, 26** and **28.** Switches **34** through **46** are connected in parallel to a source **48** of electrical energy which is connected in series to electrically conductive coating **14.**

Interposed between substrates **10** and **12** is a thin layer **50** of a nematic-smectic liquid crystal composition. It is known that smectic liquid crystals do not exhibit dynamic scattering such as is required when the liquid crystal is used in an electrooptical display cell as above described. It has been found, however, that when the smectic liquid crystal is mixed with a nematic liquid crystal, the resulting composition does exhibit dynamic scattering when an electrical potential is impressed across the composition.

Further, most of the smectic liquid crystal materials have a very low crystal-to-mesomorphic state transition temperature that is typically below room temperature. Thus, when such a smectic material is mixed with a nematic liquid crystal material, which typically has a very high crystalline-to-mesomorphic transition temperature, the resulting composition exists in the mesomorphic state at room temperatures.

For purposes of clarity, in the exploded view of Figure 4.11b the thin layer **50** is shown separated from the substrates **10** and **12**, however, in operation, the thin layer is intimately in contact with the coating **14** and the electrodes, such as **16** through **28,** formed on the surface of the substrate **12.** A light source **52** is positioned behind substrate **12** to direct a light beam at an angle toward the rear face of the substrate **12.**

FIGURE 4.11: DISPLAY DEVICES UTILIZING SMECTIC-NEMATIC LIQUID CRYSTALS

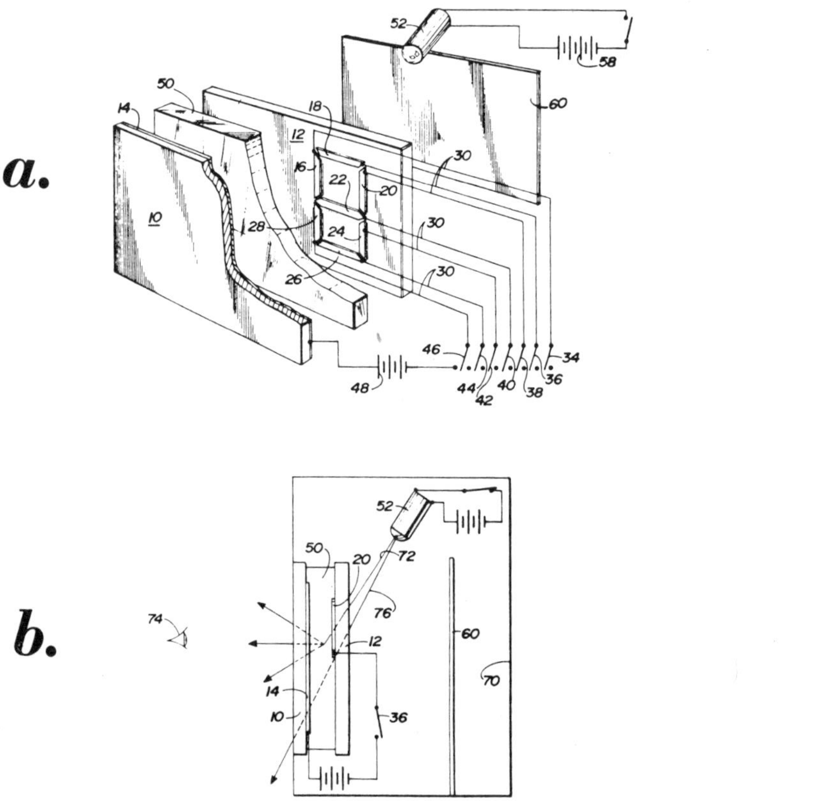

Source: L.T. Creagh, D. Jones and S. Lu; U.S. Patent 3,716,289; February 13, 1973

Since both of the substrates **10** and **12** and the electrodes and coating thereon are optically transmissive, the light beam from source **52** normally passes through the display cell. The source of light **52** is electrically connected to a suitable electrical energy source **58** and may be any conventional source including an incandescent source. A nonreflective light absorbing plate **60**, which is preferably a dull black in color, is positioned behind and spaced from the substrate **12**. Operation of the display device will be described with reference to Figure 4.11b where there is a side view of the display shown in Figure 4.11a. For clarity of description, only one electrode **20** of electrodes **16** through **28** is illustrated.

The display device is shown enclosed by a suitable casing **70**. As may be seen, the thin layer **50** of the nematic-smectic mesomorphic composition contacts the electrically conductive coating **14** and electrode **20**. When switch **36** is closed, a voltage is impressed across the layer **50** between electrode **20** and that portion of coating **14** corresponding thereto. This voltage causes the portion of the layer **50** to which the voltage is impressed to diffuse light coming from light source **52**. For purposes of illustration, light beam **72** is shown being diffused toward the eye **74** of an observer. A second representative light beam **76** passes through the portion of the nematic-smectic mesomorphic composition across which no voltage is impressed without being diffused. Thus, the observer is unable to see light from the light source in that portion of the layer. Plate **60** is provided to eliminate internal reflections from the light source **52** and also to eliminate stray room light which may interfere with the display characteristics of the display screen. With reference to Figure 4.11a, it may be seen that if switches **34** and **46** are energized, a voltage will be impressed across layer **50** in that portion of the layer corresponding to electrodes **16** and **28.** An observer, such as observer **74** of Figure 4.11b, will see a numeral 1 displayed. Similarily if all of the switches **34** through **46** are closed, the observer will see a numeral 8 displayed.

Examples of specific nematic-smectic mesomorphic compositions which may be utilized in the process are set forth in the table below. These compositions exhibit a wide transition temperature range and several of the compositions exist in the mesomorphic state at room temperature. Specific compositions were prepared by placing the appropriate weight of each component in a sealed vial. The vial was warmed until the materials were isotropic and was then placed in an ultrasonic bath at that temperature for 2 to 3 hr. The vial was allowed to cool slowly to room temperature in the ultrasonic bath.

Example Number	Composition* ---- Weight Percent-----		Crystal to Mesomorphic Temperature	Mesomorphic to Isotropic Temperature
1	20% C	80% A	72°C	111°C
2	40% C	60% A	71°C	108°C
3	60% C	40% A	58°C	85°C
4	80% C	20% A	28°C	62°C
5	20% C	80% A	58°C	112°C
6	40% B	60% A	55°C	106°C
7	60% B	40% A	59°C	95°C
8	80% B	20% A	35°C	89°C
9	90% B	10% A	30°C	78°C
10	60% B	40% D	67°C	96°C
11	80% B	20% D	43°C	90°C
12	90% B	10% D	22°C	77°C
13	10% D	90% C	19°C	49°C
14	20% D	80% C	22°C	55°C

*Examples of Materials Utilized:
A is 4,4'-bis(heptyloxy)azoxybenzene
B is p-ethoxybenzylidene-p-n-butylaniline
C is p-methoxybenzylidene-p-n-butylaniline
D is ethyl p-hexyloxybenzal-p-aminocinnamate

As noted in the above table, a composition including by weight 90% p-ethoxybenzylidene-p-n-butylaniline and 10% ethyl p-hexyloxybenzal-p-aminocinnamate (Example 12) exists in the mesomorphic state at room temperature. This composition was tested in a 1 mil electro-optical display cell at 25°C and exhibited dynamic scattering when a potential of about 22 volts was applied across the cell. Further, when the above compositions are doped with about 1 to 10% by weight of a cholesteric liquid crystal such as cholesterol nonanoate, the compositions exhibit memory, and may be erased with an alternating current signal of about 7 kHz.

Infrared Laser Beam

A process described by *F.J. Kahn; U.S. Patent 3,796,999; March 12, 1974; assigned to Bell Telephone Laboratories, Inc.* provides a liquid crystal optical display apparatus, which can be both selectively written and selectively erased. Both write-in and erase steps involve the use of localized thermal heating by means of an infrared laser, for example. The localized written state of the liquid crystal is characterized by a significantly greater optical scattering than the localized erased state of the crystal. Depending upon the conditions under which a localized area of the liquid crystal is cooled, the final state of the local heat treated area becomes the nonscattering erase state or the scattering written state, thus affording localized erase or write-in, as desired.

The factors which determine this final state (scattering vs nonscattering) include the rate of cooling. In turn, the rate of cooling can be controlled by the rate of laser scan or by the cross section of the laser beam used for the heating of the selected areas of the liquid crystal. Additionally, the final state can be affected by the presence or absence of applied fields across the liquid crystal during the cooling period.

In a specific form of the process, a smectic liquid crystal, such as N-(p-cyanobenzylidene)-p-n-octylaniline (CBOA) in the form of a thin layer, is sandwiched between two visibly transparent, but infrared absorbing, electrode layers. The CBOA is heated either to the isotropic or to the nematic state and then cooled slowly back to the smectic state in order to produce a nonscattering smectic texture over the entire layer (uniform erase or clean slate). Thereafter, a write-in infrared laser beam is allowed to scan selectively the liquid crystal at those portions thereof at which it is desired to write in a pattern of information.

Advantageously, the write-in beam temporarily heats the crystal selectively into the isotropic state. In this way a stored pattern of disordered (optically scattering) smectic regions is formed in the otherwise ordered (nonscattering) crystal, in accordance with the pattern of the selective write-in laser scan. Furthermore, this pattern of information can then be selectively erased by means of a second selective scan with the same (or another laser) beam having a scanning rate, cross section, and an optical intensity such that the rate of cooling is sufficiently slower than the cooling rate immediately subsequent to the first (write-in) scan.

This erasing beam is advantageously sufficient to produce temporarily either the nematic or the isotropic state, locally. This second (erase) scan produces again the ordered (optical nonscattering) smectic texture at those portions of the crystal to be erased, i.e., at those portions which are selectively scanned by this second scanning beam. Thus, the nonscattering smectic texture is reproduced at these erased portions, which are therefore optically transparent in substantially the same degree as the unwritten portions of the liquid crystal.

The various stored patterns of molecularly ordered and disordered states can be read out at any time by means of external optical illumination, as furnished by an ordinary light bulb, for example, or by Schlieren optical techniques. Since the ordered regions of the liquid crystal produce significantly less optical scattering than the disordered regions thereof, an optical detector such as the human eye will observe the liquid crystal as impressed with a pattern of relatively bright and dark portions.

The bright portions will be observed in those regions which are in the disordered state, that is, the optically scattering state, whereas the relatively dark portions will be observed in those regions of the liquid crystal which are in the ordered state, that is, the optically transparent state (provided of course that the detector is located off to one side obliquely with respect to the illuminated liquid crystal).

Figure 4.12a is an optical display apparatus partly in cross section and Figure 4.12b is another optical display apparatus. Referring to Figure 4.12a, an optical display device **10** includes a liquid crystal layer **11** sandwiched between a pair of visibly transparent electrodes **12** and **13.** Advantageously, one or both of these electrodes are absorptive of infrared radiation. Typically, an annular polymeric spacer **11.1** maintains proper distances between electrodes for controlling the thickness of the liquid crystal **11.** The electrodes are supported mechanically by means of transparent fused silica slabs **14** and **15,** respectively. An AC voltage source **16** is connected through a switch **17** across the transparent electrodes **12** and **13.**

An infrared laser source **18,** which can be steered by various means known in the art, provides a controlled scanning laser beam **19** incident upon the device so that this beam locally heats the electrodes **12** and **13,** and hence locally heats the liquid crystal. A readout visible optical source **20** (or ambient stray light) provides a readout incident optical beam **21,** which traverses the optical device and is scattered by the various selected portions of the liquid crystal to form an output optical wave **22.**

FIGURE 4.12: LOCALLY ERASABLE THERMOOPTIC SMECTIC LIQUID CRYSTAL STORAGE DISPLAYS

a.

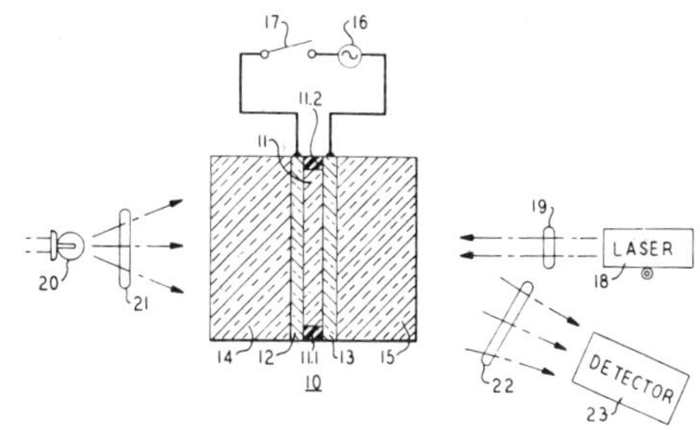

b.

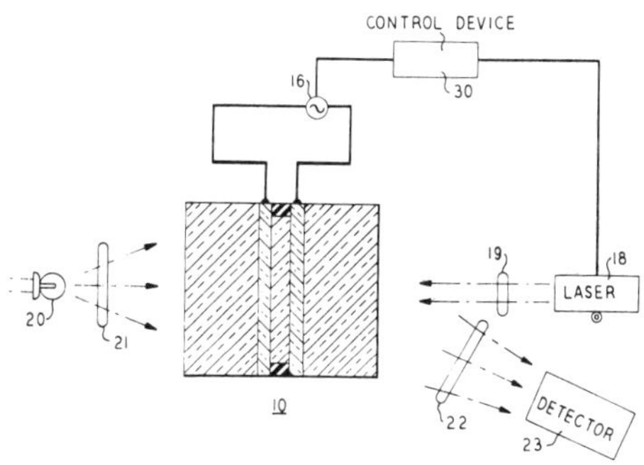

Source: F.J. Kahn; U.S. Patent 3,796,999; March 12, 1974

This output wave is incident upon the optical detector **23** (which can be the human eye) for utilization of the optical output wave. As indicated in Figure 4.12b, a control device **30** can be used to control simultaneously the laser source **18** and the AC voltage source **16**, to apply both laser beam **19** and electric fields to the device **10.**

Example: The liquid crystal **11** is a layer of N-(p-cyanobenzylidene)-p-n-octylaniline (CBOA), which is typically about 14 microns in thickness. This layer **11** is sandwiched between the two electrodes **12** and **13**, which are both coated (at their common surfaces with the liquid crystal) with a silane coupling agent, such as N,N-dimethyl-n-octadecyl-3-aminopropyltrimethoxysilylchloride (Dow Corning XZ 2-2300). The coupling agent is typically applied from water solution and then cured by drying. Thereby a preferred direction is established in the liquid crystal by means of the coupling agent. The electrodes **12** and **13** are typically made of indium-tin oxide layers having a thickness of about 0.5 to 1.0 micron each. Finally, the substrates **14** and **15** are typically made of fused silica, having a thickness of about 3 mm each.

In order to present an initially erased state which is uniform in the liuqid crystal (i.e., a clean initial state), the liquid crystal is heated uniformly into the nematic or isotropic state, by means of an auxiliary heater (not shown), such as an auxiliary electrical joule heating layer located on one or both of the exposed surfaces of the substrates **14** and **15** or by passing a heating current through one or both of the electrodes **12** and **13.** In case the liquid crystal is CBOA, a temperature in the range above about 65°C is sufficient to produce the nematic state, and a temperature above about 72°C is sufficient for the isotropic state.

The liquid crystal is thereafter cooled to the ambient temperature at a rate such that the entire crystal is brought into the ordered smectic texture. For the particular material CBOA, the operating ambient temperature at which the liquid crystal is held (after cooling), in the absence of an external beam incident thereon, should be between about 36° and 48°C. The plates **14** and **15** are maintained in this ambient temperature range in order to provide the ambient operating temperature.

In order to write a pattern of information into the crystal, the laser source **18** advantageously provides a beam **19** having a cross section of approximately 15 microns, such as furnished by a focused 20 mw YAIG infrared laser **18**. This beam **19** scans the liquid crystal at a velocity typically of about 2 cm/sec. During this scanning, the laser source **18** is directed so that the laser beam **19** selectively scans only those portions of the liquid crystal which are to be written in with the desired pattern of information. Thereby, upon illumination with the readout beam **21**, those portions which have been scanned by the laser source will scatter light toward the detector **25** in a significantly greater amount than those portions which have not been scanned by the laser source.

It should be noted that the required laser power from the source can be reduced by closing the switch **17** (Figure 4.12a) or by suitable control **30** (Figure 4.12b) during the laser scan, thereby to apply an electric field across the liquid crystal from the source **16.** Typically, an AC voltage of about 9 v RMS at 1.5 kHz is useful for this purpose of reducing the laser scan power needed for effecting the desired change in state of the liquid crystal.

In order to erase locally only certain selected portions of the previously written-in pattern of information, the liquid crystal is subjected to a second scan by the laser source **18** (or by a different laser, not shown), at least at those portions of the crystal at which it is desired to provide the local erase of the previously written portions. Of course, these portions of the crystal which are already in the ordered (nonscattering) smectic texture can also be scanned with the erasing laser scan, and these portions will thereby remain in the desired optical nonscattering texture.

In the erasure it is important that the sweeping rate of the scanning laser beam from the source **18** be significantly slower, and/or of larger cross section (but of the same optical power per unit area), or of larger power per unit area (but of same optical cross section), than during the write-in. Thereby, the cooling rate associated with the erasing scan will be slower than the rate associated with the write-in scan.

Alternatively, the switch **17** (Figure 4.12a) should be closed both during the erase scan with the laser source **18** and during the immediately subsequent cooling, thereby simultaneously applying an AC voltage from the source **16** (of about 35 v at about 1.5 kHz, for example) across the liquid crystal. This application of the electric field from the AC source **16** can also be, of course, synchronized from the erasing laser scan by suitable control **30** (Figure 4.12b). In this manner, the rate of cooling during the erasure step will be either slower, or the reordering of the liquid crystal molecules will be assisted by an electric field significantly greater, than during the write-in step.

Thereby, the portions of the previously written-in regions of the liquid crystal which have been subjected to the second (erase) scan by the laser source **18** will be returned into a nonscattering state by reason of slow cooling of the liquid crystal back to the nonscattering smectic texture, or by reason of the electric field assisting the reordering of the liquid crystal molecules back into the nonscattering smectic texture.

Whereas the above described coupling agent (XZ 2-2300) in conjunction with the above described crystal molecules (CBOA) results in an alignment of molecules which is perpendicular to the plane of the electrodes, other coupling agents or alignment methods (rubbing, for example) can also be used which result in other alignments of the liquid crystal molecules. However, it should be noted that when using a liquid whose alignment is arranged to be perpendicular to the plane of the electrodes, it is desirable that the dielectric anisotropy of the liquid crystal be strongly positive for thermal erase assisted by applied electric fields directed perpendicular to the electrodes, as is the case with CBOA described above. In addition, in order to minimize the possibility of electrical breakdown, the electrode surface facing the liquid crystal can be coated with an insulating layer such as silicon dioxide typically of a few hundred angstroms in thickness.

Variable-Tilt Compounds for Temperature Indicator

T.R. Taylor and J.L. Fergason; U.S. Patent 3,723,346; March 27, 1973; assigned to International Liquid Xtal Company have found that temperatures may be quickly and conveniently sensed and/or mapped with the use of a variable-tilt compound of the smectic C mesomorphic phase. Several organic compounds are described that are capable, within certain temperature ranges, of exhibiting such a phase, such as terephthal-bis(4-n-butylaniline) and other compounds resulting from the reaction of terephthaldehyde with amines or amides of the formula:

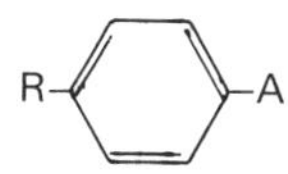

where A is $-NH_2$ or $-CONH_2$ and R is saturated alkyl or alkoxy group containing 1 to 12 carbon atoms. The compound is made, applied to a surface that has been rubbed unidirectionally, and then (while the compound is within the smectic C phase temperature range) viewed in light that is polarized planarly or circularly. The temperature indication is sensitive to changes in temperature that are quite small, and the response pattern is exceptionally rapid (being on the order of a millisecond at the greatest), since no molecular rearrangement is involved.

It has been unknown, prior to this process, that there are compounds of the smectic C phase that exhibit a temperature-dependent tilt angle. These compounds of the smectic C phase may be used for the measurement of temperature, at least within the temperature domain wherein the compound involved exhibits the smectic C phase. A prerequisite to the practice of the process is that the molecules that are within the smectic C phase are oriented in such a way that the layers are all parallel to the surfaces that contain the liquid crystal material. This parallelism may be accomplished in various ways, such as by rubbing the containing surfaces with a cotton cloth, or with paper or other material that has an orienting influence on the surface with which the liquid crystal material is in contact.

Liquid crystal material is then placed between the two rubbed surfaces in such a way that the rubbed directions of the surfaces are parallel. In some instances, a desired parallel orientation of the layers may be achieved by containing the liquid crystal material between two surfaces, only one of which has been so unidirectionally rubbed. There is thus obtained a liquid crystal unit, and when circularly polarized light is caused to become incident upon the stratum of liquid crystal material, the liquid crystal material gives a uniform retardation thereacross, provided that the temperature is constant across the sample.

If the temperature throughout the domain of the stratum of temperature-dependent variable-tilt smectic C liquid crystal material is not uniform, the liquid crystal material then does not have a tilt angle that is uniform, but rather has a tilt angle that varies in accordance with the temperature in various portions of the domain, and there is thus obtained an optical retardation that varies from one location to another within the sample, with the value of the optical retardation depending upon the tilt angle at any particular location within the sample.

Thus, if the sample is observed in transmission of a circular polarizer having a sense opposite to that of the incident light, there is observed an interference color that is characteristic of the temperature of the liquid crystal material at a particular point within the above-mentioned domain. One example of a material that exhibits the properties that are discussed above is terephthal-bis(4-n-butylaniline), which has a variable-tilt smectic C phase in the temperature range of 172.5° to 144.1°C.

In that temperature range, the tilt angle changes from 0° to 26°. The rate of change of the tilt is not linear over the temperature range. In the temperature range of 172.5° to 166°C, the tilt angle changes rather rapidly at a rate exceeding 3°/°C, and from 166° to 144.1°C, the rate of change is on the order of 0.5°/°C. This change of tilt angle is reversible with temperature, so long as the liquid crystal material is not allowed to crystallize into a different phase.

Terephthal-bis(4-n-butylaniline) may be prepared as follows. 1 mol of terephthaldehyde in 8 mols of ethanol is mixed with 2 mols butylaniline. The mixture is refluxed for 16 hr. Terephthal-bis(4-n-butylaniline) is obtained in 85% yield, after recrystallization from ethanol. It has been found, moreover, that the abovementioned tilt angle is repeatable or reproducible with temperature variations within a range of ± 0.2°C.

The reproducibility may be better than that, since the equipment used to determine such reproducibility was itself only accurate to within ± 0.2°C. Those skilled in the art will appreciate that liquid crystal materials of the variable-tilt smectic C phase comprise exceedingly good temperature indicators, because very small changes in optical retardation may easily be observed with the use of suitably (planarly or circularly) polarized light.

There are other liquid crystal materials or compounds that have exhibited properties that indicate that they possess a variable-tilt smectic C phase. Such compounds are terephthal-bis(4-n-butylaminobenzoate); terephthal-bis(4-n-hexyloxyaniline); and terephthal-bis(4-n-octyloxyaniline). Their smectic C temperature ranges are, respectively: 91° to 134°C, 175° to 230°C, and 173.7° to 231.5°C.

Memory System

According to a process described by *T.R. Taylor and J.L. Fergason; U.S. Patent 3,775,757; November 27, 1973; assigned to International Liquid Xtal Company* properties of the smectic C phase of a liquid crystal are used to store information represented by the bistable orientation directions of molecular axes. Groups of molecules contained between transparent surfaces which are treated in a particular manner to develop optical homogeneity are used as storage elements. The information state of the storage element is set by causing the molecular axes within the element to assume either a first or a second stable orientation of the axes.

It has been found that when the surfaces containing the smectic C phase liquid crystal are properly treated, as is considered in detail in the following, the long axes of the molecules of the crystal are then constrained to assume what is referred to herein as a normal tilt angle with respect to one containing surface in one stable state and what may be referred to as a complementary tilt angle with respect to the surface in a second stable state.

It has been found that the molecular axes of many smectic C compounds have normal tilt angles of approximately 45° so that the complementary tilt angle is approximately 135°, being rotated in space by approximately 90° from the tilt angle. Compounds of this type are bis(4'-n-decyloxybenzal)-2-chloro-1,4-phenylenediamine, 4-n-nonyloxybenzoic acid and 4,4'-di-n-heptyloxyazoxybenzene.

In order to simplify the explanation of the process in the following, it will be assumed that the normal tilt angle is 45° and that the complementary tilt angle is rotated 90° therefrom. It will also be assumed that the first stable state occurs when the molecular axes of a storage element are aligned along the 45° tilt angle and that this stable state occurs after erasure and represents an off or binary 0 condition.

Thus, with this convention, the second stable state becomes that during which the molecular axes of a selected storage element are aligned along the complementary tilt angle. The second state will, as a matter of convenience only, be assumed to occur in a selected storage element that is subjected to writing energy and the state will be assumed to be the on state or binary 1 condition.

It will also be assumed that the energy required to cause rotation from the first stable state to the second stable state is write energy and that the energy required to cause rotation from the second stable state to the first stable is erase energy. This terminology is selected entirely as a matter of convenience since the process may be employed to use erase energy to force the second stable state rather than the first and write energy to force the first stable state rather than the second.

In reading the state of the memory elements, use is made of the birefringent characteristic of the liquid crystal layer. According to the optical principle of birefringence, if the molecules of a particular molecular group have axes parallel to the plane of polarization of low intensity light (selected so as not to change the energy state of the molecules), the polarization of the light is rotated through some angle which depends upon the magnitude of birefringence and other factors as it passes through that particular molecular group, whereas no rotation occurs when the molecular axes of the group are normal to the polarization plane.

The important thing to note is that when the long axes of a selected molecular group have been rotated by means of write energy to align with the second stable molecular axes orientation, low intensity reading light passes through the selected molecular group without rotation whereas when erase energy sets a selected or all molecular groups to the first stable state the low intensity reading light is rotated.

From the point of view of reading, it may be stated that the two information states are represented by whether or not the liquid crystal at a particular memory element point or within a particular grouping of molecules causes birefringent rotation or not. The state of all molecular groups may then be observed and sensed electrically by viewing the light which passes through all memory elements and through a second plane of polarization which is parallel to the first. The polarization in the second plane of polarization is crossed or rotated 90° with respect to the polarization of the first plane.

Thus memory elements in the first stable state exhibiting birefringence will rotate the polarized light received so as to be viewed as a corresponding light spot through the second, crossed polarizer, whereas memory elements in the second stable state which are not birefringent will not cause rotation of the reading light and consequently will appear as dark spots as viewed through the second polarizer. In order to avoid confusion in the following, the convention followed is such that the on state of a memory element exists when the axes are aligned with the second stable axis and birefringence is absent, resulting in a dark spot in reading and that the off state exists when the axes are aligned with the first stable axis and birefringence occurs resulting in a light spot in reading.

The use of the smectic C phase liquid crystal provides a very high density of information storage which theoretically is in the order of 10^{12} information storage bits per square inch. This magnitude of storage makes it possible, in principle, to store many volumes of literature in one square inch of crystal. It can be considered that the crystal memory comprises approximately 1 million lines of information each of which includes approximately 1 million bits of information.

Suitable optical markings may then be added to the memory to designate smaller segments of memory perhaps having an information storage capacity in the order of magnitude of one page of a book where each line of print may be assumed to have up to 120 characters each of which is represented by up to 100 dots or bits. Thus, if one line of a page is assumed to require in the order of 12,000 bits of representation, and if it is further assumed that the page includes up to 100 lines, it may be seen that something in the order of 100,000 bits might be a suitable representation of a page of literature of very high density.

With this assumption then, it may be seen that on the order of 10 pages of literature can be stored on one line of the crystal memory of the process and that with 1 million lines available, the memory capacity in terms of pages is 10 million pages per memory.

Figure 4.13a shows a liquid crystal memory system according to the process and Figures 4.13b and 4.13c illustrate how the containing surfaces of the memory system of Figure 4.13a are prepared to cause proper optical homogeneity. Another variation of liquid crystal memory using prisms to avoid bending a light is shown in Figure 4.13d. A molecular group taken out of the liquid crystal layer used in the memory of Figures 4.13a and 4.13d, illustrating the formation of molecular layers is shown in Figure 4.13e. Figures 4.13f and 4.13g are views taken along line **V–V** of Figure 4.13e, showing how the molecular axes of a memory element group are aligned for the first and second stable orientation directions, respectively.

Figure 4.13h is a perspective view of the memory layer and various planes of polarization with reference to the reading, writing and erasing methods and Figure 4.14i is a block diagram of an optical memory system employing the liquid crystal memory of the process. Reference is made to Figure 4.13a where one form of liquid crystal memory constructed according to the process is shown. It will be noted that a layer of smectic C phase liquid crystal **1** is contained between transparent plates **2** and **3**. As shown in Figures 4.13b and 4.13c, inner surfaces **21** and **31** of plates **2** and **3** are rubbed along lines **23** and **33**, respectively, in order to constrain the direction of the molecular axes in a manner more fully described with reference to Figures 4.13e, 4.13f, and 4.13g.

Lines **23** and **33** are not intended to signify that actual lines may be observed after the treatment of the surfaces but rather the direction of the rubbing which must be followed to accomplish the proper alignment of the molecular axes. To insure orientation plates **2** and **3** may be glass treated with a foreign material as, for example, a dilute aqueous (1%) solution of polyvinyl alcohol. Another arrangement of the memory provided by the process is shown in Figure 4.13d where, instead of plates **2** and **3**, prisms **5** and **6** are employed. A layer of smectic C phase liquid crystal **1** is contained between surfaces **51** and **61** of prisms **5** and **6**.

FIGURE 4.13: LIQUID CRYSTAL MEMORY SYSTEM

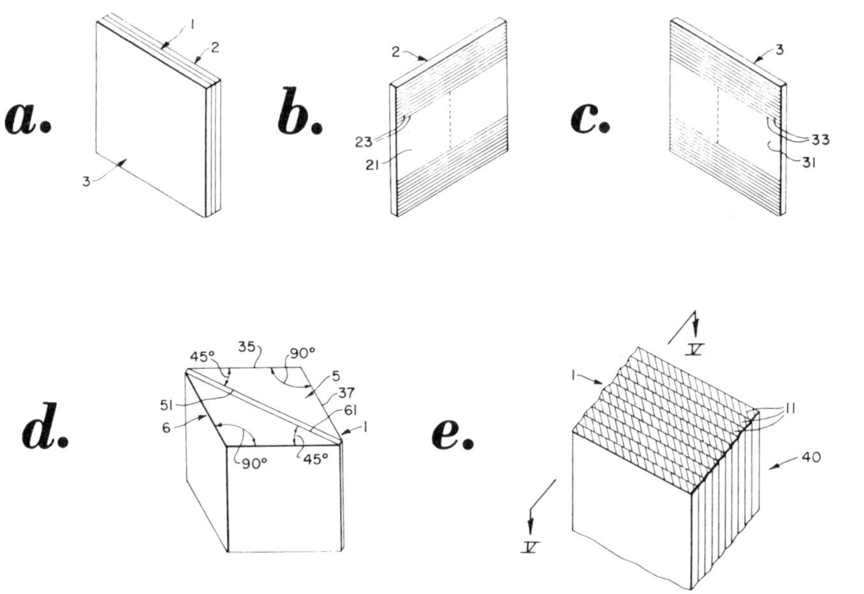

(continued)

FIGURE 4.13: (continued)

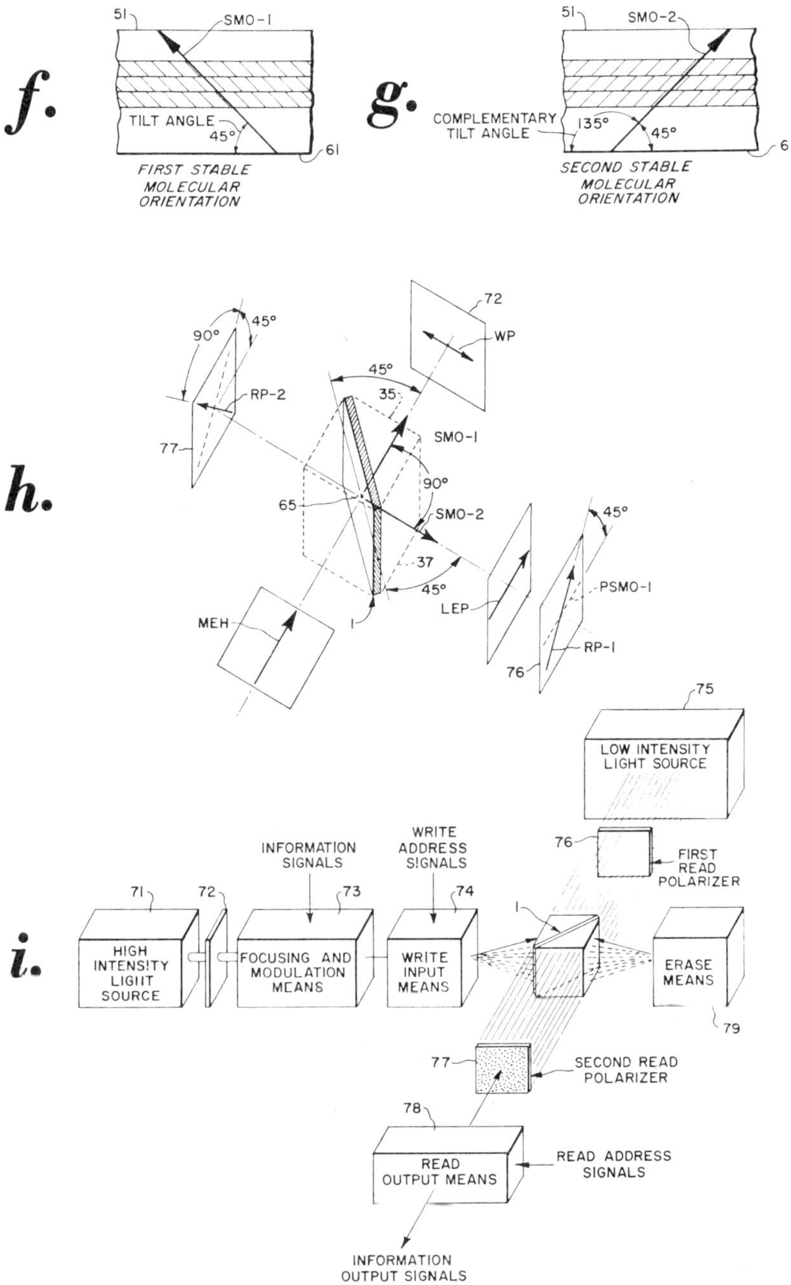

Source: T.R. Taylor and J.L. Fergason; U.S. Patent 3,775,757; November 27, 1973

Surfaces **51** and **61** are treated in the same manner as surfaces **21** and **31** shown in Figures 4.13b and 4.13c, the assumption being made that the normal tilt angle is 45°. In Figure 4.13e, a piece **40** of the liquid crystal layer **1** is shown as having a plurality of molecular layers **11** which are parallel to containing surfaces **51** and **61** (Figure 4.13d). In Figures 4.13g and 4.13f, the orientation of the molecular axes in the piece **40** shown in Figure 4.13e is represented with reference to a plane cutting through the piece perpendicular to surfaces **51** and **61** and parallel to the rubbing lines **23** and **33**.

The orientation of the molecular axes shown in Figure 4.13f is such that all of the axes lie parallel to the plane of the paper which is that representing the plane passed through piece **40** of Figure 4.13e and, within this plane, the axes are all parallel to a reference vector **SMO-1** representing the first stable molecular axis orientation abovementioned. In Figure 4.13g, the molecular axes are assumed to be parallel to the plane of the paper and to be parallel to a second stable molecular axis orientation represented as **SMO-2.** It will be noted that direction **SMO-1** has a tilt angle of 45° with respect to surface **61,** whereas direction **SMO-2** has a complementary tilt angle of 135° with respect to surface **61.**

Thus, the angle between directions **SMO-1** and **SMO-2** is 90°. While this is the preferred orientation of the two stable molecular orientations, it will be understood that substantial variations are possible without departing from the basic concept of the process. It is only important that the two stable molecular orientations exist and not that they be exactly rotated with respect to each other by 90°.

Referring to Figure 4.13h which provides a three dimensional showing of a liquid crystal memory layer **1** as it is referenced to a write electric field polarization direction **WP,** first and second molecular orientation directions **SMO-1** and **SMO-2,** a first read polarization direction referenced as **RP-1,** a second read polarization direction referenced as **RP-2,** a magnetic field erase orientation referenced as **MEH** and a light erase polarization direction referenced as **LEP.** All of the direction vectors are referenced to a memory element **65** which, as a matter of convenience, is shown as approximately in the center of the liquid crystal layer. It will be assumed that a number of molecules are included within memory element **65** and that all of the molecular axes within memory element group are oriented parallel to direction **SMO-1** or parallel to direction **SMO-2.**

In considering the operation of writing, it will be assumed that element **65** is in the first stable state with molecular axes aligned along vector **SMO-1.** A high intensity light beam with an intensity in the order of 1 w/cm^2 which may be obtained from a suitable laser source is then directed to element **65** by suitable means (not shown) after being passed through a polarizer **72** corresponding to the plane containing write electric field vector **WP.**

Since it will be assumed that vector **SMO-1** has a tilt angle of substantially 45° with layer **1,** it may be assumed that the plane of polarization containing vector **WP** is at 45° angle with layer **1.** If the prism memory of Figure 4.13d is used the write polarization plane is parallel to surface **35.** The light intensity is selected to be sufficient to cause the rotation of all those molecular axes within element **65** from the first stable direction **SMO-1** to the second stable direction **SMO-2.**

As will be more fully understood after a system using the process has been described as in Figure 4.13i, the write beam may be scanned in various modes of operation to select a plurality of memory elements such as **65** to store information. During the scanning, the beam intensity is modulated in accordance with the on-off state of the information bits to be stored.

Before continuing it will be helpful to consider the manner in which the molecular orientation is rotated by a high intensity light beam. Since most known liquid crystals with smectic C phases are optically positive, at light frequencies the dielectric constant is greater along the long molecular axis than that axis perpendicular thereto. The high intensity light, properly polarized, then causes a torque to be exerted on the selected crystal element or molecular group such as to turn the axis of highest polarizability into the direction of the electric field vector of the light. This torque is equal to $\vec{E} \times \vec{P}$, where $\vec{E}$ is the electric field vector of the light and $\vec{P}$ is the polarizability. After the high intensity write beam has been directed to element **65**, it will assume either an on or off state depending upon whether the energy of the impinging high intensity light beam was sufficient to cause the rotation from the first stable state to the second stable state. Both possibilities will now be considered with reference to the operation of the process in reading the information stored in element **65.** In reading, a source of low intensity light as may be obtained from an incandescent light bulb, is directed to the memory so that the rays of light are parallel to direction **SMO-2.**

Thus, the plane containing the first read polarization direction **RP-1** is assumed to be perpendicular to direction **SMO-2** or parallel to surface **37** where the prism of Figure 4.13b is used. The low intensity light is passed through a polarizer parallel to the plane containing vector **RP-1** with the polarization vector **RP-1** being directed to assume a 45° angle with respect to line **PSMO-1** which represents the projection of direction **SMO-1** on the plane of reference. If element **65** has remained in the first stable state after writing, in the case where the intensity of the writing beam is not sufficient to cause rotation from the first stable direction to the second stable direction, rays of the low intensity light will have their polarization rotated by birefringence.

The 45° angle between directions **PSMO-1** and **RP-1** is selected to permit maximum reading light to be rotated or not. The characteristic of the liquid crystal is such that maximum light passes in angles of 45°, 135°, 225° and 315° between **PSMO-1** and **RP-1** or every 90° starting with 45°. Minimum light passes for angles of 0°, 90°, 180° and 270°. The birefringent rotation effect can best be analyzed in terms of the light transmission between crossed polarizers. This may be represented as:

$$I = I_0 \cdot \sin^2 2\phi \cdot \sin^2 [\pi(N_2 - N_1)t/\lambda]$$

where ϕ is the angle which the direction of molecular alignment makes with the polarization axis of the first polarizer (45° to give maximum transmission); t is the thickness of the liquid crystal layers; and $(N_2 - N_1)$ represents the birefringent factor and is, effectively, the difference between the indices of refraction parallel and perpendicular to the molecular alignment. According to the above equation, when the particular molecular group alignment is in state 1 (aligned with **SMO-1**), $N_2 - N_1$ is a maximum so that maximum light passes whereas in state 2 (aligned with **SMO-2**), $N_2 - N_1$ is substantially zero so that little or no light passes.

The reading operation is completed by passing the light passing through selected element **65** through a second plane of polarization containing polarization vector **RP-2.** This may constitute a second polarizer **77.** Thus, if element **65** is in the first stable state, low intensity light will pass through the second plane of polarization. Thus, a light spot appears corresponding to element **65** if it is in the first stable state.

If, on the other hand, the write beam causes rotation of the molecular axes of element **65** to the second stable direction, the read light polarization is not rotated because the birefringent effect only operates through molecular orientations which are parallel to the planes of read polarization. Thus, the low intensity light passing through element **65** in this case has the direction of its polarization vector unchanged and at right angles to the polarization direction of the second polarizer and light therefore does not pass through the second polarizer and the stable state of element **65** is, in this case, represented by a dark spot.

It should be apparent that after writing in memory layer **1** has been completed, the layer, as viewed, appears as light and dark spots representing on and off states of information bits. It is immaterial, of course, as to whether a light spot represents an on or off state or a binary 1 or 0 and conversely whether the dark spot represents an on state or an off state or a binary 1 or 0. It is only important that the variation in light intensity passing through elements in different states be sufficient to be distinguishable. It is assumed that, after an information dot pattern has been stored in the manner described above, those skilled in the art will find appropriate means for scanning the stored information in order to translate such information into appropriate signals for various applications. One method is to cause an electron beam to scan across selected parts of the memory constituting words, lines, pages or the like. The amount of information storage which may be provided, according to the process, is then a complex function depending upon the resolution capabilities possible in reading and writing.

Two methods of performing erasing are contemplated. A magnetic field may be passed through the memory, with the field vector referenced as **MEH** in Figure 4.13h being aligned substantially with stable molecular direction **SMO-1.** In this context it should be understood that the proper definition of all directions as employed herein is to accomplish the desired result. Thus, the direction of vector **MEH** is such as to cause rotation of all of those molecular axes aligned with **SMO-1** to assume such alignment. The other method of erasing is to employ a high intensity light beam which, in the case of a small memory size, may be used to erase the entire memory as is the case of the magnetic field erase. The light beam erase, having an electric field polarization direction **LEP** (Figure 4.13h), may also be used for selective erasing of certain memory elements.

Reference is now made to Figure 4.13i which shows the general form of a system utilizing the process. In Figure 4.13i, it will be noted that a high intensity light source **71,** which may be a laser source, produces a beam which is passed through a suitable polarizer **72,** adapted to establish write electric field polarization direction **WP** (Figure 4.13h) and thence passes to focusing and modulation means **73** which responds to information signals and produces an output beam having variations in intensity corresponding to the information to be stored in the memory.

The output of means **73** is applied to write input means **74** which also receives suitable write address signals representing locations in memory which are to receive energy corresponding to the information to be stored. The write beam is directed to selected memory elements in layer **1** which may be contained between prisms in the form previously considered with respect to Figure 4.13d. Light from a low intensity source **75** is directed through a suitable first read polarizer **76,** in the manner previously considered, to illuminate layer **1.** This light passes through the layer either with birefringent rotation or not corresponding to the information state read and may be observed through a second read polarizer **77.** Suitable read output means **78** are provided which receives read address signals for controlling the scanning of the surface of polarizer **77.**

Means **78** may be similar to a television transmitter where the light and dark patterns are translated into corresponding electrical signal variations which constitute information output signals. It will be understood, however, that any means which translates the varying light intensities observed through polarizer **77** will be suitable in the practice of the process. Erase means **79** are shown for passing suitable energy through the memory to cause the rotation of the memory element molecular axes from the on representing state to the off representing state as previously considered.

Means **79** may constitute a magnetic source or a high intensity light source depending upon the particular application thereof. In the case of the high intensity light erase, means **79** could include suitable address selection means to permit selective erasing of certain memory elements to permit modification of the memory state without total erasure. It is assumed in the case where a magnetic field is used that the total memory is erased.

COMPANY INDEX

The company names listed below are given exactly as they appear in the patents, despite name changes, mergers and acquisitions which have, at times, resulted in the revision of a company name.

Bell Telephone Laboratories, Inc. - 79, 200, 221, 224, 247
Boeing Co. - 6
KK Daini Seikosha - 165, 167
E.I. du Pont de Nemours and Co. - 126
Eastman Kodak Co. - 36, 38, 189, 191
Farbwerke Hoechst AG - 191
General Electric Co. - 204
KK Hattori Tokeiten - 165
Hoffmann-La Roche, Inc. - 227, 229
Hughes Aircraft Co. - 180, 206, 232
Ing. C. Olivetti & Co., SpA - 173, 205
International Business Machines Corp. - 167, 193, 207
International Liquid Xtal Co. - 217, 251, 252
Ise Electronics Corp. - 243
Johnson & Johnson - 43
Liquid Crystal Industries - 40, 47, 92, 96
Matsushita Electric Industrial Co., Ltd. - 174
Merck Patent GmbH - 137
National Cash Register Co. - 21, 30, 32, 34, 133, 200, 212, 240
Optel Corp. - 195
Owens-Illinois, Inc. - 150, 151, 164, 165, 172
RCA Corp. - 143, 148, 162, 187, 194, 214, 217, 221, 243
KK Seikosha - 167
Siemens, AG - 182
Sperry Rand Corp. - 156, 161
KK Suwa Seikosha - 140
Temple University - 146
Texas Instruments Inc. - 188, 230, 245
Tokyo Shibaura Electric Co., Ltd. - 145
U.S. Secretary of the Navy - 105, 127
Vari-Light Corp. - 41
Westinghouse Electric Corp. - 7, 15, 16, 17, 20, 43, 47, 115, 129, 131
Xerox Corp. - 3, 10, 19, 52, 58, 62, 63, 68, 69, 79, 83, 89, 97, 105, 111, 116, 120, 123, 124, 135, 152, 197, 210

INVENTOR INDEX

Adams, J.E. - 19, 52, 58, 62, 63, 68, 69, 79, 83, 105, 116, 123, 124, 135, 152, 197
Allan, F.V. - 205
Allen, P.J. - 105
Aviram, A. - 207
Baker, C.E. - 230
Baltzer, D.H. - 41
Batesky, D.C. - 191
Becker, J.H. - 69, 79, 83, 97, 124
Berreman, D.W. - 200
Bleha, W.P., Jr. - 232
Boller, A. - 227
Braunstein, M. - 232
Cartmell, J.V. - 21, 30, 32, 34, 200, 212, 240
Castellano, J.A. - 143, 148, 162, 214
Churchill, D. - 21, 30, 32, 34, 212, 240
Cole, H.S., Jr. - 204
Creagh, L.T. - 180, 245
Dailey, J.L. - 58
Davis, F. - 40, 47, 96
Dietrich, H.J. - 150, 151, 161, 164, 165, 172
Dir, G.A. - 97
Dixon, G.D. - 20
Fergason, J.L. - 7, 15, 16, 43, 47, 115, 129, 131, 251, 252
Flam, E. - 43
Flannery, J.B., Jr. - 19, 105, 116, 152, 197
Forest, E. - 120
Franke, H.G. - 38
Goldberg, L.S. - 127
Goldberg, N.N. - 7, 15, 16, 17, 43, 47, 131
Goldmacher, J.E. - 148, 162, 195, 243
Green, D.C. - 167
Greubel, W. - 182
Haas, W.E.L. - 19, 52, 62, 68, 69, 79, 83, 105, 111, 116, 123, 124, 135, 152, 197
Haller, I. - 193, 207
Harsch, T.B. - 217
Hatfield, W.B. - 79
Heilmeier, G.H. - 148, 221, 243
Helfrich, W. - 229
Hodson, T.L. - 30, 133
Hsieh, P.Y. - 173, 205
Huggins, H.A. - 193
Hutfles, L.H. - 196
Jones, D. - 188, 245
Jones, J.W. - 30, 133
Kahn, F.J. - 200, 221, 247
Katagira, Y. - 165
Kawamoto, M. - 145
Kelker, K. - 191
Keller, C.K. - 120
Klein, R.I. - 194
Koopman, D.E. - 240
Krueger, H. - 182
Labes, M.M. - 146
Leder, L.B. - 10, 52, 63
Lodolini, M. - 189
Lu, S. - 188, 245
Madrid, R.W. - 69, 79, 89, 105
Mahler, W. - 126
Matsumoto, S. - 145
McCaffrey, M.T. - 143
Mechlowitz, B. - 19, 116, 152, 197
Meiboom, S. - 200
Melchior, H. - 224
Miller, R.D. - 207
Miller, R.E. - 32
Miyata, Y. - 165

Morikawa, M. - 243
Nagasaki, O. - 165
Oh, C.S. - 187
Ohtsuka, T. - 174
Panar, M. - 126
Pasierb, E.F. - 187
Rafuse, M.J. - 156, 161
Saeva, F.D. - 3, 68
Scala, L.C. - 20
Schadt, M. - 229
Schank, R.L. - 3
Scherrer, H. - 227, 229
Scheurle, B. - 191
Schnur, J.M. - 127
Sharpless, E.N. - 92, 96
Soree, R.A. - 161
Sorkin, H. - 194
Stahr, R.W. - 38
Steiger, E.L. - 150, 151, 161, 164, 165, 172
Steinsträsser, R. - 137
Sutton, R.C. - 36
Tayag, M.G. - 195
Taylor, T.R. - 251, 252
Tsukamoto, M. - 174
Whitaker, J.G. - 133
White, D.L. - 79, 200
Williams, R. - 217
Wolff, U. - 182
Wong, S.-Y. - 180, 206
Woodmansee, W.E. - 6
Wysocki, J.J. - 68, 69, 79, 89, 97, 105, 124, 210
Yamazaki, Y. - 140
Young, W.R. - 167, 207

U.S. PATENT NUMBER INDEX

3,322,485 - 217
3,409,404 - 129
3,441,513 - 6
3,459,839 - 196
3,499,702 - 148
3,529,156 - 7
3,533,399 - 43
3,540,796 - 162
3,575,493 - 221
3,576,761 - 47
3,578,844 - 21
3,580,864 - 16
3,580,865 - 17
3,585,381 - 30
3,594,126 - 47
3,597,044 - 148
3,600,060 - 21
3,604,930 - 105
3,617,374 - 133
3,619,254 - 40
3,620,889 - 41
3,622,224 - 89
3,627,408 - 115
3,627,699 - 15
3,642,348 - 105
3,647,279 - 96
3,650,603 - 243
3,650,608 - 230
3,652,148 - 79
3,655,270 - 180
3,655,971 - 124
3,656,834 - 193
3,656,909 - 20
3,657,538 - 131
3,661,142 - 43
3,663,390 - 131
3,666,947 - 116
3,666,948 - 116
3,669,525 - 62
3,671,231 - 123
3,674,338 - 212
3,675,987 - 156
3,679,290 - 52
3,680,950 - 105
3,687,515 - 197
3,689,525 - 191
3,690,745 - 188
3,694,053 - 221
3,697,150 - 210
3,697,152 - 63
3,697,297 - 32
3,698,449 - 194
3,700,306 - 240
3,703,329 - 214
3,704,056 - 105
3,707,322 - 79
3,711,181 - 58
3,711,713 - 79
3,716,289 - 245
3,718,380 - 69
3,720,456 - 68
3,720,623 - 34
3,720,658 - 15
3,723,346 - 251
3,726,584 - 135
3,728,008 - 205
3,732,119 - 32
3,732,429 - 232
3,734,597 - 21
3,738,942 - 145
3,742,053 - 164
3,742,067 - 172
3,743,681 - 164
3,744,920 - 135
3,749,474 - 200
3,751,467 - 151
3,756,694 - 161
3,766,061 - 126
3,768,886 - 92
3,769,313 - 161
3,769,314 - 151
3,770,653 - 165
3,770,654 - 165
3,771,065 - 127
3,772,209 - 191
3,772,210 - 189
3,773,747 - 137
3,775,631 - 243
3,775,757 - 252
3,776,927 - 131
3,779,751 - 19
3,780,307 - 68
3,781,088 - 174
3,784,513 - 150
3,785,721 - 217
3,787,110 - 200
3,789,225 - 10
3,790,251 - 97
3,790,498 - 167
3,792,915 - 187
3,795,436 - 227
3,795,516 - 38
3,795,517 - 36
3,795,529 - 34
3,796,479 - 229
3,796,999 - 247
3,799,971 - 165
3,803,050 - 197
3,804,618 - 120
3,806,230 - 111
3,809,456 - 195

3,811,180 - 232
3,814,700 - 207
3,815,972 - 173
3,816,113 - 152
3,816,334 - 140
3,816,336 - 143
3,819,531 - 3
3,821,720 - 182
3,826,757 - 180
3,827,780 - 146
3,836,230 - 83
3,836,231 - 204
3,836,243 - 224
3,836,478 - 167
3,837,730 - 79
3,838,059 - 206

NOTICE

Nothing contained in this Review shall be construed to constitute a permission or recommendation to practice any invention covered by any patent without a license from the patent owners. Further, neither the author nor the publisher assumes any liability with respect to the use of, or for damages resulting from the use of, any information, apparatus, method or process described in this Review.

PHOTOGRAPHIC EMULSIONS RECENT DEVELOPMENTS 1973

by Ronald W. James

Chemical Technology Review No. 21

The key link between the exposing and developing steps of any photographic system is the emulsion. Typically the emulsion consists of millions of tiny crystals of a silver halide salt dispersed in a water-permeable binder, such as gelatin. While these grains are responsive to all forms of electromagnetic radiation shorter than some critical wavelength in the blue region, this region can be extended to other wavelengths by the use of energy transfer chemicals named spectral sensitizers.

Color couplers are commonly added when color photographs are desired. Image dyes are designed to by cyan, magenta, and yellow which are complementary to the primary colors. Other additives are stabilizers, emulsifiers, synthetic binders, antifog and developing agents, and many more.

This book provides an in depth review of over 150 recent processes as described in the patent literature. Indicative of the high activity in this rapidly changing technology is the observation that in 1972 over 100 U.S. patents relating to photographic emulsions were issued. Process technology is extremely impressive in the direct positive and silver dye methods as well as in solvent-transfer (Polaroid-Land) systems.

The 10 chapters of the book are arranged according to the following 10 companies.

1. EASTMAN KODAK COMPANY
2. FUJI PHOTO FILM CO., LTD.
3. AGFA-GEVAERT AG & GEVAERT-AGFA NV
4. KONISHIROKU PHOTO INDUSTRY CO., LTD.
5. ILFORD LIMITED
6. CIBA LIMITED
7. E.I. DU PONT DE NEMOURS AND CO.
8. POLAROID CORPORATION
9. MINNESOTA MINING AND MANUFACTURING CO.
10. XEROX CORPORATION

To indicate the detailed information given in this book a breakdown of Chapter 1 follows here:

SPECTRAL SENSITIZERS
Tricarbocyanines and Stilbene Sulfonates
Quinoline—Substitued Cyanines + Stilbene Sulfonates
Methine Dyes with a Cyclopentanone Nucleus
2-Methylbenzimidazole Compounds
Merocyanine Dyes with a 3-Pyrrolinylalkyl Group
Cyanine Dye with Urazole Nucleus
Other Modified Cyanine Dyes
Glycol-Methanol Dye Dispersions
Multicolor Processes
Silver Dye Bleach Material

COLOR COUPLERS
Pyrazolo [3,2-c]-s-Triazoles
Substituted Thiazoline-2-thiones and Hindered Phenols as Magenta Dye Stabilizers
Cyclooxy-Substituted Acetanilides
Competing Couplers in Direct Reversal
Development Inhibitor for Releasing Couplers

DEVELOPING AGENTS
Complexed p-Phenylenediamine
Carboxylic Acid Aryl Hydrazides plus Ascorbic Acid
Hexose Reductones
Bis(Isothiuronium) Compounds

CHEMICAL SENSITIZERS
1-Phenethyl-2-Picolinium Bromide
Thioamine-Glutaraldehyde Adducts
1-(Aminophenyl)-5-Mercaptotetrazoles
Silver Ethynylcyclopentanol Complex
Other Silver Organic Complexes
Infrared Sensitizers
Electron Acceptor Desensitizers in Unfogged Systems

ANTIFOG AGENTS
Alkyl Selenols and Thiols
High Surface Concentration of Silver Halide

PREPARATION OF EMULSIONS
Acrylic Emulsion Vehicles
Manganous Salts as Emulsion Stabilizers
Acrylic Copolymers as Peptizers, Coagulants

DIRECT POSITIVE EMULSIONS
Reactive Quaternary Salts as Nucleating Agents
Dihydroaromatic Quaternary Salts
Na and K Ions with Benzotriazole
Ferricyanides as Oxidizing Agents
Applications of Nitron
Mercaptan Retarders
Electron Acceptor Desensitizers
Gold Salt Complexes
Grain Precipitation Medium
Vacuum Treatment

ISBN 0-8155-0512-4 303 pages

STORAGE BATTERIES AND RECHARGEABLE CELL TECHNOLOGY 1974

by Louis F. Martin

Chemical Technology Review No. 37

The automotive lead-sulfuric acid type storage battery, also known as accumulator in Europe, continues to account for well over 80% of the dollar volume of rechargeable or secondary batteries. Remarkable advances in chemistry and design, furthering the unique ability of the lead-acid battery to deliver large currents at relatively constant voltage, even at low temperatures, have made this type of battery an indispensable implement of the modern world.

All other commonly used rechargeable cell systems are alkaline, potassium hydroxide being the usual electrolyte. Today, rechargeable alkaline cells (nickel-cadmium, nickel-iron, or silver-zinc) are available in almost any size to fit electrically driven warehouse trucks as well as hand-held electronic calculators.

This book is organized to emphasize the advances being made in a broad range of secondary, rechargeable cells. While it is a down-to-earth, practical book, it is also intended to stimulate further research (e.g., in weight reduction). In addition to the cell systems described above, sections are also devoted to rechargeable fuel cells, reserve and delayed action batteries, and a number of batteries chemically and mechanically designed for specific end uses.

Over 230 patent-based processes related to research and production techniques. A partial and condensed table of contents follows. Numbers in parentheses indicate a plurality of processes per topic.

1. LEAD-ACID BATTERIES (47)

Lead Pastes
Dispersion of Antimony Lead Oxide
Fiberglass-Polypropylene Sheath
Lignin Sulfonate Coatings
Plumbous Oxide & Lead Acetate
Auxiliary Electrodes
Thixotropic Electrode with H_3PO_4
Lead Gel
Amorphous Silica
Titanyl Sulfate Buffer
Reserve Batteries
Lead Dioxide + Fluoboric Acid

2. ALKALINE STORAGE BATTERIES (39)

Nickel Electrodes
High Purity Divalent Nickel Hydroxide
Sintered Nickel
Platinum Gas Scavenger
Cadmium Electrodes
Chemical Control of Charging Rates

3. HIGH ENERGY CELLS (41)

Silver-Zinc
Microporous Teflon® Separators
Zirconium Oxide Additive
Zinc-Air
Reserve Batteries
Alkali Metal Coating
Zinc Powder + Mercuric Chloride
Dendrite Growth Prevention
Extended Area Electrodes
  Using Zinc Fibers

4. RECHARGEABLE ZINC CELLS (20)

Amalgamated Zinc
Amalgamation of Copper Support
Zinc-Lead Alloy
Zinc-Zinc Halide Battery
Zinc Bromide Battery
Substituted Biurea as Depolarizer
Amphoteric Metals Added
  to Zincate Electrolyte

5. RECHARGEABLE FUEL CELLS (23)

Nickel-Manganese Alloys
Silver Magnesium Alloys
Tape Fed Cells
Mobile Tape Separator
Silver Peroxide Coated Tape
Thermal Activation

6. THERMAL BATTERIES (38)

Sodium-Sulfur
Sealed Containers for
  Elemental Sodium
Lithium Cells
Nickel Fluoride Electrodes
Metal Chromate Electrodes
Bismuth Electrodes

7. OTHER CELLS AND ELECTRODES (30)

Rechargeable Manganese Alkaline Cells
Nuclear Battery
Pacemaker Battery
Seawater Batteries
Deferred Action Cells

ISBN 0-8155-0545-0 **365 pages**

DRY CELL BATTERIES 1973

Chemistry and Design

by L. F. Martin

Chemical Technology Review No. 6

Stimulated by the huge demand for dry cells required in transistorized radios, tape recorders, movie cameras, mobile communications, hearing aids, calculators, electronic watches and clocks, burglar alarms, toys, and countless other devices, the industry has spent considerable effort toward making lighter and virtually leakproof power sources of this type.

The development of new electrochemical systems, placed within the cell, also brought the additional advantage of causing portable electronic equipment to operate more uniformly and over longer periods.

The accelerating pace in battery electrochemistry and design is nowhere more evident than in the patent literature. This book describes over 130 processes relating to recent developments in dry cell batteries employing zinc, manganese, magnesium, nickel-cadmium, lithium, solid and organic electrolytes, some with thermal activation, and includes many modern manufacturing techniques.

A partial and condensed table of contents follows. Numbers in ( ) indicate the number of processes per topic. Chapter headings are given, followed by examples of important subtitles.

1. **BASIC ZINC DRY CELLS (30)**
   Electrodes
   Leached Amalgamated Zinc
   Manganese Dioxide Production
   Separators
   Intumescent Anion Exchange Compound
   Sealing Techniques
   Seal Arrangements
   Depolarizers
   Ammonium Peroxydisulfate
   Ionizable Peroxymonosulfates
   Acidified Peroxydiphosphate
   General Design and Construction
   Plastic Envelope Design

2. **BASIC MAGNESIUM DRY CELLS (6)**
   Closure Discs — Hermetical Seals
   Resilient Discs with Shaped Covers
   Flat Cells in Reinforced Composite Shells
   Multiple Anode and Cathode Arrangements
   Protective Anode Coatings
   Potassium Permanganate Magnesium Chloride Electrolyte

3. **ZINC-MERCURY-SILVER OXIDE CELLS (17)**
   Stacked Battery Construction
   Addition of Mn_2O_3 to Mercury Depolarizer Mix
   Zinc Powder-Sodium Hydroxide Electrode
   Barrier Layers for Mono- and Divalent Silver Oxide Electrodes
   Addition of Silicate Ions to Electrolyte
   Hermetically Sealed Button Cells
   Addition of Lithium Ions to Electrolyte

4. **CADMIUM CELLS (6)**
   Nickel-Cadmium
   Impregnation of Porous Electrode with Nickel Hydroxide
   Sealed Type Oxygen Cell
   Nickel Oxyhydroxide Electrode
   Other Cadmium Cells
   Porous Electrode Preparation

5. **ORGANIC ELECTROLYTES (9)**
   Electrodes for Organic Electrolytes
   Fluorinated Carbon
   Sulfur Dioxide as Cathode Depolarizer
   Cold Extension of Lithium Anode
   Tetrabutylammonium Chloride in Propylene Carbonate
   Organic Aprotic Solvent and Lewis Acid

6. **SOLID STATE CELLS (32)**
   Solid Electrolytes
   Quaternary Ammonium Silver Iodide Salts
   Vapor Deposition Technique for Silver Halides
   Contiguous Thin Film Layers of Halides
   Dispersion of Iodine Reactive Material
   Lithium Metal Coated with Oxygen-Containing Lithium Compounds
   Lithium-Organic Charge Transfer Complex
   Elemental Iodine and Cyanides of Zinc or Cadmium
   Sodium + Sulfur with Solid Electrolytes
   Sintered Polycrystalline Plate
   Alumina with Sodium Ion Conduction
   Alkali Metal Polysulfide Catholyte

7. **THERMAL AND ZINC-AIR CELLS (11)**
   Ceramic Sandwich Construction
   Cerium Oxide-Sintered Zirconium and Yttrium Oxides Disc
   Aluminum Salt Electrolyte
   Addition of Gallium or Tin to Zinc Electrode
   Gas Diffusion Cathode Containing Manganese Oxide Catalyst

8. **GENERAL BATTERY DESIGN AND CONSTRUCTION (22)**
   Flat Cells
   Seal for Water Cells
   Multicell Construction
   Other Cells and Batteries
   Disposable Cell with Encapsulated Electrolyte
   Spiral Plate Type Rechargeable Battery

ISBN 0-8155-0482-9 **320 pages**

INDUSTRIAL WATER PURIFICATION 1974

by Louis F. Martin

Pollution Technology Review No. 16

The Federal Water Pollution Control Act (now made into law) will have far-reaching effects on all industry. The pretreatment standards, about to be published as a result of this act, will exert increasing constraints on industrial process design and expansion plans in the years ahead.

The *Guidelines for the Pretreatment of Discharges of Publicly-Owned Works* prohibit the introduction of industrial pollutants which would pass through a municipal facility "inadequately treated," therefore, all existing industrial facilities and new plant designs must include specific downstream water purification processes, before the effluent is sewered or discharged into public systems.

This book describes over 160 recently devised processes for the treatment of contaminated industrial waters.

A partial and condensed table of contents follows here. Numbers in parentheses indicate the number of processes per topic, chapter headings are given, followed by examples of important subtitles.

While conventional separation processes for solid-liquid and water-oil effluents are applicable to many industrial streams, the process literature clearly highlights the areas of greatest immediate concern as shown by this table of contents:

1. SOLID-LIQUID SEPARATION PROCESSES (29)

Flocculation
Methylamine-Epichlorohydrins
Polyelectrolytes
Coagulation and Clarification
Continuous Filtration
Other Separation Techniques
Ion Exchange
Electrostatic and Electromagnetic Methods
Suspended Solids Techniques

2. OIL-WATER SEPARATIONS (18)

Special Equipment Needs
Flotation Cells
Gravity Separation
Vortex Separator
Multiphase Liquid/Liquid Separations
Use of Absorbents
Polypropylene Oil Mops
Permeable Foams for Coalescence

3. REMOVAL OF METALS (30)

Separation of Heavy Metals
Mercapto-s-Triazines
N-Acylamino Acids
Ion-Selective Membranes
Electrolytic Methods
Removal of Metal Ions
Specific Mercury Removal Methods
Other Specific Processes
Nuclear Fuel Wastes
Metal-Containing Lubricants
Organic Lead Removal
Vanadium and Cobalt Removal

4. METAL FINISHING EFFLUENTS (18)

Cyanide Removal
Conversion to Non-Toxic Biodegradable Compounds
Chromium Removal
Control of Aluminum
Other Removal Processes
Use of Chelates

5. PULP AND PAPER PROBLEMS (18)

Closed Cycle Process for Treating Waste Liquors
Hydropyrolysis and Use of Activated Carbon
Wet Air Oxidation
Flocculation
Oxyaminated Polyacrylamides
Sulfur Recovery
Ultrafiltration

6. COAL, ORE AND TAR SAND PROCESSING (14)

Reverse Osmosis + Neutralization
Use of Waste Steam
Salt-Free Condensates
Use of Nucleoproteins
Cyclone Floats

7. REFINING and CHEMICAL PROCESSES (34)

Refinery Operations
Fertilizers
Paint
Phenol
Photography
Meat Processing
Other Specific Processes

ISBN 0-8155-0554-X **300 pages**

SYNTHETIC GEM AND ALLIED CRYSTAL MANUFACTURE 1973

by D. MacInnes

The first practical and reproducible gemstone synthesis was achieved by Auguste Verneuil around 1900. He developed the flame fusion process for growing "boules" of alumina-based gems. His technique yields excellent rubies and sapphires.

The distinction between gems and industrial crystals is not a sharp one. Important laser crystals have also become popular diamond substitutes for jewelry. Today many crystals for semiconductor applications are pulled from the melt by a basic technique discovered by Czochralski. This method is also used for gem manufacture, and is especially well illustrated in this book. So are processes for making star sapphires and many more, including the synthesis of real industrial diamonds from carbon by explosive techniques.

112 patent-based processes. A partial list of contents follows. Numbers in ( ) indicate a plurality of processes per topic. Chapter headings are given and some of the more important subheadings are mentioned.

1. CORUNDUM (14)
- Crystal Growth by Verneuil Process
- Hydrothermal Corundum Synthesis
- Sapphires from Alumina Melts
- Sapphire Growth with Electron Beams
- Surface Treatment of Crystals

2. ASTERIATED GEMS (5)
- Star Rubies and Sapphires
- Improved Asterism
- Asteriated Rutile Crystals
- Producing Asterism
- Star Sapphires from Boules

3. RUTILE (8)
- Addition of Alumina to Titania
- $TiO_2 + Ga_2O_3$
- Color Change with BeO

4. TITANATES (12)
- Strontium Titanate Boules
- Barium Titanate Crystals
- Aluminum Titanate
- Calcium Titanate
- Zinc Orthotitanate
- Photochromic Titanate Material

5. GARNETS (8)
- Single Garnet Crystals
- Yttrium-Gallium Garnet Stones
- Growing Garnet on Non-Garnet Substrate
- Garnet Synthesis at High Pressure
- Yttrium-Iron Garnet (YIG)
- Yttrium-Aluminum Garnet (YAG)

6. MISCELLANEOUS CRYSTALS (9)
- Artificial Jade
- Synthetic Quartz (Cat's Eye)
- Synthetic Opals
- Synthetic Lapis Lazuli
- Lithium Niobate Crystals
- Single Crystals from Solutions
- Single Beryls (Emeralds)

7. G.E. DIAMOND SYNTHESES (21)
- The Hall Method
- Improved Hall Process
- Catalysts for Diamond Synthesis
- Diamonds with Conductive Surfaces
- Improved Diamond Crystals
- Diamond Seed Crystals
- Obtaining Uniform Growth
- Variations in Growth
- Cubic and Hexagonal Diamonds
- P-Type Diamond Crystals

8. OTHER DIAMOND SYNTHESES (35)
- Allmanna Svenska Process
- U.S. Army Electronics R & D
- Catalyst Variations
- Graphite and Fe, Ni, or Co plus a Metal Carbide
- Graphite plus Niobium
- Catalyst of V, W, Mo, Ti, Zr, Hf plus Cu, Ag or Au
- Phosphorus + Ni or Co as Solvent and Catalyst
- Diamonds by Electric Discharge Across Graphite Electrodes
- Shock Synthesis Techniques
- Shock Waves and Pressure Pulses
- Detonating Device
- Diamond Growth on Seed Crystals
- Low Pressure Processes
- Diamond Purification Methods
- Changing the Color of Diamonds

ISBN 0-8155-0469-1

221 pages

POLLUTION CONTROL IN THE ORGANIC CHEMICAL INDUSTRY 1974

by Marshall Sittig

Pollution Technology Review No. 9

Wastes from plants manufacturing identical compounds may be quite dissimilar, because of the difference in processes and raw materials. Since the organic chemical industry is now largely petrochemical, preference has been given to waste treatment from such operations.

Detailed treatability evaluation of each waste stream has become a prerequisite for any profitable product.

Physical methods for waste and by-product removal include gravity separation, air flotation, filtration, evaporation and adsorption techniques.

Chemical methods include precipitation, polyelectrolyte treatment, oxidation and other chemical conditioning such as neutralization.

Biological treatment methods comprise activated sludge and its modifications, trickling filters, aerated lagoons, and waste stabilization ponds.

This book intends to assist the chemical engineer in treating chemical waste products and effluents in conjunction with prudent raw material choices and thus bring about process economics. It presents condensed vital data from government and other sources of information that are scattered and difficult to pull together.

A partial and condensed table of contents follows here.

INTRODUCTION
Background & Economic Trends
Air Pollution
Water Pollution
Solid Wastes

WATER USE IN ORGANIC CHEMICALS MANUFACTURE

WASTEWATER SAMPLING AND SURVEYING
Sampling Station Selection
Case Studies of Wastewater Surveys

PROCESS CATEGORIES AND RAW WASTE LOADS

AQUEOUS EFFLUENT LIMITATIONS

SPECIFIC PRODUCTS — Their Manufacture and Pollution Problems

Acetaldehyde from Ethanol

Acetaldehyde from Ethylene

Acetic Acids from Acetaldehyde

Acetone from Isopropanol

Acetylene from Methane

Acrylates from Acrylic Acid

Acrylic Acid from Acetylene

Acrylonitrile from Propylene

Adipic Acid from Cyclohexane

Alachlor from Diethylaniline

Aldicarb from Isobutylene

Aldrin from Dicyclopentadiene

Allyl Chloride from Propylene

Aniline from Nitrobenzene

Atrazine from Cyanuric Chloride

BTX from Petroleum

Bisphenol-A from Phenol and Acetone

Butadiene from Butane

Caprolactam from Cyclohexane

Captan from Butadiene

Carbaryl from Naphthalene

Chlordane from Cyclopentadiene

Chlorobenzene from Benzene

Chloromethane by
Chlorination of Methane
Hydrochlorination of Methanol
Thermal Chlorination of Propane
Carbon Disulfide Chlorination

Coal Tar Distillation

Cresol from Toluene

Cyclohexane from Benzene

2, 4-D from Dichlorophenol

DDT from Chlorobenzene

Dieldrin from Aldrin

Dimethyl Terephthalate from Terephthalic Acid

Dinitrotoluene from Nitrotoluene

Disulfoton from Ethanol

Dyes and Pigments from Aromatics

Epichlorohydrin from Allyl Chloride

Ethylbenzene from Benzene

Ethyl Chloride by
Hydrochlorination of Ethylene
Chlorination of Ethane
Hydrochlorination of Ethanol

Ethylene or Propylene from Paraffins

Ethylene Dichloride from Ethylene

Ethylene Glycol from Ethylene Oxide

Ethylene Oxide from Ethylene

Formaldehyde from Methanol

Long Chain Alcohols from
Ethylene Oligomers

Malathion® from Methanol

Methanol from Natural Gas

Methylamines from Methanol + Ammonia

Methyl Bromide from Methanol

Methyl Methacrylate from
Acetone Cyanohydrin

Nitrobenzene from Benzene

Nitrochlorobenzene from Chlorobenzene

Nitroparaffins from Paraffins

Oxo Products from Olefins

Parathion from p-Nitrophenol

Perchloroethylene from Propane

Phenol from Chlorobenzene

Phenol from Cumene

Phorate from Ethanol

Phosgene from CO + Cl_2

Phthalic Anhydride from
Naphthalene or Xylene

Propylene Glycol from Propylene Oxide

Propylene Oxide from Propylene

Styrene from Ethylbenzene

SNG from Crude Oil

2,4,5-T from Trichlorophenol

Terephthalic Acid from Xylene

Tetraethyl Lead from Ethyl Chloride

Trichloroethane from Vinyl Chloride

Trichloroethylene by
Chlorination, then by
Dehydrochlorination of Acetylene
or by Oxyhydrochlorination
of Dichloroethane

Trifluralin from Perfluoromethylchloro-benzene

Vinyl Acetate from
Ethylene and Acetic Acid

Vinyl Chloride from Acetylene

Vinyl Chloride from Ethylene Dichloride

AIR POLLUTION CONTROL
Halogen Acids
Halogenated Hydrocarbons
Halogens
Hydrocarbons
Particulates
Sulfur Compounds
Coking of Coal

WATER POLLUTION CONTROL
Primary
Oil Separation
Equalization
Neutralization
Sedimentation
Flotation
Flocculation
Nutrient Addition
Secondary
Activated Sludge
Extended Aeration
Trickling Filters
Aerated Lagoons
Waste Stabilization Ponds
Chemical Oxidation
Nitrification—Denitrification

Tertiary
Chemical Precipitation
Gas Stripping
Microstraining
Carbon Adsorption
Electrodialysis
Ion Exchange
Evaporation
Reverse Osmosis
Chlorination
Rapid Sand Filtration
Sludge Handling
Aerobic Digestion
Anaerobic Digestion
Wet Oxidation
Thickening
Lagooning
Sand Drying Beds
Vacuum Filtration
Filtration
Land Disposal
Incineration
Sea Disposal
Ultimate Disposal
Thermal Oxidation
Deep Well Disposal

OVERALL WATER POLLUTION CONTROL MODELS
BPCTCA: Best Practicable Control Technology Currently Available
BATEA: Best Available Technology Economically Available
BADCT: Best Available Demonstrated Control Technology

THE ECONOMICS OF POLLUTION CONTROL

FUTURE TRENDS

ISBN 0-8155-0536-1

305 pages

CONTROLLED ACTION DRUG FORMS 1974

by J. C. Colbert

Chemical Technology Review No. 24

Speedy antacids and quick-acting analgesics have long been extolled for their fast remedial performance in all forms of advertising. Enteric coatings for specific release along the intestinal tract have been known to pharmacists since the early 1900's. Even the 12-hour cold capsule is now commonplace.

All this is a far cry from today's sophisticated drug design. Hydrophilic polymers and biological binding substances, such as select proteins and polysaccharides, enable the medicinal chemist and pharmacologist to design drug systems by which simple implants will release measured low doses of a hormone or contraceptive continuously for as long as one year. By means of skillful lamination techniques it is possible to withhold release or absorption until a certain poison enters or is generated within the body.

This review covers 211 patents issued since 1956 covering numerous methods of preparing practical dosage forms of human or veterinary drugs having some means of controlling the action or release of the active ingredients. Numbers in parentheses indicate a plurality of processes per topic. Chapter headings and some of the more important subtitles are given in the partial and condensed table of contents that follows here.

PART I: EXCIPIENTS AS CONTROL AGENTS

1. **Coatings Materials (20)**
   Shellac Coatings
   Cellulose Derivatives

2. **Wax and Lipid Matrix Materials (17)**
   Glyceryl Stearate
   Carnauba Wax

3. **Synthetic Polymer Matrix Materials (15)**
   Acrylic Copolymers
   Polyvinylpyrrolidone
   Other Polymers

4. **Other Matrixes (19)**
   CMC
   Gelatin-Sorbitol

PART II: DOSAGE FORMS FOR SUSTAINED RELEASE OR QUICK DISINTEGRATION

5. **Tablet Structure (15)**
   Laminated Tablets
   Tablet Configuration
   Active Cores
   Inlays
   Apertured Tablets
   Other Tablet Structures

6. **Capsule Technology (10)**

7. **Powders, Beads, & Pellets (17)**

8. **Other Dosage Forms**

PART III. SPECIFIC MEDICAMENTS FOR SLOW AND FAST ACTION

9. **Antibiotics (16)**
   Penicillin Tablets
   Spraycoated Penicillin Particles
   Injectable Penicillins
   Penicillin-Sulfonamide Combinations
   Acetoxymethyl Benzylpenicillinate
   Penicillanic Acid Derivatives
   Tetracycline-Novobiocin Combinations
   Slow Release Tetracycline
   Tetracycline-Chymotrypsin Combinations
   Tetracycline Fumarates
   Erythromycin-Sulfonamide Combinations
   Spiramycin Complex

10. **Blood and Cardiovascular Agents (11)**
   Heparin Formulations
   Diuretic and Saluretic Preparations
   Lipid and Cholesterol Control
   Aluminum Nicotinate
   Dextran Sulfate

11. **Analgesics and Antipyretics (9)**
   Indomethacin Microcapsules
   Acyloxyalkyl Indenylacetates

12. **Gastrointestinal Medicines (7)**
   Antacid Formulations
   HCl Release Compound
   "Digestive" Dragees

13. **Other Therapeutic Agents (24)**
   Antigens
   Viral Dosage Forms
   Narcotic Antagonists

14. **Veterinary Drugs (19)**
   Rumen Resistant Coatings
   Hormone Implants
   Avian Specifics

ISBN 0-8155-0520-5

340 pages

UTILIZATION OF WASTE HEAT FROM POWER PLANTS 1974

by David Rimberg

Pollution Technology Review No. 14

Energy Technology Review No. 3

Present-day steam-driven turbo-electric power plants in the United States discharge as waste heat an amount of energy roughly equivalent to twice their total electricity-generating capacity. This energy is most difficult to utilize, because it is degraded in temperature. The effluent water is warm, but it is far from being near the boiling point. It constitutes a necessary, but unwanted, by-product of the energy conversion process for generating electricity.

The growing quantities of waste heat discharged, and the increasing ecological anxieties about undesirable growths, energy utilization and thermal discharge problems, have stimulated an examination of methods for productively using energy presently wasted on the environment.

Part I of this book discusses the reasons for present-day ineffective utilization, but it also shows ways and means to promote more efficient energy usage.

Part II assesses the cause, magnitude, and possible effects of heat discharges into water from steam electric power plants and related condenser cooling systems. Also contained in this section is a discussion of the thermodynamics of the electric power generation cycle detailing the reasons for the "inefficiencies" in by-product heat generation.

Ultimately all the accessible waste energy appears as low temperature heat, and the term "waste heat utilization" refers to the performance of useful functions with this heat before it is discharged into the environment. Within the framework of this book, the minimum temperature of the effluent water considered for subsequent use is about 38°C or 100°F. These uses are discussed in **Part III:** Food production in agriculture, hydroponics and aquaculture (fish farming) and the use of heat in wastewater treatment.

In the past the dissipation of waste heat was accomplished by wet and dry cooling towers and cooling ponds, lakes and streams. These methods are now being challenged by some sectors of society, and industry is being forced to consider their environmental impact. In this regard **Part IV** discusses the research needs necessary to equate the complicated interactions of the physical, engineering, biological, and social aspects of this waste heat problem.

This Pollution Technology Review is based on studies conducted by industrial and engineering firms or university research teams under the auspices of various governmental agencies, e.g. The National Water Commission, Oak Ridge National Laboratory Federal Water Pollution Control Administration, Office of Water Resources Research, Environmental Protection Agency, National Science Foundation, and the Department of Commerce.

A partial and condensed table of contents follows here:

PART I

ENERGY CONSERVATION THROUGH EFFECTIVE UTILIZATION

Problems in Energy Supply
Present Uses of Energy in the U.S.
Possible Savings in Use of Energy
Summation of the Problems
Means to Promote Effective Energy Utilization

PART II

GENERATION OF WASTE HEAT BY ELECTRIC POWER PLANTS

The Waste Heat Problem
Waste Heat and Cooling Water Requirements
Effects of Steam-Electric Power Plant Operation on the Water Environment
Steam-Electric Power Plant Siting Alternatives
Thermodynamics of the Electric Power Generating Cycle
Thermal Pollution Effects
Waste Heat Treatment

PART III

WASTE HEAT UTILIZATION

Concepts and Demonstration Projects
Agricultural
Aquacultural (Fish Farms)
Wastewater Treatment

PART IV

Research Needs
Future Trends

ISBN 0-8155-0555-8 **175 pages**